Schools, Mathematics and Work

Edited by
Mary Harris

Maths in Work Project,
Department of Mathematics,
Statistics and Computing,
University of London Institute of Education

The Falmer Press

(A member of the Taylor & Francis Group)
London • New York • Philadelphia

UK	The Falmer Press, Rankine Road, Basingstoke, Hampshire, RG24 0PR
USA	The Falmer Press, Taylor & Francis Inc., 1900 Frost Road, Suite 101, Bristol, PA 19007

First published 1991

British Library Cataloguing in Publication Data
Schools, mathematics and work.
1. Education. Mathematics. Applications. Teaching.
I. Harris, Mary
510.7

ISBN 1-85000-893-0
ISBN 1-85000-894-9 pbk

Library of Congress Cataloguing-in-Publication Data

Schools, mathematics, and work/edited by Mary Harris.
p. cm
Includes bibliographical references and index.
ISBN 1-85000-8983-0 (HC): — ISBN 1-85000-894-9 (SC):
1. Mathematics. 2. Mathematics — Study and teaching. I. Harris, Mary.
QA36.S36 1991   90-43999
510Œ.71–dc20   CIP

Jacket design by Caroline Archer

Typeset in 11/13 Bembo by
Chapterhouse, The Cloisters, Formby L37 3PX

Printed in Great Britain by Burgess Science Press, Basingstoke on paper which has a specified pH value on final paper manufacture of not less than 7.5 and is therefore 'acid free'.

Contents

Foreword   *ix*
Professor Celia Hoyles

Preface   *xi*
Mary Harris

Acknowledgments   *xiii*

1 Introduction
   M. Harris

Part 1: The Context of Mathematics Education   1

   Introduction: Part 1   13
2 Ethnomathematics and its Place in the History and Pedagogy of
   Mathematics   15
   U. D'Ambrosio

3 Mathematics and Ballistics   26
   H. J. M. Bos

4 Mathematics Education in its Cultural Context   29
   A. Bishop

5 Foundations of Eurocentrism in Mathematics   42
   G. Joseph

6 Mathematics in a Social Context: Math within Education as Praxis
   versus Math within Education as Hegemony   57
   M. Fasheh

7 Folk Mathematics   62
   E. Maier

8 Hidden Messages   67
   J. Maxwell

9 Mathematics Education Post GCSE: an Industrial Viewpoint   71
   T. S. Wilkinson

10 The Computer as a Cultural Influence in Mathematical Learning   77
   R. Noss

11 The Contextualizing of Mathematics: Towards a Theoretical Map   93
   P. Dowling

Part 2: Mathematics in the Workplace — Research Views   121

   Introduction: Part 2
12 Mathematics and Workplace Research   123
   M. Harris and J. Evans

13 Looking for the Maths in Work   132
   M. Harris

14 The Role of Number in Work and Training   145
   D. Mathews

15 Skills Versus Understanding   158
   R. Strässer, G. Barr, J. Evans and A. Wolf

16 Mathematics in and out of School: A Selective Review of Studies from
   Brazil   169
   D. Carraher

17 Theories of Practice   202
   J. Evans and M. Harris

Part 3: Mathematics in the Workplace — User Views   211

   Introduction: Part 3
18 An Industry and Mathematics: A View from Courtaulds   214
   S. Ingham

19 An Industry and Mathematics: A View from a Training Board   220
   K. Pye

20 The Gendering of Work   230
   J. Holland

Part 4: School Mathematics in Context   253

Introduction: Part 4
21 The Certificate in Pre-Vocational Education   256
    S. Sullivan

22 'Maths in the Workplace': Some Issues Arising out of the Development
    of a Resource Pack   266
    P. Drake

23 Mathematics and TVEI   273
    A. Grey

24 Work Reclaimed: Status Mathematics in Non-Elitist Contexts   277
    M. Harris and C. Paechter

25 Postscript: The Maths in Work Project   284
    M. Harris

Notes on Contributors   292
Index   295

Foreword

This is an important and exciting book concerned with mapping the intricate network of relationships between school, mathematics and work.

From various perspectives, authors in this edited volume discuss the important issues of the meaning of mathematics in use — in the workplace and in everyday life — and how this meaning should transcend merely arithmetic and training in basic skills. Authors provide a crucial historical perspective on the development of mathematics and in particular on the ways in which theoretical and practical mathematics or academic and basic mathematics have become differently oriented and have acquired different status. As well as giving this historical perspective, the volume addresses contemporary issues, concerning for example TVEI and the role of the computer as an expressive medium in education.

Mary Harris has drawn together an excellent collection of papers from authors from a wide range of countries with very different experiences and very different perspectives. All too frequently issues concerned with applying mathematics education and particularly how it should meet the needs of society, are addressed in a simplistic way. This book implicitly and explicitly rejects this simplicity and provides us with a resource for making a major step towards coming to understand the complexity of the situation.

Professor Celia Hoyles
Department of Mathematics, Statistics and Computing,
University of London Institute of Education.

Preface

Schools, Mathematics and Work has been prepared as the reader for an In-Service course called 'Mathematics Outside School'. It is intended however that *Schools, Mathematics and Work* should speak to the wider readership of all involved in initial and in-service teacher education. It will also be of relevance and use to anyone involved in 'basic' mathematics or numeracy training from an industrial point of view.

Of the twenty-four papers included in *Schools, Mathematics and Work*, some have already been published in a range of journals and some, including two research reviews have been written specially for it. Much current practice in the field of mathematics and work adopts a limited approach that largely ignores the experience, expertise and research of mathematics education. Recent changes in curricula and in the social and political context of teaching have brought pressures on the education system such that it has not been able to articulate its view in a form and with a force that demands a hearing. The broad spread of papers in this collection has been chosen as a way of redressing the current imbalance by challenging some current assumptions and practices in both industry and education and opening up the field to some of its wider, mathematically richer, more positive and more imaginative potential. Themes that run through the papers are the cultural nature of mathematical thinking and action and the under-utilization in paid work and degradation in unpaid work of the intellect of half the population.

The course 'Mathematics Outside School' is published through the database NERIS, an information service operating throughout the UK to provide schools and colleges with details of curriculum materials and, in many instances, full teaching and learning resources. The service may be searched by using a modern link to the computer at the Open University which holds the NERIS database. Alternatively, it is possible to receive regularly updated copies of the entire database on compact disc. These discs may be used with suitable CD-ROM equipment. Full details of all aspects of the service offered by NERIS are available from User Support, NERIS, Maryland

College, Leighton Street, Woburn, Bedfordshire, MK17 9JD. Telephone 0525
290364.

Mary Harris
1990

Acknowledgments

A project that is run by one person necessarily make demands on a number of people, without whose goodwill it would not be able to function. All the authors who wrote papers specially for this book or updated previously published work, did so in time they made from their own work demands.

The editor is also grateful

- to her colleagues in the Department of Mathematics, Statistics and Computing at the University of London Institute of Education, whose range of talents and whose friendliness and professional support make the Department an exceptionally rewarding place in which to work,
- to colleagues Pat Drake, Declan O'Reilly and Stan Silver in particular for their criticisms, advice and support during the development of the Course and Reader,
- to Pat Drake, Jeff Evans, Carrie Paechter and Declan O'Reilly for advice, criticisms and additions to the text,
- to the Maths in Work Steering Committee, Professor Celia Hoyles, Ms. Pat Drake, Mrs. Anne Taylor, Senior Inspector for Assessment and Student Training in the London Borough of Camden and Derek Turner of the Department of Trade and Industry, for their constant and cheerful support,
- and to her family whose only complaint at the overtime worked has been that the sound of the printer running at all hours sounds like the cat in distress.

The Editor is grateful to the Editors of the following journals for permission to reprint articles from:

For the Learning of Mathematics — for D'Ambrosio and Bos
Educational Studies in Mathematics — for Bishop and Noss
Race and Class (Institute of Race Relations) — for Joseph

UNESCO *Science and Technology Education Document Series* — for Fasheh
Mathematics Teaching (Association of Teachers of Mathematics) — for Maier and Maxwell,
Bulletin of the Institute of Mathematics and its Applications — for Wilkinson
Zentralblatt für Didaktik der Mathematik — for Strässer, Barr, Evans and Wolf.

The choice and editing of the papers and the emphasis of the collection is the responsibility of the Editor.

This work was produced with the assistance of a grant from the Department of Trade and Industry. The views expressed are those of the authors and do not necessarily reflect those of the Department of Trade and Industry or any other Government Department.

1
Introduction

Mary Harris

Mathematics in Context

All three nouns in the title of this book, Schools, Mathematics and Work, are controversial. In the early 1990s, maintained schools in England and Wales are coping with this controversy as their administrative structures, their curricula and their relationships with their communities all undergo radical and rapid change.[1] One of the major changes is in mathematics, a core subject of the first compulsory national curriculum to be imposed by government. But the national curriculum is not the only vehicle in which mathematics is delivered. The increasingly organized interests of business and industry in schools include curricular influence in vocationally oriented courses in the secondary sector and the much more subtle influence of the many schools-industry organizations now active in the primary sector as well. Such organizations carry hidden, unexamined or quite explicit messages about school mathematics and education which often differ from those of educationalists. But while business and industry influence school mathematics education, however subtly and indirectly, their own training and curricula remain very largely uninfluenced by mathematics education itself; the relationship between school mathematics and work is mainly one-way (Chapter 22).

Meanwhile other demands from society bring a range of different pressures on school mathematics. Many people who do paid work work in small firms or in self-employment where demands on employees can differ from those of large employers (Chapter 13). Many people who do unpaid work use mathematics though they may not call it that (Chapters 4, 6, and 11). The academic discipline of mathematics itself undergoes constant development, for example, recently as the result of the advent of large computers in research. The discipline of mathematics education constantly raises new questions and brings new understandings of how people learn mathematics and what sort of mathematics they learn under different circumstances (Chapters 12 and

16). England and Wales is a society of people who bring differing experiences, contributions, aspirations, views of learning and views of mathematics itself. It is the duty of schools to provide an education in mathematics broad enough and deep enough for all, whatever their aspirations. There is more to the 'non-academic' end of school mathematics education than the specific training of 'basic skills' demanded by large employers.

Two themes run through the analysis of the uncomfortable relationship between mathematics education and the use of mathematics at work. One is the problem of differences in value in the distinctions made in the discipline of mathematics itself between 'pure' and 'applied'.[2] The other is the constant efforts of people who have always done mathematics as they worked (Chapter 4) or sought to learn mathematics because it was needed in their work. The number of people who learn mathematics for fun has always been extremely small.[3]

D'Ambrosio's concise review (Chapter 2) of the origins and development of the theoretical-practical split in mathematics charts its historical influence but itself suffers a little from the tendency of professional western mathematicians to interpret 'ethno-mathematics' (which D'Ambrosio defines) in their own terms (Chapter 11). A characteristic of the eurocentric historiography of mathematics, in spite of its nods towards the Middle East, is that it consciously locates mathematics *outside* the practical domain (Chapter 5). Other traditions are not so squeamish and that the Greeks themselves acknowledged their debt to them is sometimes forgotten (Bernal, 1986).

The first English, post-Roman schools were 'vocational' (Howson, 1982), the vocation being the church. Mathematics was on the curriculum; its aim was to seek underlying design in the universe, its one practical application that of calculating the dates of the movable feasts. Textbooks were variations of Boethius' 'jejune and exceedingly elementary abbreviation of earlier classics' (Boyer, 1985, p. 212). Even the mathematics curriculum of the newly founded clerical universities of the thirteenth century can be interpreted as having a vocational element, with the *quadrivium* (which included mathematics) as the post-graduate part of the liberal arts curriculum. This being a licence to teach was the reason that many clerics studied it (Howson, 1982). But as one of the liberal arts there was of course no intention that mathematics should be useful in a practical sense.

Outside academic institutions most people received their education through their employment. Mathematics was learned as it was needed and the effectiveness of technical innovation required it. The Middle Ages formed one of the great inventive eras of mankind and has strong parallels with the industrial revolution of the eighteenth and nineteenth centuries (Gimpel, 1988). Capitalist companies were formed whose shares were bought and sold, accounts were kept and accountancy itself advanced. Conditions favoured free enterprise which led to the rise of self-made men and women. The major factor in the technical innovation of the time was the development of the cam and gearing in water-powered mills,[4] including tidal mills which enabled the mechanization of a whole series of industries which until then had operated by hand or foot. Advances in the production of cloth, leather, paper, iron and coal improved the standard of living. Production was raised in an agricultural revolution which improved design in harnessing and the manufacture of iron shoes so that horses could be used for ploughing. The skill of the masons was responsible for the sophisticated, beautiful and well-plumbed buildings in which the scholastics cut themselves off from the world outside. The ramifications of the textiles industry following the mechanization of fulling[5] led to the developments in Italian banking and its vernacular manuals that were put to widespread practical use in the Renaissance. Literacy became necessary (until then it had been a church monopoly) and increasing literacy lead to the idea of education having both an investment and a consumption function, an idea which grew throughout Europe (Cipolla, 1969).

Long before Adelard's translation of Euclid (Chapter 2), medieval builders were using the manuals of Vitruvius and others, indeed they had been in continuous use since Roman times (Gimpel p. 134). The practical problem-solving of architect-engineers[6] was dependent on an inventive practical geometry developed from that of the civil engineers of classical antiquity who, being practical men, did not have the status of Euclid. In the Middle Ages many of the masons had the status accorded to the folk heroes of today, indeed some commanded such high fees as to cause jealousy among the academic practitioners of dialectic. One, Nicolas de Biard complained that although their training was merely manual, they worked 'by words only' telling their masons where and how to cut and generally behaving like intellectuals (Gimpel, *op. cit.*). People who teach and learn 'by words' leave only the results of their labour from which later generations of historians must judge their mathematical skill, something that cannot be done adequately without entering the practical task. In times when calculation was done mainly by counters and the abacus there were no recorded results other than in the artefact, for unfortunately there are 'no recorded moments in the past when great numbers of typical citizens stepped forward and recited their multiplication tables' (Cohen, 1982, p. 11). This does not mean that they could not cope with problems like calculating the diameter of a column when only a part is visible (Gimpel, *op. cit.*).

In spite of the complaints of a few academics, the Middle Ages is marked by a co-operation between scholars and craftsmen. 'The real progress of the period, quite inexplicable if its origins are sought in scholastic philosophy and science, took place in virtue of painstaking trial and error on the part of the masters of the various crafts . . . ' (Gimpel, p. 180). Indeed 'during the whole of the "high Middle Ages" from 1250 to 1550 actual practice based on empiric methods was far in advance of contemporary thought'. (Harvey, 1947). Roger Bacon was influenced both by his university training and by his study of the mechanical arts of the Greeks, Romans and Arabs in the recent translations from the Arabic. His proposal of reforms in education advocated the wider teaching of mathematics since 'he who is ignorant of it cannot know the other sciences or the things of this world'. (quoted by Boyer, p. 272). His work extended to courses in business management and other vocational skills given in Oxford as a response to the demand from the burgeoning mercantile classes (Gimpel, *op. cit.*). Although the newly arrived Hindu-Arabic numerals took time to be accepted, (Fibonacci's *Liber Abaci* was published in 1202) their use in accounting stood in efficient contrast to the Roman numerals which could only record.[7] The ideas of Bacon and others proved too revolutionary for the establishment however, who condemned their errors of belief and restricted their activities. The long term result of such a reaction separated mathematics and science from liberal humanism and launched a dualism from which education still suffers.

War is a major stimulus for technological advance (and for employment) and improvements in iron, steel and bronze techniques begun in the eleventh century culminated in the fourteenth in the improved accuracy and range of canon. It was the German canon of the Turks that ended the formal Middle Ages with their break in the indestructibly engineered walls of Constantinople that had stood for 1000 years. The original practical mathematics of canonry was concerned with problems of manufacturing maximum and accurate bore and ball within the bounds of transportability and effective functioning. By the fifteenth century, gunners were in great demand; they had to be literate and, in various gunnery schools of the sixteenth century, they were trained. (Cipolla, *op. cit.*). When mathematical theory arrived it was marked by a mis-match with the actual performance (Chapter 3). At a time when some intellectuals were again learning from the skills of 'superior artisans' (Lilley, 1958). Galileo developed his theory with the acknowledged help of the practitioners but its re-application to practice was less than successful. The inter-relations between pure and applied mathematics, between theory and utility are not simply a question of applying superior knowledge to an ignorant recipient.

By the fifteenth century, the needs of commerce including those for more sophisticated open-sea navigation techniques and instruments had increased the investment demand for both literacy and mathematics. The invention of printing, together with the tradition of merchants' manuals written in the vernacular stemming from Italy, combined to answer this. A wide range of pocket-sized business manuals for 'teaching arithmetic, book-keeping and accounts, as well as compendia of useful information about weights and measures in various countries, the drawing up of instruments of credit and exchange and conditions on the various trade routes' was published in the sixteenth century (Charlton, 1965). The need for improved navigation had been recognized in the Act for the Maintenance of the Navy in England of 1540 (Charlton, *op. cit.*). Wringing their hands at how much better things were in Europe, a delegation was sent there to recruit experts; proposals for home-based systematic technical instruction were made and opposed by vested interests — and the venture left to the private enterprise of mathematics practitioners

(Taylor, 1954). Public mathematics lectures were one of the results, the sales of reprints together with rising sales of navigation manuals both written in English, indicating the need. As in the days of the earlier Bacon there was some co-operation between individual scholar and practical craftsman though university education was concerned with Ptolemaic texts and Boethius was still in use. There was some cracking of the traditional Platonic disdain for technical instruction, and educational advice on how best to combine theoretical and practical mathematics was offered. Vives (1492–1540), the Spanish tutor hired by Henry VIII for his daughter Mary warned against the dangers of too academic a study of mathematics in that 'it leads away from the things of life, and estranges men from the perception of what conduces to the common weal' (Howson, *op. cit.*, p. 5). Again the need for reconciling the theoretical and practical in a balanced mathematics education was recognized but 'the subsequent history of mathematics education in England is largely a chronicle, on the one hand, of how this problem was ignored and, on the other hand, of how individual educators have constantly sought to effect a reconciliation.' (Howson, *op. cit.*, p. 5).

Robert Recorde's *Grounde of Artes* published in 1544 represents an attempt at such a reconciliation and was the beginning of the first attempt in English at a coherent and systematic course that combined both practical and theoretical mathematics and considered methods of teaching it. The *Grounde of Artes*, an arithmetic with mainly commercial applications was followed by a Copernican astronomy, a modified Euclid and an algebra. In effect Recorde constructed the first English mathematics curriculum (Howson, *op. cit.*) and he wrote for both scholar and practical user. During the 150 years in which the *Grounde of Artes* was in print, however, the interests of the two users diverged and by the beginning of the eighteenth century the binary system of mathematics education in England with the inherited value loadings attached to the distinction, was well and truly established. As editions multiplied and more authors wrote more arithmetics the genre became much more oriented towards commerce and omitted the explanations and questions on which Recorde had insisted, on the ground that the reader would be too ignorant to cope with them. The textbook writers simply presented material as 'rules to be remembered followed by examples drawn from business life'. (Cohen, 1982, p. 25). The effect of such books on practical arithmetic was thus to both stimulate and constrain the spread of numerical skills and though arithmetic knowledge undoubtedly spread in the seventeenth and eighteenth centuries it spread only in a manner that discouraged easy familiarity with mathematics (Cohen *op. cit.*). The splitting of commercial arithmetic from mathematics and its reduction to rote skills in sixteenth and seventeenth century England formed attitudes, even syllabi and teaching methods, available until very recently in chain bookstores in a form that would have been recognized in the sixteenth century.

Renaissance humanism, however, was mainly a literary event. It may have recognized Leonardo but it still regarded him as a craftsman, for his education had not been the liberal one of his time. In England some attempts had been made at reform, indeed the Edwardian statutes of 1549 laid down that all freshmen at Cambridge 'were to be taught mathematics as the foundation of liberal education' but the Elizabethan statutes twenty-one years later removed it again because 'its study appertains to practical life and has its place in a course of technical education rather than in the curriculum of a university'. (Howson, *op. cit.*, p. 12). The universities remained aloof and mathematics was of interest only to the connoisseur.

As a result, Samuel Pepys, old boy of St. Paul's School, graduate of Cambridge and civil servant at the Admiralty had to hire a tutor to teach him the arithmetic he needed for his work in controlling the supply of timber for shipbuilding. His diary of July 1662 records that his tutor was Mr. Cooper, Mate of the *Royal Charles*. Pepys' predicament and no doubt Mr. Cooper's experience, forced the realization that the navy as a whole also lacked the mathematics it needed. The decision was made to establish a school of navigation for officers but immediately there was the problem of finding teachers who could treat the subject practically within an academic curriculum (still of course in Latin and including Greek), for academic respectability could not be made subject to practical need. By the beginning of the eighteenth century however several mathematics schools had eventually been founded.

The initiatives in mathematics education arising from the practical needs of commerce and its carriers, that had flowered in the sixteenth and seventeenth centuries continued in the eighteenth as demands arising from increasing industrial mechanization began to lead towards the second industrial revolution. The growing feeling that mathematics and science should be included in the curriculum of educational institutions had made some headway during Cromwell's Commonwealth but reaction from the restoration church establishment effectively put the curriculum back into the traditional mould, where it flourished physically until the nineteenth century and spiritually until today. Spurred by hostile legislation, the dissenting academies of the late seventeenth century were at first as traditional as the establishment institutions they emulated, for example Timothy Jollie who founded an academy at Attercliffe, 'forbade mathematics as tending to sceptism and infidelity' (Howson, *op. cit.*, p. 47). A new definition of the content of a general education was eventually worked out, however, and put into practice with mathematics and science increasingly on the curriculum (Williams, 1961). The older urban schools and the newly founded non-conformist ones began to respond to local need and offer mathematics, and a number of vocational academies serving commerce and engineering as well as the arts joined the mathematical schools.

Literacy at the beginning of the eighteenth century was still relatively high at about 57 per cent (Cipolla, *op. cit.*, p. 62) indeed the industrial revolution was the outcome, not of 'one or two high priests of science', but of 'the daily down-to-earth experiment and tinkering on the part of a number of literate craftsmen and amateur scientists'. (Cipolla, *op. cit.*, p. 10–12). Without the scientific and mathematical knowledge acquired by the manufacturing middle classes it could not have taken place (Lawson and Silver, p. 218). As before, developments in mathematics education in these times were in response to utilitarian demand and took place outside the establishment institutions of formal education. Many of those involved contributed to both 'pure' and 'applied' mathematics, for practical mathematics arising in a job to be done encourages speculation and the possibility of more formal abstraction.

But increasing urbanization resulting from industraial and social change caused a fundamental change in education. Between 1751 and 1821 the population of Britain doubled to fourteen million and by 1871 it had reached twenty-six million. With the overall increase came an increase in the proportion of children and the proportion of people living in towns. Until the mid eighteenth century, a haphazard system of local parish and private schools in a variety of forms that included associations with shop-keeping and trade, had offered literacy and numeracy to the children of the poor. As official reports began to detail the results of urbanization, the desire to improve conditions through education, demonstrated by radicals in the tradition of Robert Owen, was countered by a mixture of fear of working-class rebellion and the widespread acceptance that the social order was divinely obtained. There was a shift in emphasis in education from moral instruction to moral rescue and the idea of a special type of education for workers' children began. Teaching them to read the Bible would save their souls but teaching arithmetic would 'produce in them a disrelish for the laborious occupations of life' (quoted by Williams, p. 156). The system based on a local social order gradually changed to a national system of social class, so that when state education did arrive eventually it was firmly class based. By the time arithmetic appeared in elementary schools it was limited and circumscribed by the tests of the Revised Code, set up not with any regard to the intellectual potential of children but for the purely administrative purpose of establishing a system of accountability for state grants in a system devised to be a cheap and efficient way of keeping working class children in their place.

The Taunton Commission, which investigated the endowed schools of the middle classes was published in 1868 and resulted in three grades of school each oriented towards a different 'level' of employment and supported by a different sort of mathematics curriculum. The curriculum for the first grade, for the sons of 'men with considerable incomes independent of their own exertions, or professional men, and men in business whose profits put them on the same level' plus 'the great majority of professional men, especially the clergy, medical men and lawyers; the poorer gentry; all in fact, who, having received a cultivated education themselves, are very anxious that their sons should not fall below them' there was the study of classical mathematics suitable for the classical curriculum. For the second grade (for the mercantile classes) the curriculum stressed the practical with arithmetic and the rudiments of mathematics beyond it. The third grade, for the sons of 'the smaller tenant farmers, the small tradesmen [and] the superior artisans, advocated 'very good arithmetic' (Cooper, 1985). The public schools with no tradition of mathematics behind them were investigated by their own commission (Clarendon) who recommended arithmetic, geometry, algebra and trigonometry for all. Practical geometry of the sort published by Recorde was associated with the mathematics needed for earning a living so Euclid was studied 'as an exercise of the intellectual powers' (Howson, *op. cit.* p. 130). Practical mathematics and practical arithmetic once the proud tool of the practitioner, then the artefacts of vulgar commerce were now the tools of social stratification and control.

In a new society forming by the growth of industry and of democracy, such curricula were the result of compromise between the demands of the public educators, the old humanists and the industrial trainers, with the latter predominating (Williams, p. 162). It was the extreme defensive reaction of the old humanists however that declared that learning with the aim of practical advantage was not real learning at all. Huxley's plea that science and the mathematics it needed should become part of a new general liberal education and culture was ignored and the relevant Act of 1889 was carefully called Technical Instruction, not Education. The mathematics and science that had been part of the curriculum of Co-operative Societies and Mechanics Institutes was now isolated in a separate formal education system inaccessible and unacceptable to the working classes whose needs had been their origin 'for it was precisely the interaction between techniques and their general living' that had given this class its new consciousness (Williams, p. 164). When middle-class technical education was finally set up in the new South Kensington institutions in the mid nineteenth century the only people capable of teaching there, that is the only people with any education in practical mathematics themselves, had received it in the army or in Germany where vocational education was, then as now, far in advance of that available in England.

Differentiated mathematics education remained the norm until the 1950s by which time the Education Act of 1944, resulting in a secondary education for all, made a mathematics curriculum for all at least debatable. Curriculum reform at the beginning of the century (described by Howson as the golden age of mathematics education) had been in the minority (in numbers if not in influence) private sector and mainly in the hands of the Mathematical Association of university and public school mathematicians. Establishment mathematics was beginning to emerge at last from the abstract mathematics of the liberal tradition that on the whole and over the centuries had stood aloof from the practical mathematics of people who needed to use it. Emerging from its cloisters with a status quite unjustified by its public performance it took the position of overall advisor on the education of the very people it had disdained for so long. The foundation of the Association for Teaching Aids in Mathematics (later to become the Association of Teachers of Mathematics) was essentially a reaction to academic curricula and their irrelevance to secondary modern schools whose curricula were built on elementary school arithmetic. Arising from the philosophy not of arithmetic for some and mathematics for others but mathematics for all, the attempted reforms of the ATAM tried to establish a new tradition instead of re-working the hierarchical compromises of the old one.

The legacy of unbalanced distinction of class and status between the separate sorts of mathematics still lingers at an almost unconscious level that reifies it as something fundamental and immutable. It still dominates both the historical view of mathematics as a discipline and the deliberations of curriculum planners and it still has nothing to do with intellectual ability. The old tensions remain, as their extensive history ensures that they will, and they are clearly visible in the new national curriculum.

In spite of rhetoric to the contrary, the tri-partite, class-based, differentiated mathematics curriculum is maintained by employers' recruitment practices. A good mathematics degree is still regarded as the best mark of a maximally educated brain and one of the best recruitment tools for high status jobs whether or not the graduate will be required to do the mathematics of the university course. 'A' level is a mark of numeracy high enough for an engineering apprenticeship or 'good arithmetic and the rudiments of mathematics beyond arithmetic' for people taken on by building societies and banks and respectable chain stores with management trainee programmes. The arithmetic of the sixteen-year-old school leaver is that demanded by industry in evident and often complacent ignorance of research into the mathematics done at work (see Chapter 11 and the papers in Part 2 of this volume) and with time-honoured reluctance either to train or to modify training methods by consulting people whose business is learning and teaching (Wolf, 1984, p. 11). The relationship between industry and education has been argued hotly over the past century (Reeder, 1981) and the performance of industry on the whole has been neither creditable nor consistent (Barnett, 1986). Current image-polishing as a priority in much industry-education work in the late 80s and early 1990s quickly tarnishes on investigation of training records and interests. Wilkinson's justification (Chapter 9) echoes formal reports[8] and the action of the government in setting up a voucher system in the hopes that it will force firms to train if they are to recruit.

The adoption of the hierarchical form of both mathematics and employment and the continuing use of the former as a filter for the latter maintain echoes of the mores of the nineteenth century, the period of their formalization. Both mathematics and work are heavily gendered, schools a little less so, for the same social order that defined the place of working class children, defined the place of women.[9] The low participation of women in mathematics and science is bewailed at times of national shortage and the reasons for it are the subject of much research and debate. The traditional equation of low participation with low attainment, of low attainment with low ability and low ability with some female defect, has led to a situation in which 'girls and women are said to be lacking, while boys' and men's "mastery" of mathematics, their claim to superior rationality and scientific truth, is unchallenged, as though their "attaining higher eminence" were proof enough' (Walkerdine, 1989, p. 1). Over the past decades, since the 'problem' became an issue and before, there have been different interpretations. The Victorian belief (supported by Darwin) that women do not have the mental capacity, provided justification for girls not doing mathematics at all or doing the minimum arithmetic needed for household budgeting and then switching to needlework while the boys went on with more. More recently the question shifted from asking why girls cannot do mathematics to why girls do not do mathematics[10] and to theories of lack of motivation or fear of success (Willis, p. 3). As Walkerdine (1989) points out, all these approaches operate on a deficit model, on the assumption that there is something wrong with the girls. Walkerdine does not deny that there is a problem but there is no evidence that girls as a whole fail in mathematics. The same evidence about girls' and boys' attainment is interpreted differently however, so that even when girls do attain highly in mathematics there is assumed to be something wrong with their femininity.

After a long history of discrimination, women are still widely discriminated against at work (Chapter 20) and their work at home is still widely assumed to be lacking in intellectual content (Chapter 6). The textiles industry throughout the world is the major employer of women as cheap labour and, as the industry becomes mechanized it is men who take over the technical work (Cockburn, 1985). The stereotypical childhood activity that indicates a potential engineer or mathematician is the boy playing with a constructional toy or mending his bicycle. Yet the activity of dressing a doll, with all the spatial skill it contains including the handling of symmetries of garment construction, and the skills of categorizing in tidying up, and the budgeting skills learned at mother's knee are not regarded as mathematical or, when they are, are decreed as low level. The reasons for this are not mathematical in the sense of the mathematical attainment or potential of individuals, but lie in the history of some of the habits, attitudes and prejudices discussed above. For these reasons, industrial and practical examples throughout this book are taken not from engineering which is usual when discussing issues at the interface of school, work and low-level mathematics, but from the textile industry, the third largest industry in Britain and the highest employer of women in the lowliest jobs. By drawing attention to the unexploited mathematical nature of textiles work both at home and in the factory, it is intended to draw attention to some of the extreme limitations in attitude to both mathematics and work that currrently dominate thinking at the schools-industry interface.

Meanwhile, outside the debates of content and pedagogy, people go about their daily business relatively unaffected by either (Chapter 7), or by academic opinions as to whether or not the mathematics they do in their work is 'proper' mathematics or not (Chapter 11 and Part 2). All peoples of all cultures invent and use mathematics as it is needed, as they have always done (Chapter 4). It may not have always been very sophisticated but as part of the culture that generated it, it was valued and if something is valued then it can be studied because it is valuable. It is ironic that when such mathematics is brought under the 'liberal' tradition of education it becomes devalued as un-intellectual and that when the tradition is taken over by the state, politicized as well (Chapter 8).

There is however an alternative to retaining the debate within the parameter of the current top-down thinking of both mathematics education and industry. The original demand-led mathematics of workplaces is now being investigated by techniques deriving from social anthropology in the field that is broadly known as ethnomathematics. The raw material for the research is the mathematical activity of people in the context of their daily work. The field is relatively new but already it is restoring the dignity of 'low-level' work as a resource for learning far-from-trivial mathematics, one that has implications for the school-work curriculum. Such an approach takes the total culture of the learner as the starting point so that, for example, the use of computers in schools is not seen from the programming and keyboard skills orientation of a skills and applications curriculum, but as part of the culture and learning resources of the child (Chapter 10).

As we suggested in the beginning of this book there is more to mathematics education at the lower end of the curriculum than the learning of the arithmetic requested by a few employers dictated by the limited demands of particular and often short-term interests. A sample of quite how much more forms the detail of its contents.

Notes

1 It is noteworthy that during the space of the year in which this book was edited, all the authors whose papers were concerned with vocationally-oriented curricula and with industry found it necessary to make substantial modifications to their original papers. This fact gives a reasonable estimate of the pressures on schools and training establishments due to the very rapid, modifications and changes in government policies and practices.

2 A very broad distinction is meant here so that contrast between 'pure' and 'applied' also covers that between 'theoretical' and 'practical', 'academic' and 'vocational' and any other contrastive pair that express a similar meaning.

3 An apparent lack of interest may in fact be a lack of access. The popularity of the ITV mathematics programme 'Fun and Games' indicates the interest of up to 9 million people a week.

4 Domesday book lists 5634 water mills (Gimpel, p. 229).

5 Fulling is the process by which newly woven cloth is cleaned and thickened by pounding it in water. The improvement in water mills meant that the laborious business of hand (or foot) fulling could be done by power-driven hammers. But power is control and the cottage industry was transformed by monopolist mill owners insisting on their tenants bringing their cloth to the mill to be fulled. Developments in fulling were as decisive as the mechanism of spinning and weaving were in the eighteenth century (Gimpel, p. 16).

6 Of course the terms 'architect' and 'engineer' have different connotations today. The terms here are to be interpreted as people with both the combined necessary mathematical knowledge and

practical skill to raise buildings which command the respect of the architects and engineers of today.

7 As Cohen (1982) reminds us Roman numerals have survived for this purpose to this day. We write 'Henry VIII' for example. 'Henry 8' looks decidedly odd.

8 The report 'Challenge to Complacency: Changing Attitudes to Training' by Coopers and Lybrand Associates for the Manpower Services Commission and the National Economic Development Office (1985) remarks (page 4) 'few employers think training sufficiently central to their business for it to be a main component in their corporate strategy'.

9 Ethnic minorities have joined working-class women at the bottom of the pile.

10 It is notable how these questions disappear in times of war when women suddenly do have the strength to 'man' factories and the intellect to 'man' research, only to return to their positions of physical and mental incapacity when the war is over and the men want their jobs back.

References

BARNETT, C. (1986) 'Education for Industrial Decline', in *The Audit of War*, London, Macmillan.

BERNAL, M. (1986) *Black Athena: the Afroasiatic Roots of Classical Civilisation*. London, Free Association Books.

BOYER, Carl C. (1985) *A History of Mathematics*, Princeton University Press. Paperback edition John Wiley (First edition 1968).

CHARLTON, K. (1965) *Education in Renaissance England*, London, Routledge and Kegan Paul.

CIPOLLA, C. (1969) *Literacy and Development in the West*, London, Pelican.

COCKBURN, C. (1985) *Machinery of Dominance: Women, Men and Technical Know-How*, London, Pluto Press.

COHEN, P.C. (1982) *A Calculating People. The Spread of Numeracy in Early America*, Chicago, University of Chicago Press.

COOPER, B. (1985) *Renegotiating Secondary School Mathematics*, Lewes, Falmer Press.

GIMPEL, J. (1988) *The Medieval Machine: The Industrial Revolution of the Middle Ages*, England, Wildwood House, Second edition.

HARVEY, J. (1947) *Gothic England. A Survey of National Culture 1300–1550*, London, Batsford.

HOWSON, G. (1982) *A History of Mathematics Education in England*, Cambridge, Cambridge University Press.

LAWSON, J. and SILVER, H. (1973) *A Social History of Education in England*, London, Methuen.

LILLEY, S. (1958) 'Robert Recorde and the Idea of Progress: A Hypothesis and Verification', *Renaissance and Modern Studies*, 2, 3–37.

REEDER, D. (1981) 'A Recurring Debate: Education and Industry' in Dale, D., Esland, G., Fergusson, R. and MacDonald, M. (1981) *Education and the State, Volume 1, Schooling and the National Interest*, Falmer Press in Association with Open University Press.

TAYLOR, E.G.R. (1954) *The Mathematical Practitioners of Tudor and Stuart England*, Oxford, University Press.

WALKERDINE, V. (1989) *Counting Girls Out*, London, Virago in Association with University of London Institute of Education.

WILLIAMS, R. (1961) *The Long Revolution*, Harmondsworth, Penguin Books.

WILLIS, S. (1989) *Real Girls Don't do Maths: Gender and the Construction of Privilege*, Australia, Deakin University.

WOLF, A. (1984) *Practical Mathematics at Work: Learning Through YTS*, Sheffield, Manpower Services Commission.

Part 1
Introduction

The collection of papers in Part 1 is aimed at setting the social and historical scene of what is currently regarded as low level mathematics and its relationship to work; at questioning it and at suggesting a broader, more humane, more culturally and personally oriented approach.

The collection begins with a classic paper from D'Ambrosio in which he proposes the subject of ethnomathematics, one which lies between the development of mathematical ideas and cultural anthropology. Ethnomathematics is based on a broad conceptualization of mathematics which allows it to identify and analyze 'informal' and work practices which are essentially mathematical in nature. The author has up-dated his original paper with some additional remarks about the development of the concept of ethnomathematics which by its nature has relevance to an anthropological analysis of all work practices.

The article by Bos is taken from a longer one that discusses the social and historical context of mathematics. It discusses the interplay between mathematical theory and practice in a particular case and is a reminder that the relationship is a complex one. Bishop's paper is a summary and review of fifteen years work on mathematics as a cultural activity. He develops a pancultural concept of mathematics which challenges formal values of standard western mathematics and discusses implications for teacher education. Joseph too challenges establishment views of the history and development of mathematics, pointing to the danger of the eurocentric approach in a multi-cultural society. Fasheh gives a vivid account of how an acceptance of the hegemony of standard western mathematics was challenged and overthrown when circumstances forced an analysis of the nature and purpose of mathematics education in relationship to the status of the people in receipt of it.

'Folk maths' is a phrase coined by Maier to describe the 'mathematics that folks do' and the relevance of formal mathematics education to it. First published in 1980, Maier's paper still challenges the relevance of much of school mathematics to the world outside school. An examination of school texts can reveal more than simple irrelevance. The cases noted by Maxwell illustrate clearly that the examples through which school arithmetic is practised are not, and probably cannot be politically neutral.

Wilkinson, a mathematician working in an engineering company, presents some views from industry. In certain respects they are traditional views but they express clearly the points of view of industry as 'end-users' of school mathematics. Mathematics education has to be able to justify itself outside mathematics education as well as in it. Wilkinson's view of computers at work is of sophisticated tools. Noss's is a much wider one of computers as part of total culture including that of the mathematics class. Used suitably they can expand what it is possible to do, to learn and to teach in mathematics.

In the final paper Dowling, critically reviews some recent practices and issues in mathematics education and its relationship to work and develops a theoretical model in which they can be further analyzed. His theoretical perspective ends Part 1 and leads the way in to the next one.

2
Ethnomathematics and its Place in the History and Pedagogy of Mathematics

Ubiritan D'Ambrosio

This paper first appeared in *For the Learning of Mathematics* 5, 1 (February 1985). It has been updated for this volume with additions by the author.

Introductory Remarks

In this paper we will discuss some basic issues which may lay the ground for an historical approach to the teaching of mathematics in a novel way. Our project relies primarily on developing the concept of *ethnomathematics*.

Our subject lies on the borderline between the history of mathematics and cultural anthropology. We may conceptualize ethnoscience as the study of scientific and, by extension, technological phenomena in direct relation to their social, economic and cultural backgrounds (D'Ambrosio, 1977). There has been much research already on ethno-astronomy, ethno-botany, ethno-chemistry and so on. Not much has been done in ethnomathematics, perhaps because people believe in the universality of mathematics. This seems harder to sustain as recent research, mainly carried on by anthropologists, shows evidence of practices which are typically mathematical, such as counting, ordering, sorting, measuring and weighing, done in radically different ways to those which are commonly taught in the school system. This has encouraged a few studies on the evolution of the concepts of mathematics in a cultural and anthropological framework. But we consider this direction to have been pursued only to a very limited and — we might say — timid extent. A basic book by Wilder (1981) which takes this approach, and a recent comment on Wilder's approach by Smorinski (1983) seem to be the most important attempts by mathematicians. On the other hand, there is a reasonable amount of literature on this by anthropologists. Making a bridge between anthropologists and historians of culture and mathematicians is an important step towards recognizing that different modes of thought may lead to different forms of mathematics; this is the field which we may call ethnomathematics.

Anton Dimitriu's extensive history of logic (Dimitriu, 1977) briefly describes Indian and Chinese logics merely as background for his general historical study of the logics that originated from Greek thought. We know from other sources, that for example, the concept of the 'number one' is a quite different concept in the Nyaya-Vaisesika epistemology; 'the number one is eternal in eternal substances, whereas two, etc. are always non-eternal' and from this proceeds an arithmetic, (Potter, 1977). Practically nothing is known about the logic underlying the Inca treatment of numbers though what is known through the study of quipus suggests that they used a mixed qualitative-quantitative language (Ascher and Ascher, 1981).

These remarks invite us to look at the history of mathematics in a broader context so as to incorporate in it other possible forms of mathematics. But we will go further than these considerations in saying that this is not a mere academic exercise, since its implications for the pedagogy of mathematics are clear. We refer to recent advances in theories of cognition which show how strongly culture and cognition are related. Although for a long time there have been indications of a close connection between cognitive mechanisms and cultural environment, a reductionist tendency, which goes back to Descartes and has to a certain extent grown in parallel with the development of mathematics, tended to dominate education until recently, implying a culture-free cognition. Recently a holistic recognition of the interpenetration of biology and culture has opened up a fertile ground of research on culture and mathematical cognition (e.g., Lancy, 1983). This has clear implications for mathematics education, as has been amply discussed in D'Ambrosio 1985 and 1986a.

An Historical Overview of Mathematics Education

Let us look briefly into some aspects of mathematics education throughout history. We need some sort of periodization for this overview which corresponds, to a certain extent, to major turns in the socio-cultural composition of western history. (We disregard for this purpose other cultures and civilizations).

Up to the time of Plato, our reference is the beginning and growth of mathematics in two clearly distinct branches: what we might call 'scholarly' mathematics, which was incorporated in the ideal education of Greeks, and another, which we may call 'practical' mathematics, reserved mainly for manual workers. In the Egyptian origins of mathematical practice there was the space reserved for 'practical' mathematics behind it, which was taught to workers. This distinction was carried on into Greek times and Plato clearly says that 'all these studies (ciphering and arithmetic, mensurations, relations of planetary orbits) into their minute details is not for the masses but for the selected few', (Plato's *Laws* VII, 818) and 'we should induce those who are to share the highest functions of State to enter upon that study of calculation and take hold of it . . . not for the purpose of buying and selling, as if they were preparing to be merchants or hucksters', (*Republic* VII, 525b). This distinction between scholarly and practical mathematics, reserved for different social classes, is carried on by the Romans with the *trivium* and *quadrivium* and a separate practical training for labourers. In the Middle Ages we begin to see a convergence of both in one direction that is, practical mathematics begins to use some ideas from scholarly mathematics in the field of geometry. Practical geometry is a subject in its own right in the Middle Ages. This approximation of practical theoretical geometry follows the translation from the Arabic of Euclid's *Elements* by Adelard of Bath (early 12th century). Dominicus Gomdissalinus, in his classification of sciences, says that 'it would be disgraceful for someone to exercise any art and not know what it is, and what subject matter it has, and the other things that are premised of it' as cited in Victor (1979). With respect to ciphering and counting, changes start to take place with the introduction of Arabic numerals; the treatise of Fibonacci (D'Ambrosio, 1980) is probably the first to begin this mixing of the practical and theoretical aspects of arithmetic.

The next step in our periodization is the Renaissance when a new labour structure emerges; changes take place in the domain of architecture since drawing makes plans accessible to bricklayers, and machinery can be drawn and reproduced by others than the inventors. In painting, schools are found to be more efficient and treatises become available. The approximation is felt by scholars who start to use the vernacular for their scholarly works, sometimes writing in non-technical language and in a style accessible to non-scholars. The best known examples may be Galileo, and Newton with this *Optiks*.

The approximation of practical mathematics to scholarly mathematics increases in pace in the industrial era, not only for reasons of necessity in dealing with increasingly complex machinery and instruction manuals, but also for social reasons. Exclusively scholarly training would not suffice for the children of an aristocracy which had to be prepared to keep its social and economical predominance in a new order (D'Ambrosio, 1980a). The approximation of scholarly mathematics to practical mathematics begins to enter the school system, if we may so call education in these ages.

Finally, we reach a last step in this rough periodization in attaining the twentieth century and the widespread concept of mass education. More urgently than for Plato the question of *what* mathematics should be taught in mass educational systems is posed. The answer has been that it should be a mathematics that maintains the economic and social structure, reminiscent of that given to the aristocracy when a good training in mathematics was essential for preparing the élite (as advocated by Plato), and at the same time allows this élite to assume effective management of the productive sector. Mathematics is adapted and given a place as 'scholarly practical' mathematics which we will call, from now on, 'academic mathematics' i.e. the mathematics which is taught and learned in the schools.

In contrast to this we will call *ethnomathematics* the mathematics which is practised among identifiable cultural groups, such as national-tribal societies, labour groups, children of a certain age bracket, professional classes and so on. Its identity depends largely on focuses of interest, on motivation, and on certain codes and jargons which do not belong to the realm of academic mathematics. We may go a step further in this concept of ethnomathematics to include much of the mathematics which is currently practised by engineers, mainly calculus, which does not respond to the concept of rigour and formalism developed in academic courses of calculus. As an example, the Sylvanus Thompson approach to calculus may fit better into this category of ethnomathematics. Builders, well-diggers and shack-raisers in the slums also use examples of ethnomathematics.

Of course this concept asks for a broader interpretation of what mathematics is. Now we include as mathematics, apart from the platonic ciphering and arithmetic, mensuration and relations of planetary orbits, the capabilities of classifying, ordering, inferring and modelling. This is a very broad range of human activities which, throughout history, has been expropriated by the scholarly establishment, formalized and codified and incorporated into what we call academic mathematics but which remains alive in culturally identified groups and constitutes routines in their practices.

Ethnomathematics in History and Pedagogy and the Relations between them

We should like to insist on the broad conceptualization of mathematics which allows us to identify several practices which are essentially mathematical in their nature. And we also presuppose a broad concept of *ethno*, to include all culturally identifiable groups with their jargons, codes, symbols, myths and even specific ways of reasoning and inferring. Of course, this comes from a concept of culture as the result of an hierarchization of behaviour, from individual behaviour through social behaviour to cultural behaviour.

The concept relies on a model of individual behaviour based on the cycle: reality → individual → action → reality, schematically shown in Figure 2.1. In this holistic model we will not enter into a discussion of what is reality or what is an individual or what is action. We refer to D'Ambrosio, 1980b. We simply assume reality in a broad sense, natural, material, social and psycho-emotional. Now, we observe that links are possible through the mechanism of information (which includes both sensory and memory, genetic and acquired) which produces stimuli in the individual. Through a mechanism of reification these. stimuli give rise to strategies (based on codes and

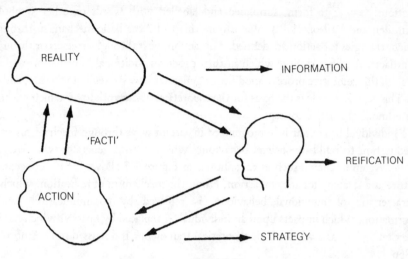

Figure 2.1.

models) which allow for action. Action impacts upon reality by introducing *facti* into this reality, both artefacts and *mentifacts*. (We have introduced this neologism to mean all the results of intellectual action which do not materialize, such as ideas, concepts, theories, reflections and thoughts.) These are added to reality, in the broad sense mentioned above, and clearly modify it. The concept of reification has been used by sociobiologists as 'the mental activity in which hazily perceived and relatively intangible phenomena, such as complex arrays of objects or activities, are given a

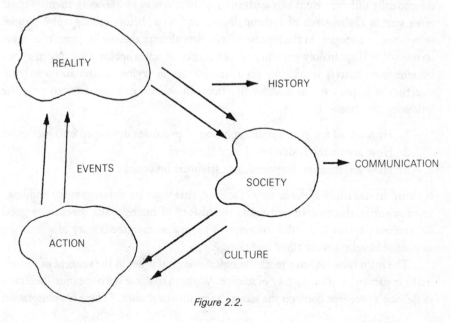

Figure 2.2.

factitiously concrete form, simplified and labelled with words or other symbols' (Lumsden and Wilson, 1981). We assume this to be the basic mechanism through which strategies for action are defined. This action, be it through artefacts or through mentifacts, modifies reality, which in turn produces additional information which, through this reificative process, modifies or generates new strategies for action, and so on. This ceaseless cycle is the basis for the theoretical framework upon which we base our ethnomathematics concept.

Individual behaviour is homogenized in certain ways through mechanisms such as education to build up societal behaviour, which in turn generates what we call *culture*. Again a scheme such as that shown in Figure 2.2 allows for the concept of culture as a strategy for societal action. Now, the mechanism of reification, which is characteristic of individual behaviour, is replaced by communication, while information, which impacts upon an individual, is replaced by history, which has its effect on society as a whole. This theoretical framework is discussed in D'Ambrosio 1986b.

As we have mentioned above, culture manifests itself through jargons, codes, myths, symbols, utopias, and ways of reasoning and inferring. Associated with these we have practices such as ciphering and counting, measuring, classifying, ordering, inferring, modelling and so on, which constitute ethnomathematics.

The major question we are then posed is the following: how *theoretical* can ethnomathematics be? It has long been recognized that mathematical practices, such as those mentioned at the end of the previous paragraph, are known to several culturally differentiated groups; and when we say *known* we mean in a way which is substantially different from the western or academic way of knowing them. This is often seen in the research of anthropologists and, even before ethnography became recognized as a science, in the reports of travellers all over the world. Interest in these accounts has been mainly curiosity or the source of anthropological concern about learning how natives think. We go a step further in trying to find an underlying structure of inquiry in these *ad hoc* practices. In other terms, we have to pose the following questions:

1. How are *ad hoc* practices and solution of problems developed into methods?
2. How are methods developed into theories?
3. How are theories developed into scientific invention?

It seems, from a study of the history of science, that these are the steps in the building-up of scientific theories. In particular, the history of mathematics gives quite good illustrations of steps 1, 2 and 3 and research programmes in the history of science are in essence based on these three questions.

The main issue is then a methodological one, and it lies in the concept of history itself, in particular in the history of science. We have to agree with the initial sentence in Bellone's excellent book on the second scientific revolution: 'There is a temptation hidden in the pages of the history of science — the temptation to derive the birth and death of theories, the formalization and growth of concepts, from a scheme (either logical or philosophical) always valid and everywhere applicable . . . Instead of dealing with real problems, history would then become a learned review of edifying tales for the benefit of one philosophical school or another' (Bellone, 1980). This tendency permeates the analysis of popular practices such as ethnoscience, and in particular ethnomathematics, depriving it of any history. As a consequence, it deprives it of the status of knowledge.

It is appropriate at this moment to make a few remarks about the nature of science nowadays, regarded as a large scale professional activity. As we have already mentioned, it developed into this position only since the early nineteenth century. Although scientists communicated among themselves and scientific periodicals, meetings and associations were known, the activity of scientists in earlier centuries did not receive any reward as such. What reward there was came more as the result of patronage. Universities were little concerned with preparing scientists or training individuals for scientific work. Only in the nineteenth century did becoming a scientist start to be regarded as a professional activity. And out of this change, the differentiation of science into scientific fields became almost unavoidable. The training of a scientist, now a professional with specific qualifications, was done in his subject, in universitites or similar institutions and mechanisms to qualify him for professional activity were developed. Standards of evaluation of his credentials were developed. Knowledge, particularly scientific knowledge, was granted a status which allowed it to bestow upon individuals the required credentials for their professional activity. This same knowledge, practised in many strata of society at different levels of sophistication and depth was expropriated by those who had the responsibility and power to provide professional accreditation.

We may look for examples in mathematics of the parallel development of the scientific discipline outside the established and accepted model of the profession. One such example is Dirac's delta function which, about twenty years after being in full use among physicists, was expropriated and became a mathematical object, structured by the theory of distributions. This process is an aspect of the internal dynamics of knowledge *vis-à-vis* society.

There is unquestionably a time lag between the appearance of new ideas in mathematics outside the circle of its practitioners and the recognition of these ideas as 'theorizable' into mathematics, endowed with the appropriate codes of the discipline, until the expropriation of the idea and its formalization as mathematics. During this period of time the idea is put to use and practised: it is an example of what we call ethnomathematics in its broad sense. Eventually it may become mathematics in the style or mode of thought recognized as such. In many cases it never gets formalized and the practice continues, restricted to the culturally differentiated group which originated it. The mechanism of schooling replaces these practices by other equivalent practices which have acquired the status of mathematics, which have been expropriated in the original forms and returned in a codified version.

We claim a status for these practices, ethnomathematics, which do not reach the level of mathematization in the usual, traditional sense. Paraphrasing the terminology of T. S. Kuhn, we say they are not 'normal mathematics' and it is very unlikely they will generate 'revolutionary mathematics'. Ethnomathematics keeps its own life, evolving as a result of societal change, but the new forms simply replace the former ones, which go into oblivion. The cumulative character of this form of knowledge cannot be recognized and its status as a scientific discipline becomes questionable. The internal revolutions in ethnomathematics, which result from societal changes as a whole, are not sufficiently linked to 'normal mathematics'. The chain of historical development, which is the spine of a body of knowledge structured as a discipline, is not recognizable. Consequently ethnomathematics is not recognized as a structured body of knowledge, but rather as a set of *ad hoc* practices.

Conclusion

For effective educational action not only an intense experience in curriculum development is required, but also investigative and research methods that can absorb and understand ethnomathematics. And this clearly requires the development of quite difficult anthropological research methods relating to mathematics, a field of study as yet poorly cultivated. Together with the social history of mathematics, which aims at understanding the mutual influence of socio-cultural, economic and political factors in the development of mathematics, anthropological mathematics, if we may coin a name for this speciality, is a topic which we believe constitutes an essential research theme in the third world countries, not as a mere academic exercise, as it now draws interest in the developed countries, but as the underlying ground on which we can develop curricula in a relevant way.

Curriculum development in third world countries requires a more global, clearly holistic approach, not only by considering methods, objectives and contents in solidarity, but mainly by incorporating the results of anthropological findings into the three-dimensional space which we have used to characterize curriculum. This is quite different to what has frequently and mistakenly been done, which is to incorporate these findings individually in each coordinate or component of curriculum.

This approach has many implications for research priorities in mathematics education for third world countries and has an obvious counterpart in the development of mathematics as a science. Clearly the distinction between pure and applied mathematics has to be interpreted in a different way. What has been labelled Pure Mathematics, and continues to be called such, is the natural result of the evolution of the discipline within a social, economic and cultural atmosphere which cannot be disengaged from the main expectations of a certain historical moment. It cannot be disregarded that L. Kronecker ('God created the integers — the rest is the work of men'), Karl Marx and Charles Darwin were contemporaries. Pure Mathematics as opposed to Mathematics, came into consideration at about the same time, with obvious political and philosophical undertones. For third world countries this distinction is highly artificial and ideologically dangerous. Clearly, to revise curriculum and research priorities in such a way as to incorporate national development priorities into the scholarly practices which characterize university research is a most difficult thing to do. But all the difficulties should not disguise the increasing necessity of pooling human resources for the more urgent and immediate goals of our countries. This poses a practical problem for the development of mathematics and science in third world countries. The problem leads naturally to a close for the theme of this paper, that is the relation between science and ideology.

Ideology, implicit in dress, housing, titles, so superbly denounced by Aimé Cesaire in *La Tragédie du Roi Christophe*, takes a more damaging turn, with even longer and more disrupting effects, when built into the formation of the cadres and intellectual classes of former colonies, which constitute the majority of so-called third world countries. We should not forget that colonialism grew together in symbiotic relationship with modern science, in particular with mathematics and technology.

Additional Remarks

Since this paper was originally published, several new aspects of ethnomathematics have been developed. We have ample evidence of the fact that when mathematics programmes in schools all over the world have been associated with European thought, it has been a hindrance to the learning of mathematics by children and adults from diverse cultural backgrounds. Our efforts in education have always had, as a main focus, the intention of raising the level of cultural consciousness and developing self-esteem through the unbiased promotion of the use of diversified modes of coping with, managing and explaining reality. This has lead to an ambitious research programme in the generation, transmission and diffusion of knowledge with obvious pedagogical implications and also clear interfaces with religion, art and societal and political behaviour in its most general sense. Of course, this adds up to a historiography and philosophy of science. An etymological exercise speaks in favour of calling this very broad programme ethnomathematics. We are essentially aiming at the study of the art or technique (*techné* = *tics*) of explaining, understanding, coping with and managing reality (*mathema*) in different cultural environments (*ethno*). One of these *tics* of *mathema*, associated with the expansion of mediterranean culture through Europe and later through the entire globe is called mathematics — as it is universally accepted in educated circles. Several other of these *tics* of *mathema* are found in the most diverse cultural environments or *ethnos*. Of course, according to the complexity of their entire cultural history, every group (*ethnos*) explains and copes with their very diverse realities in the most diverse ways, through sophisticated decorative and musical schemes, through divinatory and ritual practices, or through ways of measuring, sorting and ciphering. These and many other ways all come from the pursuit of explanation, of understanding, of coping with mysteries and difficulties posed by life, of finding one's way in everyday life, summing up, or surviving and transcending one's existence. These things are inherent in our species. Diverse modes of surviving and transcending lead to different modes of thought. Thus, it is preposterously eurocentric to try to identify mathematics or zoology or other disciplines as compartmentalized pieces of knowledge in different cultures, just as it is preposterously adult-centric to impose on young children these structured modes of explanation and of dealing with situations. And it is preposterously scholarly to expect that adults submitting themselves to a late education will easily modify their modes of thought to accept disciplinary approaches to explaining and dealing with situations. Each individual uses the ethnomathematics of his group to provide explanations and to cope with his environment. Eventually, he may command several other ethnomathematics.

References

ASCHER, M. and ASCHER, R. (1981) *Code of the Quipu*. Ann Arbor, University of Michigan Press.

BELLONE, E. (1980) *A World on Paper* Cambridge, Mass., The MIT Press, (original edition 1976)

D'AMBROSIO, U. (1977) 'Science and Technology in Latin America during Discovery', *Impact of Science on Society*, 27, 3, pp. 267–274.

D'AMBROSIO, U. (1980b) 'Uniting Reality and Action: a Holistic approach to Mathematics Education', in Steen, L. A. and Albers, D. J. (eds.) *Teaching Teachers, Teaching Students*, Boston, Birkhäuser.

D'AMBROSIO, U. (1980a) 'Mathematics and Society: Some historical considerations and pedagogical implications'. *Int. J. Math. Educ. Sci. Technol*, II, pp. 478–488.

D'AMBROSIO, U. (1985) *Socio-Cultural Bases for Mathematics Education*. Campinas, Brazil, UNICAMP.

D'AMBROSIO, U. (1986a) 'Cultural, Cognition and Science Learning Science' in Gallagher, J. and Dawson, George. *Education and Cultural Environments in the Americas*. Proceedings of an Interamerican Seminar on Science Education, Panama, Dec. 10–14 1984, Washington DC. NSTA/NSF/OAS.

D'AMBROSIO (1986b) *Da Realidade ä Ação: Reflexões sobre Educação (e) Matemätica* (From Reality to Action: Reflections on Education (and) Mathematics), Sao Paulo, Summus Editorial.

DIMITRIU, A. (1977) *History of Logic*, 4 vols., Kent, Abacus Press.

LANCY, D. F. (1983) *Cross-cultural Studies in Cognition and Mathematics*, New York, Academic Press.

LUMSDEN, C. J. and WILSON, E. G. (1981) *Genes, Mind and Culture*, Cambridge, Mass., Harvard University Press.

POTTER, K. H. (1977) *Indian Metaphysics and Epistemology*, Princeton, N.J., Princeton University Press.

SMORINSKI, C. (1983) 'Mathematics as a Cultural System', *Mathematical Intelligence*, 5, 1, p. 9–15.

THOMPSON, S. (1986) (many editions since 1910) *Calculus Made Easy*, London, Macmillan.

VICTOR, S. K. (1979) *Practical Geometry in the High Middle Ages*, Philadelphia, The American Philosophical Society.

WILDER, R. L. (1981) *Mathematics as a Cultural System*, Oxford, Pergamon.

3
Mathematics and Ballistics

H. J. M. Bos

This extract is from the paper 'Mathematics and its social context: a dialogue in the staff room, with historical episodes' by H. J. M. Bos published in the journal *For the Learning of Mathematics* 4, 3 (November, 1984). The article is the text of a lecture to a conference on 'The Use of Historical Topics in the Teaching of Mathematics' held at Vingsted, (Denmark) January, 1983.

The first studies on the paths of projectiles fired from guns date from the sixteenth century but one can safely say that ballistics (or rather exterior ballistics, i.e. what happens to the projectile outside the gun) became a theory only after Galileo's discovery (published 1638) that projectiles describe parabolas. The theory was worked out; it provided tables and instruments by which the range of the projectile could be calculated from the angle of elevation and the range of one test shot. The theory became a teachable subject called *parabolic ballistics*; it was taught in military schools in the seventeenth and eighteenth centuries. Throughout that time it was of virtually no use to the practice of artillery and all artillery men knew this. The reasons were obvious; the theory took no account of air resistance and assumed that in a series of shots from the same gun the initial velocity would be constant. Neither assumption held in practice.

By the seventeenth century scientists were already turning their attention to the influence of air resistance on the path of projectiles. Huygens, Newton and others derived differential equations for the trajectories on the basis of various suppositions about the relation of the resistance to the velocity of the projectiles. The most fruitful supposition (in that it led to differential equations that were solvable and interesting) was to assume the resistance to be proportional to the square of the velocity. Thus a new theory called quadratic ballistics was developed; it became a teachable subject at the end of the eighteenth century. This theory also provided tables. It was taught to military engineers and artillery technicians. It still was of little or no practical use because the guns, the projectiles and the powder that were used were not standardized in such a way that the initial velocity of the projectile could be considered constant in a series of shots. By this time, the second half of the eighteenth century, the top mathematicians had lost interest in the subject. The theory was later worked out mainly within the army by the artillery engineers themselves.

During the nineteenth century, ballistics theory gradually became useful for artillery in the field. First of all, bullet-shaped projectiles were introduced that were fired from guns with spiral grooves in their barrels. The motion of such projectiles is much more stable and therefore more predictable. Moreover, the projectiles and guns were standardized so that the initial velocity of the projectile was reasonably constant. It is noteworthy that this process of standardization of the materials (one might call it mathematization of the material) was as necessary for the success of the theory as the improvement of the theory itself. Experimental research on air resistance led to a better understanding of resistance as a function of velocity. Thus more accurate differential equations could be set up and after some re-adjustment these turned out to be integrable. Thus tables could be calculated and by the end of the century these proved to be fairly reliable, at least for projectiles with rather flat trajectories.

World War I (1914–1918) made new demands on ballistic theory; there were the long-distance guns (with highly curved trajectories) and there was ground-to-air artillery. In both cases the density of the air along the trajectory could no longer be considered constant and this led to complications in the differential equations, which meant that they were no longer directly solvable. Now the compilers of tables had to resort to numerical integration. By then this was a well-known method for dealing with otherwise intractable differential equations; it had been developed in astronomy. The disadvantage of the method was that it was very time-consuming. Still, it was the only possible way and so, in the period between the world wars, ballistic tables were produced in what can be called ballistic computational laboratories, where groups of calculators, with the help of simple adding and multiplication machines, produced the entries in the ballistic tables. The need for the automation of this calculating process was keenly felt and this led to the development of the electronic computer. Indeed the two most important motivating forces behind the construction of the first computer in the USA during World War II were the need to automate ballistic computation and the need to perform scientific calculations in connection with the development of the atomic bomb.

So much for the ballistics episode, which is a clear example of the way in which mathematics is inter-related with (at least one part of) society. But in which direction does the influence operate? In the early stages one can speak of inspiration; mathematicians were inspired by the practice of artillery to develop theories of ballistics. But there was hardly any influence in the other direction; the theories were of no use in practice. This was, indeed a very common feature of mathematical physics in the seventeenth and eighteenth centuries; however brilliant, deep and theoretically fruitful the theories (of mechanics, hydro-mechanics, electricity, heat, etc.) were, they could hardly be applied in practice. The situation is well illustrated in a quotation from a pamphlet published by Arbuthnot in 1701:

> The great objection that is made against the necessity of mathematics in the . . . great affairs of navigation, the military art, etc. is that we see those affairs carry'd out and managed by those who are not great mathematicians; as seamen, engineers, surveyors, gaugers, clock-makers, glass-grinders etc., and that the mathematicians are commonly speculative, retir'd, studious men, that are not for an active life and business, but content themselves to sit in their studies and pore over a scheme of calculations.   (Taylor 1954)

The non-effectiveness of the early theory in practice is one of the main obstacles to accepting Hessen's picture of the development of science[1]; it seems difficult to rely exclusively on economic explanations in a case where science could develop for such a long time without evident practical results.

It is noteworthy that in the case of ballistics, the theory became effective only after a period of rather strong antithesis between theory and practice. The synthesis which brough success came about in the nineteenth century when theorists were working within artillery and when guns and ammunition were standardized and thus adjusted to the requirements of theory.

The episode may serve as a reminder that inter-relations between theory and practice, between pure mathematics and applications, and between theoretical developments and social utility, can be very complex and one-sided.

Notes

1 The author refers here to a previous passage in the same article in which he mentions the argument of B. Hessen 'in a famous article "The Social and Economic Roots of Newton's Principia" that science, and hence also mathematics, is strongly determined by economic factors. Ever since it was presented, as a lecture by one of the Soviet delegates to the second international congress on the history of science and technology in London in 1931, the article has been a constant source of inspiration, and a constant cause of dissent and debate'.

References

HESSEN, B. (1931) 'The Social and Economic Roots of Newton's Principia', in Wersky P. G. (ed) 1971. *Science at the Crossroads*, London, Cass.

TAYLOR, E. G. R. (1954) *The Mathematical Practitioners of Tudor and Stuart England*, Cambridge, University Press.

4
Mathematics Education in its Cultural Context

Alan J. Bishop

An earlier version of this paper appeared in a Special Issue of *Educational Studies in Mathematics*. The issue was Vol. 19 No. 2, and was called *Mathematics, Education and Culture*. It was published in May 1988 and was compiled to coincide with a unique event, a whole day's meeting on 'Mathematics, Education and Society' held during the Sixth International Congress on Mathematical Education at Budapest, Hungary. The issue of 'culture' was high on the agenda of that day and the whole proceedings have appeared in a publication edited by Keitel *et al.* (1989).

Introduction

In this article I shall summarize the results of the analyses and investigations which have engaged me over the past fifteen years. There have been two major and related areas of concern in that time, and both seem to have important implications for research, for theory development and for classroom practice.

Cultural Interfaces in Mathematics Education

The first concern is with what I think of as 'cultural interfaces'. In some countries like the UK, pressure has mounted to reflect in the school curriculum the multicultural nature of their societies, and there has been widespread recognition of the need to re-evaluate the total school experience in the face of the education difficulties of many children from ethnic minority communities. In other countries like Papua New Guinea, Mozambique and Iran, there is criticism of the 'colonial' or 'Western' educational experience, and a desire to create instead an education which is in tune with the 'home' culture of that society. The same concern emerges in other debates about the formal education of Aborigines, Amerindians, of the Lapps and of Eskimos. In all of these cases, a culture-conflict situation is recognized and curricula are being examined.

One particular version of this problem relates to the mathematics curriculum and its relationship with the home culture of the child. Mathematics curricula though, have been slow to change, due primarily to a popular and widespread misconception. Up to five or so years ago, the conventional wisdom was that mathematics was 'culture-free' knowledge. After all, the argument went, 'a negative times a negative equals a positive' wherever you are, and triangles the world over have angles which add up to 180 degrees. This view though, confuses the 'universality of truth' of mathematical ideas with the cultural basis of that knowledge. The mathematical ideas are decontextualized and abstracted in such a way that 'obviously' they can apply everywhere. In that sense they are clearly universal. But as soon as one begins to focus on the particulars of these statements, one's belief in the universality becomes challenged. Why is it 180 degrees and not, say, 100 or 150? Where does the idea of negative number come from? Authoritative writers on mathematical history have given answers to these kinds of questions of course, and they demonstrate quite clearly that mathematics has a cultural history.

But whose cultural history are we referring to? Recently, research evidence from anthropological and cross-cultural studies has emerged which not only supports the idea that mathematics has a cultural history, but also that from different cultural histories have come what can only be described as different mathematics. One can cite the work of Zaslavsky (1973), who has shown in her book *Africa Counts*, the range of mathematical ideas existing in indigenous African cultures. Van Sertima's *Blacks in Science* (1986), is another African source as is Gerdes (1985). On other continents, the research of Lancy (1983), Lean (1986) and Bishop (1979) in Papua New Guinea, Harris (1980) and Lewis (1976) in Aboriginal Australia, and Pinxten (1983) and Closs (1986) with the Amerindians, has also added fuel to this debate. The term 'ethnomathematics' has been revived (d'Ambrosio, 1985) to describe some of these ideas, and even if the term itself is still not well defined, there is no doubting the sentiment that the ideas are indeed *mathematical* ideas. The thesis is therefore that mathematics must now be understood as a kind of cultural knowledge, which all cultures generate but which need not necessarily 'look' the same from one cultural group to another. Just as all human cultures generate language, religious beliefs, rituals, food-producing techniques etc., so it seems do all human cultures generate mathematics. Mathematics is a pan-human phenomenon. Moreover, just as each cultural group generates its own language, religious belief etc., so it seems that each cultural group is capable of generating its own mathematics. Clearly this kind of thinking will necessitate some fundamental re-examination of many of our traditional beliefs about the theory and practice of mathematics education, and I will outline some of these issues below.

Values in Mathematics Education

The second area of concern to me is our ignorance about 'values' in mathematics education. In the same way that mathematics has been considered for many years to be culture-free, so it has also been considered to be value-free. How could it be concerned with values, the argument goes, when it is about indisputable facts concerning triangles, fractions or multiplication? Once again anthropologically-oriented researchers like Pinxten (1983), Horton (1971), Lewis (1976) and Lean (1973) have presented us with plenty of evidence with which to challenge that traditional view. Moreover any mathematics educators who work in cultural-interface situations soon become acutely aware of the influence of value-conflicts on the mathematical learning experience of the children for whom they are responsible.[1] Furthermore one can argue that a mathematical education is no education at all if it does not have anything to contribute to values development. Perhaps that is a crucial difference between a mathematical training and a mathematical education?

Indeed it would seem to me to be thoroughly appropriate to conceptualize much current mathematics teaching as merely mathematical training, in that generally there is no explicit attention paid to values. I am not saying that values are not learned — clearly they are — but implicitly, covertly and without much awareness of conscious choice. Surely a mathematical *education*, on the other hand, should make the values explicit and overt, in order to develop the learner's awareness and capacity for choosing?

There is a pressing need today to consider values because of the increasing presence of the computer and the calculator in our societies. These devices can perform many mathematical techniques for us even now and the arguments in favour of a purely mathematical training for our future citizens are surely weakened. Society will only be able to harness the mathematical power of these devices for appropriate use if its citizens have been made to consider values as part of their education. For some pessimists however, like Ellul (1980) the situation is far too out of control in any case for education to be able to do anything constructive at this stage. Nevertheless the ideas of other analysts such as Skovsmose (1985) do offer, in my view, the potential for developing strategies for change. My own perspective on this area of values has been stimulated by the culture-conflict research mentioned earlier and it is this perspective which I propose to enlarge on here. The fundamental task for my work was to find a rich way to conceptualize mathematics as a cultural phenomenon.

Mathematics as a Cultural Phenomenon

The most productive starting point was provided by White (1959) in his book *The Evolution of Culture* in which he argues, as others have done, that 'the functions of culture are to relate man to his environment on the one hand, and to relate man to man, on the other'. (p. 8). White though, went further and divided the components of culture into four categories:

- ideological: composed of beliefs, dependent on symbols, philosophies;
- sociological: the customs, institutions, rules and patterns of interpersonal behaviour;
- sentimental: attitudes, feelings concerning people, behaviour;
- technological: manufacture and use of tools and implements

Moreover whilst showing that these four components are inter-related White argues strongly that 'the technological factor is the basic one; all others are dependent upon it. Furthermore, the technological factor determines, in a general way at least, the form and content of the social, philosophic and sentimental factors'. (p. 19).

Writers such as Bruner (1964) and Vygotsky (1978) have also shown us the significance of written language, and one of its particular conceptual 'tools', mathematical symbolism. Mathematics, as an example of a cultural phenomenon, has an important 'technological' component, to use White's terminology. But White's schema also offers an opportunity to explore the ideology, sentiment and sociology driven by this symbolic technology, and therefore to attend to values as well. Mathematics in this context is conceived of as a cultural *product*, which has developed as a result of various activities. These I have described in other writings (Bishop 1986; Bishop, 1988) so I will just briefly summarise them here. There are, from my analyses, six fundamental activities which I argue are both universal, in that they appear to be carried out by every cultural group ever studied, and also necessary and sufficient for the development of mathematical knowledge.

They are as follows:

Counting: the use of a systematic way to compare and order discrete objects. It may involve body or finger counting, tallying or using objects, or string to record, or special number names. Calculation can also be done with the numbers, with magical and predictive properties associated with some of them. (See for example Lean 1986; Menninger 1969; Ascher and Ascher 1981; Closs 1986; Ronan 1981; Zaslavsky 1973.)

Locating: exploring one's spatial environment and conceptualizing and symbolizing that environment, with models, maps, drawing and other devices. This is the aspect of geometry where orientation, navigation, astronomy and geography play a strong role. (See for example Pinxten 1983; Lewis 1976; Harris 1980; Ronan 1986.)

Measuring: quantifying qualities like length and weight for the purposes of comparing and ordering objects. Usually measuring is used where phenomena cannot be counted (e.g. water, rice) but money is also a unit of measure of economic growth. (See for example Menninger 1969; Gay and Cole 1967; Jones 1974; Harris 1980; Zaslavsky 1973).

Designing: creating a shape or a design for an object or for any part of one's spatial environment. It may involve making the object, as a copyable 'template', or drawing it in some conventional way. The object can be designed for technological or spiritual use and 'shape' is a fundamental geometrical concept. (See for example Gerdes 1986; Temple 1986; Ronan 1981; Bourgoin 1973; Faegre 1979; Ostwald 1976).

Playing: devising and engaging in games and pastimes with more or less formalized rules that all players must abide by. Games frequently model a significant aspect of social reality, and often involve hypothetical reasoning. (See for example Huizinga 1949; Lancy 1983; Jayne 1974; Roth 1902; Falkener 1961; Zaslavsky 1973).

Explaining: finding ways to represent the relationships between phenomena. In particular, exploring the 'patterns' of number, location, measure and design, which create an 'inner world' or mathematical relationships which model and thereby explain the outer world of reality. (See for example Lancy 1983; Horton 1971; Pinxten 1983; Gay and Cole 1967).

Mathematics as cultural knowledge, derives from humans engaging in these six universal activities in a sustained and conscious manner. The activities can either be performed in a mutually exclusive way or perhaps, more significantly, by interacting together as in 'playing with numbers' which is likely to have developed number patterns and magic squares, and which arguably contributed to the development of algebra. I would argue that, in the Western mathematics which I and many other have learnt, these activities have contributed at least the following highly significant ideas.

Counting: Numbers. Number patterns. Number relationships. Developments of number systems. Algebraic representation. Infinitely large and small. Events, probabilities, frequencies. Numerical methods. Iteration. Combinatorics. Limits.

Locating. Position. Orientation. Development of coordinates — rectangular, polar, spherical. Latitude/longitude. Bearings. Angles. Lines. Networks. Journey. Change of position. Loci (circle, ellipse, polygon . . .). Change of orientation. Rotation. Reflection.

Measuring. Comparing. Ordering. Length. Area. Volume. Time. Temperature. Weight. Development of units — conventional, standard, metric system. Measuring instruments. Estimation. Approximation. Error.

Designing. Properties of objects. Shape. Pattern. Design. Geometric shapes (figure and solids). Properties of shapes. Similarity. Congruence. Ratios (internal and external).

Playing. Puzzles. Paradoxes. Models. Games. Rules. Procedures. Strategies. Prediction. Guessing. Chance. Hypothetical reasoning. Games analysis.

Explaining. Classifications. Conventions. Generalizations. Linguistic explanation — arguments, logical connections, proof. Symbolic explanations — equations, formulae, algorithms, functions. Figural explanations — diagrams, graphs, charts, matrices. (Mathematical structure — axioms, theorems, analysis, consistency.) (Mathematical models — assumptions, analogies, generalizability, predictability).

From these basic notions, the rest of 'Western' mathematical knowledge can be derived[2], while in this structure can also be located the evidence of the 'other mathematics' developed by other cultures. Indeed we ought to re-examine labels such as 'Western Mathematics' since we know that many different cultures contributed to the knowledge encapsulated by that particular label.[3]

However, I must now admit to what might be seen as a conceptual weakness. There is no real prospect of *my* being able to test whether or not this 'universal' structure will be adequate for describing the mathematical ideas of other cultural groups. On the contrary, I would maintain that it must be for others from these cultural groups to determine this. Far from my inability being interpreted as a weakness, I believe it is important to recognize that in this kind of analysis one must be constantly aware of the dangers of culturo-centrism. It may well be the case that my analysis will not hold up under cross-cultural scrutiny — it is my hope that it may in fact stimulate some other analytic developments which again could be tested crossculturally.

This kind of culturo-centrism is well explained by Lancy (1983) who has proposed a 'universal' stage theory of cognitive development. Lancy shows that his Stage 1 corresponds to Piaget's sensory-motor and pre-operational stages 'the accomplishments of this stage are shared by all human beings' (p. 203). Stage 2 is where the enculturation begins: 'What happens to cognition during Stage 2 then, has much to do with culture and environment and less to do with genesis' (p. 205). This, for me, is the stage where different cultures develop different mathematics. However Lancy also has a Stage 3 in his theory, which concerns the metacognitive level: 'In addition to developing cognitive and linguistic strategies, individuals acquire "theories" of language and cognition' (p. 208). For Lancy, therefore, the 'formal operation' stage of Piaget's theory represents the particular theory of knowledge which the 'Western' cultural group emphasizes. Other cultural groups can, and do, emphasize other theories of knowledge.

This idea gives a useful cultural entree into the area of values, linking as it does with White's idea that the technology of a culture (in our case the symbolic technology of mathematics) not only relates humans to their environment in a particular way, but also 'drives' the other cultural components — the sentimental, the ideological and the sociological. It is these that are the heart of the values associated with mathematics as a cultural phenomenon. Before turning to examine these in more detail it is necessary once again to point out that my own cultural predisposition makes it very difficult to attempt any more at this stage than merely outlining the values which I feel are associated with the 'Western mathematics' with which I am familiar. I do know that enough evidence exists to suggest that White's schema does have some credibility in 'Western' culture. I am in no position however to argue that for any other culture. Once again that verification must be left to those in other cultural groups.

The three value components of culture — White's sentimental, ideological and sociological components — appear to me to have pairs of complementary values associated with mathematics, which give rise to certain balances and tensions. If we consider first the 'sentimental' component we can see that so much of the power of mathematics in our society comes from the feelings of security and *control* that it offers. Mathematics, through science and technology, has given Western culture the sense of security in knowledge — so much so that people can become very frustrated at natural or human-made disasters which they feel should not have happened. The inconsistency of a mathematical argument is a strong motive for uncovering the error and getting the answer 'right'. The mathematical valuing of 'right' answers informs society which also looks (in vain of course) for right answers to its societal problems. Western culture is fast becoming a mathematico-technological culture.

Where control and security are sentiments about things remaining predictable, the complementary value relates to *progress*. A method of solution for one mathematical problem is able, by the abstract nature of mathematics, to be generalized to other problems. The unknown can become known. Knowledge can develop. Progress, though, can become its own reward and change becomes inevitable. Alternativism is strongly upheld in Western culture and as with all the values described here contains within itself the seeds of destruction. It is therefore important to recognize that it is the interactions and tensions between those values of control and progress which allow cultures to survive and to grow.

If those are the twin sentiments driven by the mathematical symbolic technology then the principle *ideology* associated with Western mathematics must be *rationalism*. If one were searching for only one identifiable value, it would be this one. It is logic, rationalism, and reason which has guaranteed the significance of mathematics within Western culture. It is not tradition, not status, not experience, not seniority, but logic which offers the main criterion of mathematical knowledge. With the advent of computers this ideology is extended even further, if that is possible.

The Indo-European languages appear to have rich vocabularies for logic —

Gardner (1977) in his (English) tests used over 800 different logical connectives. The rise of physical technology has also helped this development, in that 'causation', one of the roots of rational argument, seems to be developed much more easily through physical technology than through nature — the time-scales of natural processes are often too fast or too slow. It was simple physical technological devices which enabled humans to experiment with process, and to develop the formidable concept of 'direct causation'. However there is also a complementary ideology which is clearly identifiable in Western culture, and that is *objectivism*. Western culture's world-view appears to be dominated by materials, objects and physical technology. Where rationalism is concerned with the relationship between ideas, objectivism is about the genesis of those ideas. One of the ways Western mathematics has gained its power is through the activity of objectivizing the abstractions from reality. Through its symbols (letters, numerals, figures) mathematics has enabled people to deal with abstract entities, *as if they were* objects.

The final two complementary values concern White's *sociological* component, the relationships between people and mathematical knowledge. The first I call *openness* and concerns the fact that mathematical truths are open to examination by all, provided of course that they have the necessary knowledge to do the examining. Proof grew from the desire for articulation and demonstration, so well practised by the early Greeks, and although the criteria for the acceptability of proof have changed, the value of opening the knowledge has remained as strong as ever. And there is a complementary sociological value which I call *mystery*. Despite that openness, there is a mysterious quality about mathematical ideas. Certainly everyone who has learned mathematics knows this intuitively, whether it is through the meaningless symbol-pushing which many children still unfortunately experience, or whether it is in the surprising discovery of an unexpected connection. The basis of the mystery again lies in the abstract nature of mathematics — abstraction can literally be meaningless. Of course mathematical ideas offer their own kind of context so it is very possible to develop meaning within mathematics itself.

These then are the three pairs of values relating to Western mathematics which are shaped by, and also have helped to shape, a particular set of symbolic conceptual structures. Together with those structures they constitute the cultural phenomenon which is often labelled as 'Western mathematics'. We certainly know that different symbolizations have been developed in different cultures and it is very likely that there are differences in values also, although detailed evidence on this is not readily available at the present time. How unique these values are, or how separable a technology is from its values must also remain open questions.

Some Issues Arising from this Analysis

White's (1959) view of culture has enabled us to create a conception of mathematics different from that normally drawn. It is a conception which enables mathematics to be understood as a pan-cultural phenomenon. It seems that what I have been referring to as 'Western mathematics' must be recognized as being similar to, yet different from the mathematics developed by other cultural groups. There appear to be differences in symbolization and also difference in values. Just how great those differences are will have to be revealed by further analyses of the available anthropological and cross-cultural evidence. Hopefully the analysis here will help to structure the search for that evidence.

But what educational issues has this analysis revealed to us? From an anthropological perspective, mathematical education is a process of inducting the young into part of their culture, and there appear anthropologically to be two distinct kinds of process. On the one hand there is *enculturation*, which concerns the induction of the young child into the home or local culture, while on the other there is *acculturation*, which is to do with the induction of the person into a culture which is in some sense alien, and different from that of the home background. Appealing as this simple dichotomy is, the real educational situation is rather more complex. Consider inducting a child in 'Western' mathematics — for which children is enculturation the appropriate model? Is it really part of anyone's *home* and *local* culture? It is certainly not the product of any one culture and therefore no-one could claim it as theirs exclusively. Moreover there are plenty of practising mathematicians in universities all over the world who would object fundamentally (and rightly so in my opinion) to the suggestion that they were engaged in developing Western cultural knowledge.

So what culturally, is the mathematics which until a few years ago was generally thought to represent the only mathematics? Is it better to think of it as a kind of internationalized mathematics which all can learn to speak and understand? Or is it more a sort of Esperanto of the mathematical world, an artificial, pragmatic solution to a multi-cultural situation? That doesn't seem to be a good analogy because of the strong cultural values associated with it and the fact that it was not deliberately created in the way Esperanto was. Perhaps it is more appropriate educationally to recognize that different societies are influenced to different degrees by this international mathematico-technological culture and that the greater the degree of influence the more appropriate would be the idea of enculturation? What of acculturation? That clearly does raise other educational issues. Whilst acculturation is a natural kind of cultural development when cultures meet, there is something very contentious to me about an education which is *intentionally* acculturative. There is a clear intention implied in that notion to induct the child into an alien culture without any concern for the ultimate preservation of that child's home culture. Those whose children are being so acculturated have a perfectly understandable right to be concerned. Moreover I would stress that only those people for whom Western mathematics is an alien cultural product should decide what to do in the culture-conflict situation so created. It might be possible to develop a bi-cultural strategy, but that should not be for 'aliens' like me to decide. In my view the same questions arise over choice of teacher, and choice of educational environment. I would argue that in general in a culture-conflict situation it is better in the long run for the teacher to be from the 'home' culture, and for that teacher to be closely linked with the local community. If culture-conflict is to be handled sensitively, then schooling and the teacher, should stay close to the people affected, in my view.[4]

Another set of issues relate to the mathematics curriculum in schools, particularly in those societies where there are several ethnic minority groups. What ideas should we be introducing the children to? To what extent should any of the mathematical ideas from other cultures be used? And how is it possible to structure the mathematics curriculum to allow this to happen? It certainly would seem valuable to use mathematical ideas from the child's home culture within the overall mathematical experience, if only to enable the child to make good contact with the construct of *mathematics per se*. We know only too well some of the negative effects of insisting on children experiencing only alien cultural products — the meaningless, the rote-learning syndrome, the general attitude of irrelevance and purposelessness. So how can we overcome this? One possible way is to use as a structural framework the six *activities* which I described earlier. If those activities are universal, and if they are both necessary and sufficient for mathematical development, then a curriculum which is structured around those activities *would* allow the mathematical ideas from different cultural groups to be introduced sensibly. Is it indeed possible by this means to create a culturally-fair mathematics curriculum — a curriculum which would allow all cultural groups to involve their own mathematical ideas whilst also permitting the 'international' mathematical ideas to be developed?

Finally, what about the education of values? One implication of the values analysis earlier could be a consideration of the emphasis given in present mathematics education to certain values. I do not think it would be too cynical to suggest that a great deal of current mathematics teaching leans more towards control than to progress, to objectism rather than to rationalism, and to mystery rather than to openness. Perhaps a greater use of such teaching activities as group work, discussion, project work and investigations could help to redress the balances in each of the complementary pairs. We may then move our mathematical education more towards 'progress', 'rationalism' and 'openness', a goal with which other educators appear to agree. Certainly I believe that we should educate our children *about* values and not just train them into adopting certain values, although I realize that different societies may desire different approaches. (Nevertheless I cannot imagine how, or why, one would *train* a child to adopt a value like openness!). Again it seems to me to depend on the extent to which the particular society is influenced by those mathematico-technological cultural values, and relates once again to the enculturation/acculturation issues I described earlier.

In Conclusion

Perhaps the most significant implication for mathematics education of this whole analysis lies in teacher education. It is clear that teacher educators can no longer ignore these kinds of issues. Mathematics education in practice is, and always should be, mediated by human teachers. Inducting a young child into part of its culture is necessarily an inter-personal affair, and therefore teachers must be made fully aware of this aspect of their role. More than that, they need to know about the values inherent in the subject they are responsible for, they need to know about the cultural history of their subject, they need to reflect on their relationship with those values, and they need to be aware of how their teaching contributes not just to the mathematical development of their pupils, but also to the development of mathematics in their culture. Teacher education is the key to cultural preservation and development.

Notes

1 Fasheh (1982) and Kothari (1978) are educators who also express the values-conflict clearly.
2 As mathematical ideas develop, of course, they become part of the environment also, ready to be acted on as with any other part of the environment.
3 Kline (1962) and Wilder (1981) are two authors who have explored the cultural history of 'Western' mathematics and Joseph (1987) and in this volume has challenged and up-dated that knowledge.
4 Taft (1977) in a wide-ranging article describes many of the complex issues surrounding people in culture-conflict situations, and also indicates just how widespread a phenomenon it is.

References

ASCHER, M. and ASCHER, R. (1981) *Code of the Quipu*, Chicago, University of Michigan Press.
BISHOP, A. J. (1979). 'Visualising and mathematics in a pre-technological culture' *Educational Studies in Mathematics* 10, 2, pp. 135–46.
BISHOP, A. J. (1986) 'Mathematics Education as Cultural Induction', *Nieuwe Wiskrant*, October, pp. 27–32.
BISHOP, A. J. (1988) *Mathematical Enculturation: A Cultural Perspective on Mathematics Education*, Dordrecht, Kluwer.
BOURGOIN, J. (1973) *Arabic Geometrical Pattern and Design*, New York, Dover.
BRUNER, J. S. (1964) 'The Course of Cultural Growth', *American Psychologist*, 19, pp. 1–15.
CLOSS, M. P. (ed) (1986) *Native American Mathematics*, Austin, Texas, University of Texas Press.
D'AMBROSIO, U. (1985) 'Ethnomathematics and its Place in the History and Pedagogy of Mathematics', *For the Learning of Mathematics*, 5, 1, pp. 44–48.

ELLUL, J. (1980) *The Technological System*, New York, Continuum Publishing.

FAEGRE, T. (1979) *Tents: Architecture of the Nomads*, London, John Murray.

FALKENER, E. (1961) *Games Ancient and Oriental — How to Play Them*, New York, Dover.

FASHEH, M. (1982) 'Mathematics, Culture and Authority', *For the Learning of Mathematics*, 3, 2, pp. 2–8.

GARDNER, P.L. (1977) *Logical Connectives in Science*, Melbourne, Australia, Monash University, Faculty of Education.

GAY, J. and COLE, M. (1967) *The New Mathematics in an Old Culture*, New York, Holt, Rinehart and Winston.

GERDES, P. (1985) 'Conditions and Strategies for Emancipatory Mathematics Education in Underdeveloped Countries', *For the Learning of Mathematics*, 5, 1, pp. 15–20.

GERDES, P. (1986) 'How to Recognise the Hidden Geometrical Thinking: A Contribution to the Development of Anthropological Mathematics', *For the Learning of Mathematics*, 6, 2, pp. 10–17.

HARRIS, P. (1980) *Measurement in Tribal Aboriginal Communities*, Northern Territory Department of Education, Australia.

HORTON, R. (1971) 'African Traditional Thought and Western Science', *Africa* also in Young, M.F.D. (ed) *Knowledge and Control*, pp. 208–226, London, Collier-MacMillan.

HUIZINGA, J. (1949) *Homo Ludens*, London, Routledge and Kegan Paul.

JAYNE, C.F. (1974) *String Figures and How to Make Them*, New York, Dover. (first published as *String Figures* by Scribner 1906).

JONES, J. (1974) *Cognitive Studies with Students in Papua New Guinea*. (Working Paper No. 10) University of Papua, New Guinea, Education Research Unit.

JOSEPH, G.G. (1987) 'Foundations of Eurocentrism', *Race and Class* 28, pp. 13–28, also reprinted in this volume.

KEITEL, C., DAMEROW, P., BISHOP, A.J. and GERDES, P. (1989) *Mathematics Education and Society*. Science and Technology Education Document Series No. 35. Paris, Unesco.

KLINE, M. (1962) *Mathematics: A Cultural Approach*, Mass., Addison Wesley.

KOTHARI, D.S. (1978) *Keynote Address* in Proceedings of Asian Regional Seminar of the Commonwealth Association of Science and Mathematics Educators. (New Delhi) British Council, London.

LANCY, D.F. (1983). *Cross-cultural Studies in Cognition and Mathematics*, New York, Academic Press.

LEACH, E. (1973) 'Some Anthropological Observations on Number, Time and Common-Sense', in Howson, A.G. (1986) *Developments in Mathematical Education*, Cambridge. Cambridge University Press.

LEAN, G.A. (1986) *Counting Systems of Papua New Guinea*. Research Bibliography, 3rd Edition, Department of Mathematics, Papua New Guinea University of Technology, Lae, Papua New Guinea.

LEWIS, D. (1972) *We the Navigators*, Hawaii, University Press of Hawaii.

LEWIS, D. (1976) 'Observations on Route-finding and Spatial Orientation among the Aboriginal Peoples of the Western Desert Region of Central Australia', *Oceania* XLVI, 4, pp. 249–282.

MENNINGER, K. (1969) *Number Words and Number Symbols — A Cultural History of Numbers*, Cambridge, Mass., MIT Press.

OSTWALD, W.H. (1976) *An Anthropological Analysis of Food-getting Technology*, New York, Wiley.

PINXTEN, R. van DOOREN, I. and HARVEY, F. (1983) *The Anthropology of Space*, Philadelphia, University of Pennsylvania Press.

RONAN, C.A. (1981) *The Shorter Science and Civilisation in China*, 2, Cambridge, Cambridge University Pres.

RONAN, C. A. (1983) *The Cambridge Illustrated History of the World's Science*, Cambridge, Cambridge University Press.

ROTH, W. E. (1902) *Games, Sports and Amusements*, North Queensland Ethnographic Bulletin, 4, pp. 7–14.

VAN SERTIMA, I. (1986) *Black in Science*, New Brunswick, Transaction Books.

SKOVSMOSE, O. (1985) 'Mathematical Education versus Critical Education', *Educational Studies in Mathematics*. 16, 4, pp. 337–354.

TAFT, R. (1977) 'Coping with Unfamiliar Cultures', in Warren, N. (ed) *Studies in Cross-Cultural Psychology*, 1, London, Academic Press.

TEMPLE, R. K. G. (1986) *China. Land of Discovery and Invention*. Wellingborough, UK. Stephens.

VYGOTSKY, L. S. (1978) *Mind in Society*, Cambridge, Mass., MIT Press.

WHITE, L. A. (1959) *The Evolution of Culture*, New York, McGraw-Hill.

WILDER, R. L. (1981) *Mathematics as a Cultural System*, Oxford, Pergamon Press.

ZASLAVSKY, C. (1973) *Africa Counts*, Boston, Mass., Prindle and Schmidt, Inc.

5
Foundations of Eurocentrism in Mathematics

George Ghevarghese Joseph

This article first appeared in *Race and Class* **XXVIII** 3 1987: published by the *Institute of Race Relations*. The issues raised in it are elaborated further in *The Crest of the Peacock: Non-European Roots of Mathematics*, published by Penguin 1990.

Introduction

There exists a widespread Eurocentric bias in the production, dissemination and evaluation of scientific knowledge. And this is in part a result of the way many people perceive the development of science over the ages. For many Third World societies, still in the grip of an intellectual dependence promoted by European dominance during the past two or three centuries, the indigenous scientific base which may have been innovative and self-sufficient during pre-colonial times is neglected or often treated with a contempt that it does not deserve. An understanding of the dynamics of precolonial science and technology in these societies and an identification of the nature of the base on which the superstructure rested are essential in formulating a strategy of meaningful adaptation of the indigenous forms that remain to present-day scientific and technological requirements.

Now an important area of concern for anti-racists is the manner in which European scholarship has represented the past and potentialities of non-white societies with respect to their achievement and capabilities in promoting science and technology. The progress of Europe and its cultural dependencies[1] during the last 400 years is perceived by many as inextricably — and even causally — linked with the rapid growth of science and technology. So that in the minds of many, scientific progress becomes a uniquely European phenomenon, to be emulated only by following the European path of social and scientific development.

Such misrepresentation of the history and culture of societies outside the

European tradition raises a number of issues which are worth exploring. First, there are certain implications for the nature of the relationship between knowledge and power which was indicated at the beginning of this article. Second, there is the issue of who 'makes' science and technology. In a material and non-élitist sense, people from all continents have contributed to the growth of knowledge in general and of science in particular. Third, if one is imprisoned within the ethnocentricity of a particular place/time location, then non-European reality may only impinge marginally either as an unchanging residual experience to be contrasted with the dynamism and creativity of Europe, or as a rationale for the creation of disciplines congealed in subjects such as development studies, anthropology, orientalism, sinology and indology. These subjects then serve as the basis from which theories of social development and history can be developed.

The shaky foundations of these 'adjunct' disciplines are being increasingly exposed by scholars, mainly from those countries which provide the 'raw materials' of these disciplines. In a recent contribution to the journal *Race and Class* Edward Said (1985) points to a number of examples of 'subversive' analyses, inspired by similar impulses as his seminal anti-orientalism critique (1978), which are aimed at nothing less than the destruction of the existing Eurocentric paradigmatic norms. For example the growing movement towards promoting a form of indigenous anthropology which sees its primary task as questioning, redefining and, if necessary rejecting particular concepts which grew out of colonial experience in western anthropology, is well examined in Fahim (1982). In a similar vein, I propose to show that the standard treatment of the history of non-European mathematics is a product of a historiographical bias (conscious or otherwise) in the selection and interpretation of facts which, as a consequence, results in ignoring, devaluing or distorting contributions arising outside European mathematical traditions.[2]

The Historical Development of Mathematical Knowledge

Most histories of mathematics which were to become standards for later work were written in the late nineteenth or early twentieth centuries. During that period, two contrasting developments were taking place which would have an impact both on the content and the balance of the books produced on both sides of the Atlantic. On the one hand, exciting discoveries of ancient mathematics written on papyri in Egypt and clay tablets in Mesopotamia, dating back from the second millenium BC, had pushed back the origins of written mathematical records by at least 1,500 years.

A far stronger counter-influence was the political climate of the day, when the same period saw the culmination of European domination in the shape of a 'scramble for Africa' and the final subjugation of Asia by imperialist powers. As an adjunct to imperial domination arose the ideology of racism and white superiority which spread over a wide range of social and economic activities, including the writing of histories of science. These histories emphasized the unique role of Europe as providing the soil and spirit for scientific discovery. The contributions of the colonized were ignored and devalued as part of the rationale for subjugation and dominance. And the developments in mathematics before the Greeks — notably in Egypt and Mesopotamia — suffered a similar fate, being dismissed as of little importance to the future of the subject.

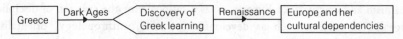

Figure 5.1   The 'classical' Eurocentric approach.

Figure 5.1 represents a 'classical' Eurocentric view of how mathematics developed over the ages. This development is seen as taking place in two areas separated by a period of inactivity lasting for a thousand years; Greece (from about 600BC to 300AD) and post-Renaissance Europe from the fifteenth century to the present day. The intervening period of inactivity constituted the 'Dark Ages' — a convenient label which was both an expression of post-Renaissance prejudices about its immediate past and of the true inheritors of the 'Greek miracle' which was supposed to have sprung up spontaneously from the Ionian soil 2,000 years earlier. (A detailed analysis of the construction of the idea of the 'Arian Greek' is given in Martin Bernal's *Black Athena*, Free Association Books 1987.)

Two passages, one by a well-known historian of mathematics writing at the turn of the century and the second by a contemporary writer whose books are still widely referred to on both sides of the Atlantic, show how impervious is Eurocentric scholarship to new evidence and sources.

> The history of mathematics cannot with certainty be traced back to any school or period before that of the Ionian Greeks. (Ball, 1960)

> [Mathematics] finally secured a firm grip on life in the highly congenial soil of Greece and waxed strongly for a short period . . . With the decline of Greek civilisation, the plant remained dormant for a thousand years . . . when the plant was transported to Europe proper and once more imbedded in fertile soil. (Kline, 1953).

The first statement is a fair summary of what was generally known at the turn of this century, except for the intriguing omission of early Indian mathematics contained in the *Sulbasutras* (c800–c500 BC), translated by Thibaut between 1874 and 1877, which were at least contemporaneous with the earliest known Greek mathematics. The second statement ignores a substantial body of research evidence pointing to significant development in mathematics in Mesopotamia, Egypt, China and pre-columbian America. Mathematics is perceived as an exclusive product of white men and European civilisations. And that is the central message of the Eurocentric trajectory described in Figure 5.1.

But this comforting rationale for an imperialist/racist ideology of dominance became increasingly untenable for a number of reasons. First, there was the fulsome acknowledgment by the ancient Greeks themselves of the intellectual debt they owed to the Egyptians and Babylonians. There are scattered references from Plato (c380 BC) to Plutarch (c100 AD) to the early knowledge acquired from the Egyptians in various fields, including astronomy, mathematics and surveying, with a number of commentators considering the priests of Memphis as founders of science. Both Thales (cd546 BC), the legendary founder of Greek mathematics, and Pythagoras (cd 500BC), one of the earliest and greatest of Greek mathematicians, were reported to have travelled widely in Egypt and Babylonia and learned much of their mathematics from these areas. Some sources even credit Pythagoras with having travelled as far as India in search of knowledge, which may explain some of the close parallels between Indian and Pythagorean philosophy and geometry.[3]

A second reason why the trajectory described in Figure 5.1 is untenable arose from the findings of the combined efforts or archaeologists, translators and interpreters who unearthed evidence of a high level of mathematics practised in Mesopotamia and to a lesser extent in Egypt from the beginning of the second millenium BC, which provided further confirmation of Greek reports on the nature of such mathematics. In particular, the Babylonian mathematicians had invented a place value number system, understood (but not proved) the so-called Pythagorean Theorem[4] and evolved an iterative method of solving quadratic equations which would only be improved upon in the sixteenth century AD.

Third, the significance of the Arab contribution to the development of European intellectual life could no longer be ignored. The course of European cultural history and the history of European thought are inseparably tied up with the achievement of Arab scholars during the Middle Ages (or the Dark Ages as they came to be known by post-Renaissance Europe) on account of their seminal contributions in mathematics, natural sciences, medicine and philosophy. In particular, we owe to the Arabs in the field of mathematics the bringing together of the technique of measurement, evolved from its Egyptian and Babylonian roots to its final form in the hands of Greeks and Alexandrians, with the remarkable instrument of computation (our number system), which originated in India, and finally supplementing these strands with a systematic and consistent language of calculation which came to be known by its Arabic name, algebra. A grudging acknowledgment of this debt by certain books contrast sharply with a general neglect when it came to recognising other Arab contributions.[5]

Fourth, there was some recognition that in talking about the Greek contribution one should separate the classical period of Greek civilisation (i.e. from the sixth century to the third century BC) from the post-Alexandrian dynasties (ie: from the third century BC to the third century AD). In early Eurocentric scholarship, the Greeks of the ancient world were perceived as ethnically homogeneous and originating from areas which were mainly within the geographical boundaries of present-day Greece. It was part of the Eurocentric mythology that from the mainland of Europe had emerged a group of people who had created out of virtually nothing the most impressive of all civilisations of ancient times. And from that civilisation had sprung not only the cherished institutions of the present-day western culture but also the main-spring of modern science and technology. The reality is, however, more complex and problematic.

Before the appearance of Alexander (356–323BC), the term 'Greek' did encompass a number of independent city states, often at war with one another, but exhibiting close ethnic and cultural bonds, and above all sharing a common language — whose alphabet was borrowed from the Phoenicians of North Africa. The conquests of Alexander changed the situation dramatically, for at his death his empire was divided among his generals who established separate dynasties. The two notable dynasties from the point of view of mathematics were the Ptolemaic Dynasty of Egypt and the Selucid dynasty, which encompassed the earlier site of Mesopotamian civilization. The most famous centre of learning and trade became Alexandria in Egypt, established in 332 BC and named after the conqueror. From its foundation, one of its most striking features was its cosmopolitanism — part Egyptian, part Greek and a liberal sprinkling of Jews, Persians and Phoenicians, and even attracting scholars and traders from as far away as India. A lively contact was maintained with the Selucid dynasty. It thus became the meeting-place for ideas and different traditions, and over the period the character of Greek mathematics changed mainly as a result of the continuing cross-fertilization between different mathematical traditions, notably the algebraic and empirical traditions of Babylonia and Egypt interacting with the geometric and anti-empirical traditions of classical Greece. And from this mixture came some of the greatest mathematicians of antiquity — notably Euclid, Archimedes, Apollonius and Diophantus. It is, therefore, misleading to speak of Alexandrian mathematics as Greek, except in so far as the term indicates the Greek cultural traditions served as the main inspiration and the Greek language as the medium of instruction and writing in Alexandria. In that sense, the use of the term 'Greek' is closely analogous to the use of the term 'Arab' to describe a civilisation which encompasses a number of ethnic and religious groups, but all of whom were imbued with the Arab culture and language.

Figure 5.2 describes a 'modified' Eurocentric trajectory which takes a limited account of the contributions made by other cultural areas to the development of mathematical knowledge. There is some awareness of the existence of mathematics before the Greeks and the debt to these earlier mathematical traditions in Babylonia and Egypt. But this awareness is likely to be tempered with dismissive rejections of their importance compared to Greek mathematics — 'the scrawling of children just learning to write as opposed to great literature' (Kline 1962).

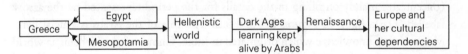

Figure 5.2   The modified Eurocentric tragectory.

The differences in character of the Greek contribution before and after Alexander are recognized to a limited extent in Figure 5.2 by a chronological separation of Greece from the Hellenistic world (where the Ptolemaic and Selucid dynasties were the crucial instrument of mathematical creation for that period). There is also a recognition of the Arabs, but merely as custodians of Greek learning during the so-called Dark Ages in Europe.[6] Their role as transmitters and creators of knowledge is ignored. So are the contributions of other civilisations — notably those of China and India — which are perceived as borrowers from Greek sources, as having made only minor contributions or as having an insignificant role in mainstream mathematical developments (i.e.; developments culminating in European dominance).[7] More recently, histories of mathematics carry separate chapters, serving as 'residual' dumps, entitled 'Oriental' mathematics or 'Indian/Chinese' mathematics, which are of marginal relevance to the mainstream themes pursued in these books. (see for example Boyer 1978; Eves, 1976; Scott, 1958; Struik, 1967) This marginalization of non-European mathematics is reflected in the nature of the scholarship that characterizes the treatment of these subjects in successive text books. An openness to more recent research findings, especially in the case of Indian and Chinese mathematics, is sadly missing. As a consequence, paraphrases of the contents of earlier texts or quotes from individuals whose scholarship or impartiality have been seriously questioned are reproduced in each succeeding generation of textbooks.[8]

Figure 5.2 therefore remains a flawed representation of how mathematics developed over time. It encompasses a series of biases and remains impervious to new evidence and arguments. With minor modifications, it presents the model to which most books on the history of mathematics conform. While I propose in the next section to explore the nature and sources of the biases that such a representation reflects, it should be noted that similar Europe-centred bias exists in other disciplines as well. For example, diffusion theories in anthropology and social geography indicate that 'civilization' spreads from the centre (i.e., 'greater' Europe) to the periphery (i.e., to the rest of the world). Again, theories of modernization or evolutionary schemes developed within the framework of certain brands of Marxism are characterized by a similar type of Eurocentrism. In all such conceptual schemes, the development of Europe is seen to serve as a precedent for the way in which Third

World societies will develop in the future — a trajectory whose spirit is not dissimilar to the one suggested in Figure 5.2.

Figure 5.3 offers an alternative trajectory of mathematical development, but concentrates mainly on filling in the details for the period represented by the arrow labelled 'Dark Ages' in Figures 5.1 and 5.2. The role of the Arabs is crucial here. Mathematical knowledge which originated in India, China and the Hellenistic world was sought out by Arab scholars and then translated, refined, synthesized and augmented at different centres of learning, starting with Jundishapur in South-east Persia and then moving to Baghdad, Cairo and finally to Toledo and Cordoba in Spain. Considerable resources were made available to the scholars through the benevolent patronage of the Abbasid Caliphs (i.e., the rulers of the eastern Arab empire with its capital at Baghdad) and Ummayid Caliphs (i.e., the rulers of the western Arab empire with its capital first at Damascus, then moving to Cairo and finally to Cordoba).

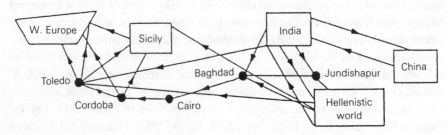

Figure 5.3   An alternative trajectory (from 8th to 15th century)

The Abbasid Caliphs, notably al-Mansur (754–775 AD), Harun al-Rashid (786–806 AD) and al-Mamun (813–833 AD), were in the forefront of promoting the study of astronomy and mathematics in Baghdad. Indian scientists were invited to settle in Baghdad. At the closure of Plato's academy in 529 AD by the Roman Emperor Justinian to placate Christian bigotry, many of its scholars found refuge in Jundishapur in Persia, which a century later became part of the Arab world. Greek manuscripts from the Byzantine empire, the translations of the Syrian schools of Antioch and Damascus, the remains of the Alexandrian library in the hands of the Nestorian Christians at Edessa were all sought out eagerly by Arab scholars, aided and abetted by the rulers who had control over or access to men and materials from the Byzantine Empire, Persia, Egypt, Syria and places as far east as India and China. Caliph al-Mansur built at Baghdad a *Bait al-Hikma* (translated as *House of Wisdom*) which contained a large library or stock of the manuscripts that had been collected from various sources, an observatory which became a meeting place of Indian, Chinese, Hellenistic and Babylonian astronomical traditions, and a university where scientific research continued apace. A notable member of this institution Mohamed ibn-Musa al-Khwarizmi (c825 AD) wrote two books which were of crucial importance to the future development of mathematics. One of the books, the Arabic text of which is extant, is entitled *Hisab al-djabr wa-al Muqabala* (which may be translated as the 'Science of Reduction and Cancellation' or probably 'Science of Equations') which introduced the word *al-djabr* (or algebra) for the first time. His second book of which only a Latin translation is extant, is called *Algorithmi de Numero Indorum*. It explained the Indian system of numeration, and was based on the work of Brahmagupta (c628 AD), an Indian mathematician-astronomer, entitled *Brahmasputa Siddhanta*. While al-Khwarizmi was at pains to point out the Indian origin of the numeration system, subsequent translations of the book attributed not only the book but the numerals to the author. Hence in Europe any scheme using these numerals came to be known as an algorism or later algorithm (i.e., a corruption of the name of al-Khwarizmi) and the numerals became Arabic numerals.

Other great Arab mathematicians continued the work begun by al-Khwarizmi and they included Thabit ibn-Qurra (826–901), Abu-Kamil (c900), Abul-Wefa (940–998), Ibn al-Haytham (c965–1039), al-Biruni (973–1048), Omar Khayam (c1050–1122), better known in the west as a poet and hedonist and Nasir Eddin al-Tusi (120–1274). The last named mathematician was no longer in the service of Arab rulers; he was an astronomer to the Mongol, Hulagu Khan, grandson of Genghis Khan. His contributions to non-Euclidian geometry, which formed the starting-point of the work of Saccheri of Italy five centuries later, show that Arab geometry had truly come of age after being tied initially to the apron strings of Hellenistic geometry.

Figure 5.3 highlights the importance of two areas of southern Europe in the transmission of mathematical knowledge to western Europe. Spain and Sicily were the nearest points of contact with Arab science and had been under Arab hegemony, with Cordoba succeeding Cairo as the capital and centre of learning of the Ummayid caliphate during the ninth and tenth centuries. Scholars from different parts of western Europe congregated in Cordoba and Toledo in search of both ancient and contemporary knowledge. As an illustration of this great thirst for knowledge, it is reported that Gherardo of Cremona (c1114–1187) went to Toledo after its recapture by the Christians, in search of Ptolemy's *Almagest*, an astronomical work of great importance produced in Alexandria during the second century AD. He was so taken by the intellectual activity here that he remained for a period of twenty years, during which he is reported to have copied or translated eighty manuscripts of Arab mathematics or Greek classics, which he then proceeded to take back to his homeland. Gherardo was one of a number of European scholars, including Plato of Tivoli, Adelard of Bath and Robert of Chester, who flocked to Spain in search of knowledge.

There are two additional features of mathematical knowledge that Figure 5.3 serves to highlight. First, it is not generally recognized that practically all topics taught in school mathematics today are directly derived from the work of mathematicians originating outside western Europe before the twelfth century AD. The failure to recognize this fact is partly a function of the heavily Eurocentred nature of school curricula and partly due to the unwarranted neglect of history (and particularly non-Eurocentric history) of mathematics in a typical mathematicians education. Second, Figure 5.3 shows the one-way traffic of mathematical knowledge into western Europe up to the fifteenth century. Thus the Arab mathematical renaissance between the eighth and twelfth centuries shaped and determined the pace of developments in the subject for the next five hundred years.

The Anatomy of Eurocentric Bias

The Eurocentric historiography of mathematics exhibits certain features which may explain the biases that result. First, there is a general disinclination to locate mathematics in a materialist base and thus link its development with economic, political and cultural changes. Second, there is a tendencey to perceive mathematical pursuits as confined to an élite, a select few who possess the requisite qualities or gifts denied to the vast majority of humanity. This is a view prevalent even today in the classroom and thus determines what is taught and who benefits from learning mathematics. Third, there is a widespread acceptance of the view that mathematical discovery can only follow from a rigorous application of a form of deductive axiomatic logic, which is perceived as a unique product of Greek mathematics. As a consequence 'intuitive' or empirical methods are dismissed as of little relevance in mathematics. Finally, the presentation of mathematical results must conform to the formal and didactic style following the pattern set by the Greeks over 2,000 years ago. And, as a corollary, the validation of new additions to mathematical knowledge can only be undertaken by a small, self-selecting coterie whose control over the acquisition and dissemination of such knowledge through journals has a highly Eurocentric character today.

As an illustration of how the features listed above can create Eurocentric bias, let us examine the status ascribed to mathematical pursuits which do not conform to the criteria mentioned in the last paragraph, notably in Egypt and Mesopotamia before the emergence of Greek mathematics.

A commonly expressed view is that, before the Greeks, there was no mathematics in the sense of the characteristic intellectual activity which goes under that name today. The argument goes: pre-Greek mathematics had neither a well-defined concept of 'proof' nor any perception of the need for proof. Where the Egyptians or Mesopotamians were involved in activities which could be described as 'mathematics' these activities were purely utilitarian, such as the construction of calendars, parcelling out land, administration of harvests, organization of public works (e.g., irrigation or flood control) or collection of taxes. Empirical rules were devised to help undertake these activities, but there is no evidence of any overt concern with abstractions and proofs which form the core of mathematics. In any case, the argument continues, the only evidence that we have to assess the mathematics of these two civilizations amounts to little more than the exercises that school children of today are expected to work out, which merely involve the application of certain rules or procedures; they are hardly 'proofs' or results which have universal application.

The word 'proof' has different meanings, depending on its context and the state of development of the subject. To suggest that because existing documentary evidence does not exhibit the deductive, axiomatic, logical inference characteristic of much of modern mathematics, these cultures did not have a concept of proof, would be misleading. Generalizations about the area of a circle and the volume of a truncated pyramid are found in Egyptian mathematics. Checking the correctness of a division by a subsequent multiplication or verifying the solutions of different types of equation by the method of substitution are found in Babylonian mathematics. A method in common use in Europe until about a hundred years ago for solving linear equations is generally known as the method of false position.[9] This method was in common use to solve practical problems such as determining the potency of beer or obtaining optimal feed mixtures for cattle and poultry in Egyptian and Babylonian mathematics. As Gillings (1972) has argued, Egyptian 'proofs' are rigorous without being symbolic, so that typical values of a variable are used and generalization to any other value is immediate. Or again, generalizations of the methods used in solving problems contained in the Ahmes papyrus (c1650) and the Moscow papyrus (c1850 BC) two of the most important mathematical documents from Egypt, involve applications of the same procedure to one example after another. To illustrate, consider one of the 'lesson texts' dating back to the time of the first Babylonian Dynasty of Hammurabi (c. 1700 BC), translated and interpreted by Neugebauer (1935). For the sake of simplicity, I have converted the quantities expressed in base 60 (i.e., sexagesimal) system to our base 10 (i.e., decimal system).

Problem
Length (*us*), width (*sag*). I have multiplied length and breadth, thus obtaining the area. Then I added to the area the excess of length over width: 183 (was the result). Then I added length and width: 27. Required (to obtain) length, width and area.

Solution
Given 27 and 183, the Sums.

Result
15 length, 12 width, 180 area.

Method
One follows this method: [step 1] $27 + 183 = 210$; $2 + 27 = 29$

Take one half of 29 and square it: [step 2] $(14.5)^2 = 210.25$
Subtract 210 from the result: [step 3] $210.25 - 210 = 0.25$
Take the square root of 0.25 [step 4] Square root of $0.25 = 0.5$
Then, length $= 14.5 + 0.5 = 15$
breadth $= (14.5 - 0.5) - 2 = 12$
area $= 15 \times 12 = 180$

Present-day method
Let length (*us*) = x and width (*sag*) = y. Then the problem is solved by evaluating the two equations;
$$xy + x - y = 183$$
$$x + y = 27 \qquad\qquad (1)$$
Now define a new variable y' such that $y' = y + 2$
Then (1) can be re-written as:
$$xy' = 27 + 183 = 210$$
$$x + y' = 2 + 27 = 29 \qquad\qquad (2)$$
(*Note.* The transformation from (1) to (2) is indicated by Step 1)
The general system of equations of which (2) is a particular case may be expressed thus;
$$xy' = P$$
$$x + y' = s$$
So that the solution is:
$$x = 1/2s + w$$
$$y' = 1/2s - w \qquad\qquad (3)$$
where
$$w = \text{square root of } [(1/2s)^2 - P]$$
Substitution of P = 210, s = 29 gives w = 0.5, which can then be used to evaluate x = 15, $y = y' - 2 = 12$ and area = 180.

What the Babylonian method involved was the application step by step of the general formula, expressed in modern algebraic symbolism, given in (3) to numbers. The Sumerian symbols *us* and *sag*, for length and width respectively, serve the same purpose as our algebraic symbols x and y. And instead of providing a formula for the solutions of this type of problem, the Babylonians gave one example after another, just as an elementary school textbook may do today to ensure that the method is correctly applied. Such a demonstration may very well be as effective as formal 'proofs' in problems of this nature. This problem is also indicative of the level of sophistication reached by Babylonian mathematics. To dismiss such a work as 'scrawlings of children just learning to write' (Kline 1962) is more a reflection of the author's prejudices than an objective assessment of the real quality of such mathematics.

A further criticism levelled against Egyptian and Babylonian mathematics is that their mathematics was more a practical tool than an intellectual pursuit. This criticism is symptomatic of a widespread attitude, again originating with the Greeks[10], that mathematics devoid of a utilitarian bent is in some sense a finer or better mathematics. This attitude has even percolated right across the mathematics curriculum in schools and colleges.[11] As a consequence, there is both a sense of remoteness and irrelevance associated with the subject among many who study it, and an ingrained élitism among those who teach it. This élitism is translated at a classroom level into a view, often implicit and not spoken, that real mathematics as opposed to 'doing sums' is an activity suited for a select few — which when extended provides the broader argument that mathematics is a unique product of European culture. Thus, élitism in the classroom is ultimately linked to the form of intellectual racism which I have described as Eurocentrism.

Countering Eurocentrism in the Classroom

The foregoing analysis illustrates the need to confront and then counter Eurocentrism in mathematics. A commonly expressed view of the educational establishment in this country is that while a correction of the Eurocentric bias in history may be a worthwhile exercise, it has little relevance to mathematical activities within the classroom. I have stated elsewhere why I think this is a misconceived view and how an unbiased historical perspective can enrich the quality of mathematical activity in the classroom as well as provide a valuable input into anti-racist education generally. (Joseph, 1984; 1985 and 1986) It would be useful to restate these arguments in the context of the themes explored here.

First, mathematics is shown to have flourished all over the wolrd, with its internal logic providing a point of convergence for different mathematical traditions, without being constrained by geography, gender[12] (see Osen, 1974) or race. Yet within this unity there is an interesting diversity which could serve to entertain and educate at the same time. By bringing to the attention of the students differences in the language and structure of counting systems found across the world, by showing how different calendars and eras operate or by examining different spatial relations contained in, say, traditional African designs, Indian *rangoli* patterns and Islamic art, they could serve both as useful examples of applied mathematics as well as increase their awareness of cultural diversity.[13]

Second, a historical approach may, if handled carefully, provide a useful materialistic perspective in evaluating contributions made by different societies. The implied myth of the 'Greek' miracle in explaining the origins of mathematics will give way to a more balanced assessment of the nature of early mathematical accomplishments. Thus, the Ishango bone, found on a fishing site by the banks of Lake Edward in Zaire dating back about 8,000 years, was first thought of as a permanent numerical record of unknown objects. A closer study of the notches on it revealed that it may have been a six-month calendar of the phases of the moon[14]. Similarly, an American *quipi* found in Peru was first thought of as an art object consisting of an intricate pattern of woven knots. But it was later recognized that the artefact contained the record of a whole population census taken about 2,000 years ago, where the knots of varying sizes stood for different numerical magnitudes and different colour coding used to show characteristics such as sex and age. In a predominantly pastoral or simple agricultural economy such ingenious devices were invented to satisfy the main mathematical requirements — the recording and preservation of such information as was required to keep track of the passage of time or predict seasons for planting seed or the coming of rains. But as societies evolved, mathematical demands became more varied and sophisticated, leading, for example to the discovery of the place value notation by Babylonians (c2000 BC) for more complex computations, and the eventual adoption 3,000 years later, when mechanical contrivances such as the abacus or rod numerals were no longer sufficient, of our number system (developed by the Indians about 2,000 years ago), when written calculations became absolutely essential for trade and commerce. Both the Babylonian invention and the Indian numerals were momentous discoveries at the time, but are taken for granted today.

Finally, if we accept the principle that teaching should be tailored to children's experience of the social and physical environment in which they live, mathematics should also draw on these experiences, which would include in contemporary Britain the presence of different ethnic minorities with their own mathematical heritage. Drawing on the mathematical traditions of these groups, indicating that these cultures are recognized and valued, would also help to counter the entrenched historical devaluation of them. Again, by promoting such an approach, mathematics is brought into contact with a wide range of disciplines, including art and design, history and social studies, which it conventionally ignores. Such a holistic approach would serve to augment, rather than fragment, a child's understanding and imagination.

Notes

1 The term 'cultural dependencies' is used here to describe those countries — notably the United States, Canada, Australia and New Zealand — which are mainly inhabited by populations of European origin and share similar historical and cultural roots as Europe. For the sake of brevity, the term 'Europe' is used hereafter to include these areas as well.

2 A concise and meaningful definition of mathematics is virtually impossible. In the context of this article, the following aspects of the subject are particularly relevant. Mathematics can be looked at as an international language, with a particular kind of logical structure. it contains a body of knowledge relating to number and space. It prescribes a set of methods for obtaining conclusions

about the physical world. And it is an intellectual activity using both intuition and imagination to arrive at proofs and conclusions which may carry a high sense of aesthetic satisfaction for the creator.

3 These parallels are found in the following areas: a) a belief in transmigration of souls; b) the theory of five elements constituting matter; c) the reasons for prohibiting consumption of beans; d) the structure of the religio-philosophical character of the Pythagorean fraternity which shared certain similarities with Buddhist monasteries and; e) the contents of the mystical speculations of the Pythagorean school which bear a remarkable resemblance to *Upanishads*. The statement of the Pythagorean theorem in geometry is found in *Sulbasutras* (the oldest extant documents containing Indian geometry). According to Greek tradition, Pythagoras, Thales, Empedocles, Anaxagoras, Democritus and others undertook journeys to the East to study philosophy and science. While it is far fetched to assume that all these individuals reached India, there is a strong historical possibility that some of them became aware of Indian thought and science through Persia.

4 The statement and demonstration of the so-called Pythagorean theorem is found in varying degrees of detail all over the world. A variety of evidence is at present available on the widespread practical use of the theorem among the Babylonians (c1800–1600 BC). The Chinese provided a proof of the theorem in their oldest extant mathematical text entitled *Chou Pei* (500 BC). As mentioned earlier, the *Sulbasutras* (c600–800 BC) contained the earliest known general proof of the theorem. It is also worth noting that even though the theorem is universally associated with the name of Pythagoras, there is no evidence that Pythagoras had either stated or proved the theorem. The earliest Greek proof, which is still to be found in school geometry texts was given by Euclid (c300 BC).

5 The unacknowledged debt to Arab science includes: a) an earlier description of pulmonary circulation of the blood by Ibn al-Nafis, usually attributed to Harvey; b) the first known statement of the principle of the refraction of light by Ibn al-Haytham, usually attributed to Newton: c) the first known statement of the law of gravity by al-Khazin, again attributed to Newton; d) the first clear statement of the theory of evolution by Ibn Miskawayh, usually attributed to Darwin and; e) the first explanation of the rationale underlying the 'scientific method' which is found in the works of Ibn Sina, Ibn al-Haytham and al-Biruni but usually credited to Bacon.

6 In a review article Nisbet (1973) has pointed out how much the myth of a Renaissance occurring in Europe between the fifteenth and sixteenth centuries has persisted, in spite of overwhelming evidence to indicate that there was continuous intellectual development taking place in Europe from the twelfth century.

7 Chinese, Mayan or Japanese mathematics are often ignored on the grounds that they fall outside the main line of mathematical development that culminated in the European advance of the subject.

8 One individual who is frequently quoted by historians as an authority on Indian mathematics is G. R. Kaye who was in the service of the raj at the turn of the century. His interpretations both with regard to dating certain mathematical documents (notably *Sulbasutras* and the *Bakhshali Manuscript*) which he generally tended to put much later than other scholars, usually on fairly flimsy grounds, as well as his tendency to attribute anything significant in Indian mathematics to a Greek origin, have been criticized by notable scholars of ancient Indian mathematics (see for example Datta and Singh, 1935 and 1938, Sarasvati Amma 1979, Srinivasiengar 1967) without apparently making much impression on those who continue to write histories of mathematics in Europe and her cultural dependencies.

9 To solve for x in the equation $x + x/5 = 24$, the method of false position involves arguing that if $x = 5$, then $x + x/5$ will equal 6. So to obtain the required 24, we need to multiply 6 by 4. Or the correct x value is 20.

10 An important distinction running right across Greek thought has been *arithmetica*, the study of the properties of pure numbers, and *logistica* the use of numbers in practical applications. The cultivation of the latter discipline was to be left to the slaves. A legend has it that when Euclid (c300 BC) was asked what was to be gained from studying geometry, he told his slave to toss a coin at the inquirer.

11 There is, however, a discernable movement towards 'utilitarian' mathematics in the modern classroom. So the tide may be turning.

12 The contribution of women mathematicians has also been neglected in standard histories, except for the occasional mention of Hypatia (cd 415) whose cruel death at the hands of a Christian mob is taken by some to represent the end of Alexandrian mathematics.

13 It is not my intention here to enter into the controversy regarding the precise meaning of culture. The relationship between a people who possess a culture and the culture itself is highly complex and very germane to the point under discussion. The term 'culture' is used here in an anthropological sense to describe a collection of customs, rituals, beliefs, tools, mores etc., possessed by a group of people who may be related to one another by factors such as a common language, geographical contiguity or class.

14 Marshack (1972) has argued, on the basis of a close fit observed between the numbers in each group of notches and the astronomical lunar periods, that the Ishango bone offers possible evidence of one of man's earliest intellectual activities, devising sequential notation based on a six-month lunar calendar for activities such as tattooing, decorating, gaming or ceremonial festivities.

References

BALL, W. W. R. (1960) *A Short Account of the History of Mathematics*. New York, Dover reprint.

BOYER, C. B. (1978) *A History of Mathematics*, New Jersey, Princeton University Press.

DATTA, B. B. and SINGH, A. N. (1962) *History of Hindu Mathematics*, 2 volumes, Bombay, Asia Publishing House.

EVES, H. (1983) *An Introduction to the History of Mathematics*, Fifth edition, Philadelphia, Holt, Rinehart and Winston.

FAHIM, H. (1982) *Indigenous Anthropology in Non-western Countries*, North Carolina, University Press.

GILLINGS, R. J. (1972) *Mathematics in the Time of the Pharoahs*, Cambridge, Mass., MIT Press.

JOSEPH, G. (1984) *The Multicultural Dimension*, Times Educational Supplement, 5th October.

JOSEPH, G. (1985) *A Historical Perspective*, Times Educational Supplement, 11th October.

JOSEPH, G. (1986) 'A non-Eurocentric Approach to School Mathematics', *Multi-cultural Teaching*, 2, 4.

KLINE, M. (1962) *Mathematics: A Cultural Approach* Reading, Mass., Addison Wesley.

KLINE, M. (1953) *Mathematics in Western Culture*, New York, Oxford University Press.

MARSHACK, A. (1972) *The Roots of Civilisation*, London, Weidenfeld and Nicolson.

NEUGEBAUER, O. (1935) *Mathematics Keilschrift-Text*. Vol. 2, Berlin, Springer Verlag.

NISBET, R. (1973) 'The Myth of the Renaissance', *Comparative Studies in Society and History*, 15, 4 October.

OSEN, L. M. (1974) *Women in Mathematics*, Cambridge Mass., MIT Press.

SAID, Edward. (1985) 'Orientalism Reconsidered', *Race and Class*, Vol. XXVI No. 2.

SAID, Edward. (1978) *Orientalism*, New York and London, Vintage Books.

SARASVATI AMMA, T. A. (1979) *Geometry in Ancient and Medieval India*, Delhi. Motilal Banarsidass.

SCOTT, J. F. (1958) *A History of Mathematics from Antiquity to the Beginning of the Nineteenth Century*, London, Taylor and Francis.

SRINIVASIENGAR, C. N. (1967) *The History of Ancient Indian Mathematics*, Calcutta, World Press.

STRUIK, D. J. (1967) *A Concise History of Mathematics*, Dover Press, New York.

6
Mathematics in a Social Context: Math within Education as Praxis versus Math within Education as Hegemony

Munir Fasheh

This paper was first read at the Sixth International Congress on Mathematical Education held in Budapest, Hungary July 27 to August 3 1988 and was published in the collection *Mathematics Education and Society* by UNESCO, Sciences and Technology Education document Series No 35, Paris 1989. It is adapted from the author's dissertation 'Education as Praxis for Liberation: Birzeit University and the Community Work Program' March 1988, Graduate School of Education, Harvard University, Cambridge, Mass., USA.

When the 1967 Israeli-Arab war broke out, I was 26 years old, already with a Masters degree and four years experience of teaching math at various levels. The war shook the foundations of my small, comfortable and seemingly consistent and meaningful world, a world created by formal institutionalized education. The war revealed how little we — the formally educated — knew. Almost none of our conceptions matched what was going on. Although I started questioning education in general almost immediately after the war, I did not question at that time the possible relation of the maths and physics (which I had studied and was teaching) to what was going on in the world nor did I question the fundamental assumptions upon which math and science were based. In fact, as a result of the war I became more convinced that one task I had as an educator was to expand the use of logic and science in the world through teaching. I thought that all that we needed was more math, a 'New Math', as well as better and more diversified ways of teaching it. Actually, I was formally involved for six years (1972–1978) in math instruction at several levels and in different ways in the schools of the West Bank. The 'New Math' I was in charge of introducing into schools was fundamentally alien, dry and abstract. In order to overcome this, I encouraged the incorporation of cultural concepts, independent avenues of exploration and personal feelings into the work. In addition to classroom teaching, I established and used math clubs, magazines, general discussion meetings and in-service course (Fasheh, 1982). This revitalized the teaching, introduced both structure and logic, and was important in developing creativity and enthusiasm among both teachers and students. It did not lead me, however to question hegemonic assumptions behind the math itself.

While I was using math to help empower other people, it was not empowering for me. It was, however, for my mother whose theoretical awareness of math was completely undeveloped. Math was necessary for her in a much more profound and real sense than it was for me. My illiterate mother routinely took rectangles of fabric and, with few measurements and no patterns, cut them and turned them into beautiful, perfectly fitted clothing for people. In 1976 it struck me that the math she was using was beyond my comprehension; moreover, while math for me was the subject matter I studied and taught, for her it was basic to the operation of her understanding. In addition, mistakes in her work entailed practical consequences completely different from mistakes in my math.

Moreover, the value of her math and its relationship to the world around her was drastically different from mine. My math had no power connection with anything in the community and no power connection with the western hegemonic culture which had engendered it. It was connected solely to symbolic power. Without the official ideological support system, no one would have 'needed' my math; its value was derived from a set of symbols created by hegemony and the world of education. In contrast, my mother's math was so deeply embedded in the culture that it was invisible through eyes trained by formal education. Her math had no symbols of power. Its value was connected to concrete and immediate needs and actions.

Seeing my mother's math in context helped make me see my math in context; the context of power. What kept her craft from being fully a praxis and limited her empowerment was a social context which discredited her as a woman and uneducated and paid her extremely poorly for her work. Like most of us, she never understood that social context and was vulnerable to its hegemonic assertions. She never wanted any of her children to learn her profession; instead, she and my father worked very hard to see that we were educated and did not work with our hands. In the face of this, it was a shock to me to realize the complexity and richness of my mother's relationship to mathematics. Mathematics was integrated into her world as it never was into mine.

I was attracted to math and physics because of what I felt to be their role in making the world more intelligible; because they could explain phenomena and predict events; and because they could make absolute statements and accomplish concrete things such as build bridges, make planes, and facilitate surgery. I was also attracted to math and physics because of such claims about them as neutrality, universality, precision and bettering human conditions. I was aware of their role in producing bombs and pollution but was convinced that the ethical, moral and humanitarian dimension was the norm and the fundamental role.

What I started to see in 1967 were the practical limits of the education I had been given. The 1976 war started a process which made the real environment and power relations more visible. My sense of the intellectual, moral and humanitarian dimensions of science and math gradually gave way to a sense of the central functions of science and math, creating power and generating hegemony. The stunning Israeli military victory in 1967 was a victory of superior science, math and technology and not a result of moral superiority or superior personal courage. The message of the highly sophisticated warplanes and bombs (mainly the prohibited napalm) used was loud and clear. Math, science and technology generate a system of control because they perform material functions successfully.

My mother's sewing demonstrated another way of conceptualizing and doing mathematics, another kind of knowledge, and its place in the world. The value of my mother's tradition, of her kind of mathematics and knowledge, while not intrinsically disempowering however, was continuously discredited by the world around her, by the culture of silence and cultural hegemony.

The most crucial issue that the above discussion raises is that of the relation of education to the world it inhabits. The education I had been given fitted me to live in a world created by education and hegemony. It left me blind to its ideological dimension, to the relationship between the knowledge transmitted to me and power. This blindness, characteristic of hegemonically educated Third World people, left me unfit to live in the real world, in the real environment. Hegemony is not only characterized by what it includes but also by what it excludes: by what it renders marginal, deems inferior and makes invisible. The effect of hegemonic education is such that it is almost possible to define the real environment by what formal education marginalizes or excludes.

Because ideology is a world-view which embodies certain conceptions, values, languages, relations and interests, which are translated into daily practices and which produce certain consciousness, the role of intellectuals and institutions is of primary importance; the reproduction of a hegemonic ideology is achieved through them (among other apparatuses). Intellectual development in the colonial hegemonic context is designed to provide ideology without a basis in power. It allows intellectuals to participate vicariously in the moral, intellectual, humanitarian and technical aspects of western culture as well as in educational, scholarly and research activities. Individuals are sometimes acknowledged and rewarded for their participation and contribution. The training of colonial intellectuals directs them to derive their sense of worth and status from this vicarious participation, alienating them from their own culture, history and people. The indigenous population often supports this tendency by giving status to such intellectuals. Hegemonic education produces, generally speaking, intellectuals who have lost their power base in their own culture and society and who have been provided with a foreign culture and ideology, but without a power base in the hegemonic society. (An intellectual at Berkeley, MIT or Harvard, for example, may actually participate in decision making concerning weapons production and in weapons production itself; or be involved in the policy decisions of multinational corporations; or with US government policy concerning other countries.) Lacking a power base at both ends, Third World intellectuals tend to sharply overvalue symbolic power and tokens (titles, degrees, access to prestigious institutions, awards).

As with individuals, the institutional hegemonic connection does give an illusion of power. Moreover, in some Third World countries, these institutions do produce people who govern the country, compete in international business and build bombs. In that sense the institutions have become a part of the community of the oppressors rather than the oppressed; they have moved from the Third to the first or second world. It is precisely this possibility which directs Third World institutions to ally themselves with hegemony and with the dominant trend in education.

In the final analysis, the power of western hegemony rests on the claims of superiority, universality and ethical neutrality of westen math and (positivist) science and technology extended into claims of western superiority in the social, cultural, moral political and intellectual spheres. Continuing to accept western math and science as universal and authoritative is detrimental to creating a healthier and more humane world. Like other activities, math and science need to be treated in a critical way, not only at the implementation or application stage but also, and more importantly, at the level of the basic premises and values governing conceptions, practices and production.

Education perceived as praxis is the opposite of hegemonic education. Praxis is the combination of concrete conditions (social, cultural and material), reflection and action in constant interplay. Education perceived as praxis, implies the existence of two inter-related concepts, the contest of authentic dialogue and the real, concrete context of facts, the social reality in which people exist. The primacy of the learners' experience in this type of pedagogy is obvious. Within this perception of education, the essential functions of intellectuals and institutions within the Third World are to make sense of the world, of our experience and of our culture; to respond to the challenges posed by the real environment; to empower people; to overcome the culture of silence; to remove obstacles to learning; and to release the human capacity to understand and act in order to help transform consciousness and society. The purpose of both structure and teachers is to facilitate that praxis and that empowerment. Crucial to achieving this are: gaining a power base in one's own culture and society; starting with the perspective and experience of the indigenous population; creating an education relevant to the community's needs and geared towards production (not only in the material sense but also in literary, artistic fields etc.); and having the community as a primary ally of institutions and intellectuals. It is also important to stress here that empowerment includes acquiring the means to critically appropriate knowledge existing outside of our immediate experience in order to broaden our understanding of ourselves, the world, and the possibilities for transforming the taken-for-granted assumptions about the way we live.

The study of and interest in ethnomath should not be an end in itself. We should avoid the pitfall of reversing the previous attitude. Indigenous cultures and ways were generally ignored, marginalized or simply considered backward in educational activities and thinking. We should avoid treating them as precious things to be preserved at any cost and celebrated at such occasions as international conferences and we should avoid glorifying them out of context, out of the realities, activities, needs and consciousness of the moment. I do believe that cultures should play a revitalizing role in math education; culture here refers to 'the particular ways in which a social group lives out and makes sense of its given' circumstances and conditions of life. In order for culture to play a positive role, however, education perceived as praxis has to be stressed; linking education with the real environment is central.

Within the perspective of education perceived as praxis, the study of ethnomath becomes an avenue to recognize the fallacies and dangers of a universal conception of life and the world; to question established beliefs and convictions; to free our imagination from ready answers and solutions; to help us gain concreteness and new meaning of words and ideas through real life situations and personal experiences; to provide us with diversified ways of seeing the world and our environment and dealing with the fundamental problems we face; and to help us transform our assumptions and structures.

In this sense, the linear concept of progress has to be abandoned. The belief in a best and universal path is detrimental to the richness which is manifest in life and in communities as well as to the production of conceptions, practices and attitudes (including those in math and science) which are meaningful and relevant to the majority of people. The strongest case against hegemonic formal math, in my opinion, is that it is useless and meaningless to the majority of students.

In conclusion, the stress in math education should not be so much on the cultural determinants or the differences in the conditions in industrialized and in developing countries, rather whether we perceive education as praxis or not. It is probably easier to do this in Third World communities because the differences between the ideological environment created by formal education and the real environment are easier to see and because other ways of doing math are still practised. Thus, the basic challenge we face in math education, in my opinion, is how to teach math within the perspective of math in practice.

References

FASHEH, M. 'Mathematics, Culture and Authority', *For the Learning of Mathematics*, 3, 2, (November, 1982)

7
Folk Mathematics

Eugene Maier

This article is an edited version of the one that first appeared in *Mathematics Teaching* 93, December, 1980.

Consider the following question from the 'consumer mathematics' section of the first National Assessment of Educational Progress, the nation-wide testing program that has surveyed the 'educational attainments' of more than 90,000 Americans.

> A parking lot charges 35 cents for the first hour and 25 cents for each additional hour or fraction of an hour. For a car parked from 10.45 in the morning until 3.05 in the afternoon, how much money should be charged?

This question was answered correctly by only 47 per cent of the 34,000 17-year-olds tested, a result widely cited as an example of Americans' poor mathematical skills. But does the 'parking-lot' exercise have any validity? Does it actually measure ability to handle real parking-lot arithmetic?

Obviously in a parking lot the problem of figuring one's bill is never so clearly or explicitly stated. One is not handed a paper on which all necessary data are neatly arranged. Instead, information must be gathered from a variety of sources — a sign, a wristwatch, a parking-lot attendant. Paper and pencil are seldom available for doing computations. One is unlikely to go through the laborious arithmetical algorithms or procedures taught in schools and used in tests. One is more likely to do some quick mental figuring.

And a few people compute an exact bill, even mentally. It is easier and more efficient to figure an approximate answer: 'I've parked here for less than six hours and the rates are slightly higher than 25 cents an hour on average, so my bill shouldn't be more than $1.50'. Unfortunately, approximation is little taught in schools, and most people do not feel comfortable enough with numbers to try it. Most people might have a very rough notion of what the bill should be, intuitively or based on prior experience. They rely on the parking attendant to compute it. The attendant, of course, almost certainly relies on some mechanical or electronic device.

There may be some mathematics done in parking lots, but probably very little of the sort measured by the National Assessment. The situations are radically different. People do parking lot arithmetic in parking lots using methods appropriate to parking lots, not in classroom using paper-and-pencil methods.

But testers are not alone in producing specious measures of 'real world' mathematical skills. The average elementary mathematics textbook is full of 'story problems' like the following:

> Mr. Brown made 3 gallons of garden spray. He put the spray into bottles holding 1 quart each. How many bottles did he fill?

It is hard to imagine anyone facing such a problem outside a classroom. Certainly few elementary children have ever manufactured garden spray, or ever witnessed adults doing what Mr. Brown was purported to do.

Yet such 'problems' give the appearance of being from the 'real world'. Supposedly they are intended to relate school experiences to life outside school. But they have little in common with that life. They are school problems, coated with a thin veneer of 'real-world' associations. The mathematics involved in solving them is school mathematics, of little use anywhere but in school.

Much school mathematics consists of abstract exercises unrelated to anything outside school. This is why it is disliked and rejected by many. School children recognize that school mathematics is not a part of the world outside school, the world most important to most people. Yet if school is to be preparatory for life outside school, the school world ought to be as much like the non-school world as possible. In particular, young people in classrooms ought to do mathematics as it is done by folk in other parts of the world. School mathematics ought to emulate folk mathematics.

Woody Guthrie defined folk music as the 'music folks sing'. In the same way, folk mathematics is mathematics that folks do. Like folklore, folk mathematics is largely ignored by the purveyors of academic culture — professors and teachers — yet it is the repository of much useful and ingenious popular wisdom. Folk mathematics is the way people handle the mathematics-related problems arising in everyday life. Folk mathematics consists of a wide and probably infinite variety of problem-solving strategies and computation techniques that people use. I believe the first goal of mathematics education should be to assist children to cultivate and enlarge their inherent affinities and abilities for folk maths.

Attempts have been made in recent years to teach 'real-world' mathematics, a phrase betraying the curious notion that school is somehow unreal. These attempts usually have failed, producing jumbles of 'story problems' like 'Mr. Brown made 3 gallons of garden spray . . . ' Such 'real-world' mathematics describes no place I have ever been or wanted to be. Such curricula only serve to make school mathematics more meaningless and absurd.

Surveys 'in the field' to determine what mathematics is used in various occupations have produced lists of topics that read much like the table of contents of an arithmetic test: addition, subtraction, multiplication, division, fractions and decimals, ratio, proportion and percentages. Teachers conclude they are already teaching those things and return to the security of the text. What is overlooked is *how* and *why* mathematics is done outside school. School maths and folk maths range over much the same mathematics topics. But folk mathematicians — that is, all of us when we are facing a mathematics-related problem in everyday life — do mathematics for reasons and with methods different from those commonly involved in school maths.

I have heard a college teacher tell how she became aware of the vast differences between school mathematics and the mathematics of her kitchen. She noted that in halving a recipe that called for 1/3 cup of shortening, she simply filled a 1/3 cup measure until it looked half full. In school this would be presented as a 'story problem' with the correct answer '$(1/3) \div 2 = 1/6$'. Solving the problem would require paper and pencil, certain reading and writing skills, and the ability to divide fractions. But in the kitchen, paper and pencil are seldom handy. Neither the problem nor solution is written. And dividing fractions is not really necessary.

I remember eavesdropping on a friend of mine, a building contractor as he related his train of thought in computing 85 per cent of 26. '10 per cent of 26 if 2.6, and half of that is 1.3' he said. 'So that's 3.9, and 3.9 from 26 is — let's see, 4 from 26 is 22 — 22.1 is 85 per cent of 26'. He computed 15 per cent of 26 and then subtracted, using some slick mental arithmetic in the process. I was struck by what I had heard, knowing that the method of finding percentages taught in school involves a complicated algorithm requiring paper and pencil. I asked him whether his school experience had anything to do with how he handled the percentage problem. 'Didn't you know? I quit school in the sixth grade to help out on the farm,' he said.

I once watched a crew of workmen replace the metal gutters and downspouts on an old three-story building. With only a few metal-working tools and measuring tapes, they quickly cut, bent and fitted the gutters and downspouts, probably unaware of how nicely they were dealing with three-dimensional space. I wondered what would happen if a crew of mathematics educators wrote a textbook on mathematics for gutter and downspout installers. I imagined all the standard school geometry and trigonometry the textbook would contain, and how the crew I was watching would see no relationship between what they were doing and the contexts of the text. Such are the differences between school and folk maths.

The differences do not only exist in the adult world. Watch children playing Monopoly. One lands on Park Lane and owes rental on two houses. 'That's £390', demands the owner. A £500 bill is offered, correct change made, and the game proceeds. But translate the same problem into school mathematics. 'Mr. Jones sent

Acme Estates a cheque for £500. However, he owed them only £390. How much should be refunded?' The same children scurry for pencil and paper, ask a flurry of questions, worry over correct procedures, and hurry to get on to other things.

Consider an incident the father of an eight-year-old related to me. His son brought home a teacher-made drill sheet on subtraction. The child had completed all the exercises, save one which asked for the difference 8 − 13. That one the child had crossed out. The father knew his son had computed such differences in trinominoes, a game in which it is possible to 'go into the hole'. The father asked the boy if he knew the answer.

'Yes', he said, 'it's − 5'.
'Well, why did you cross that exercise out?' the father asked.
'In school, we can't do that problem, so the teacher said to cross it out'.

Further discussion revealed that since negative numbers had not yet been introduced in class, the teacher had said that one cannot subtract a larger number from a smaller number. Hearing this, the father's older son, aged eleven, asked, 'Why do teachers lie?'

For the eight-year-old, school mathematics had already become different from folk maths. Knowledge and skills acquired outside school no longer seemed to apply inside, a most confusing development. To the 11-year-old, teachers were no longer trustworthy. The older boy had begun to realize that school maths is less authentic and less reliable than folk maths.

Some of the general differences between school maths and folk maths are clear. One is that school maths is largely paper-and-pencil mathematics while folk maths is not. Folk mathematicians rely more on mental computations and estimations and on algorithms that lend themselves to mental use. When computations become too difficult or complicated to perform mentally, more and more folk mathematicians are turning to calculators and computers. In folk maths, paper and pencil are a last resort. Yet they are the mainstay of school maths.

Another difference is in the way problems are formulated. In school almost all problems are presented to students pre-formulated and accompanied by the requisite data. For folk outside school, problems are seldom clearly defined to begin with, and the information necessary for solving them must be actively sought from a variety of sources. While talking with a trainer of apprentice electricians, I realized that an industrial electrician is much more likely to be asked 'What's the problem?' than be told 'Do this problem.' Yet technical mathematics courses for electricians are filled with the latter statement, whereas the former question seldom, if ever, occurs.

And the problems themselves differ between school maths and folk maths. Many so-called 'problems' in school maths are nothing more than computation exercises. They focus on correct procedures for pushing symbols around on paper. Folk mathematicians compute too, but for them it is not an end in itself. Pages of long division exercises are not part of folk maths. Problems in folk maths deal with what it will cost, how long it will take, what the score is, how much is needed. Problems in folk maths deal with a part of one's world in a mathematical way.

How can school maths be made more like folk maths? School maths should provide school children with the opportunity to deal with the mathematics in their own environments in the same way as proficient folk mathematicians do. School children should be encouraged to formulate, attempt to solve, and communicate their discoveries about mathematical questions arising in their classrooms, their playgrounds, their homes. All are rich with questions to explore: *How many tiles are in the ceiling? How big is the playground? How old are you — in seconds?* Children should be encouraged to develop their own solutions and ways of computing, building on their previous knowledge. Schools should be mathematically rich environments providing many opportunities to develop and exercise mathematical talent. In this setting, the role of the teacher is to bring mathematical questions to the attention of children, encouraging them to seek answers, and talk with them about possible solutions while allowing them to grope, err and discover for themselves.

All this is terribly idealistic and difficult to achieve. But teachers, administrators, curriculum developers and especially those of us who call ourselves mathematics educators should be seeking ways of making school maths more like folk maths. There may be something inherent in schools, in the constraints and demands placed upon them, that will prevent school maths and folk maths from ever being the same. But the gap between the two need not be a chasm.

8
Hidden Messages

Jenny Maxwell

This article first appeared in *Mathematics Teaching* Number 111, March, 1985.

There is an almost unchallenged assumption that mathematics education, for both teacher and taught, occurs in a political vacuum. This I cannot accept: it seems impossible that such a central part, mathematics, of such a political institution, education, should really be politically neutral.

It is easier to be objective about, and therefore to recognize, the social bias of mathematics questions from abroad. Chinese examples stress agricultural and military applications; a Cuban textbook asks children to find 'the average monthly number of violations of Cuban air-space by North American airplanes': Russian children are asked about collective farms: East German children have a similarity problem about a tower in Berlin and a bigger one in Moscow. All the following problems contain assumptions, some explicit and some implicit, about the society from which they (or in some cases their enemies) come.

> Twenty-three peasants are working in a field. At midday six guerilla fighters arrive to help them from a military base near to their village. How many people are working in the field.   (Mozambique)

> Once upon a time a ship was caught in a storm. In order to save it and its crew the captain decided that half of the passengers would have to be thrown overboard. There were fifteen Christians and fifteen Turks aboard the ship and the captain was a Christian. He announced that he would count the passengers and that every ninth one would be thrown overboard. How could the passengers be placed in a circle so that all the Turks would be thrown overboard and all the Christian saved?   (USA)

> A Freedom Fighter fires a bullet in to an enemy group consisting of twelve soldiers and three civilians all equally exposed to the bullet. Assuming one

person is hit by the bullet find the probability that the person is a) a soldier, and b) a civilian.   (Tanzania)

When worker Tung was six years old his family was poverty-stricken and starving. They were compelled to borrow five *dou* of maize from a landlord. The wolfish landlord used this chance to demand the usurious compound interest of 50% for three years. Please calculate how much grain the landlord demanded from the Tung family at the end of the third year.
                                                                                                         (China)

These examples coming from foreign cultures strike most of us as blatantly political, a part of the indoctrination of the young into the currently dominant values in these societies. But I feel that we are less aware of this same process when it occurs in British schools.

To find out more about this I recently made up a small collection of questions on percentages. They were based on textbook questions, some almost as printed, but slightly altered to make political points. I then asked some twenty-five teachers (in schools and further education) for their reactions to these questions.

One question concerned Mr. Jones who owned a factory employing 100 people. He drew a salary of £15,000 p.a., paid each of the 20 supervisory staff £10,000 p.a. and the other 80 employees £6,000 p.a. The question asked for the total wage bill and went on:

The company has done well in the last year so Mr. Jones decides to the give himself a 15% rise, the supervisory staff a 10% rise and the non-supervisory staff a 5% rise.

There were then questions about the new wages bill. Only ten out of my twenty-five teachers found the social and political assumptions here worth commenting on:

'There's a bit to talk about there . . . we might have a little talk about industrial justice'.
'This is the shocking one . . . I'd have to have a laugh about it . . . I couldn't resist a comment'.

It was surprising that there was divided opinion, even amongst teachers from the same institution, about whether or not the students would notice the inequalities of this industrial situation.

Another question was ridiculous, about a spider's weight increased to 500 gm. Sixteen teachers commented on this absurd situation, six more than had mentioned the social context of the previous question. Was this because the first consolidated their own view of the world?

One question was about a man who won £2,000 in a competition and the way in which he shared the prize money, his wife getting nothing and each of his sons more than their (older) sister. No one mentioned the latter point and only four teachers the former. Again, did they not notice or was it that this is how they know things are? One teacher made a different point: 'I don't believe in competition . . . I usually say so.'

It was alarming that only about half the teachers responded adversely to the use of the word 'alien' in the following question:

Assuming that the number of aliens in the UK is 1/3% of the population, how many aliens would you expect to find in a crowd of 60,000?

'I would avoid [it] . . . [it is] likely to cause embarrassment to certain people and give rise to the nastier feelings of one or two members of our society'
'Calling people 'aliens' smacks of racialism'.
'We'd be on very dangerous ground . . . to use words like "alien" or even to draw to children's attention that our society is a racial mix. We've got quite a few pupils in our school who will seize on anything like this as a means of causing friction between the various groups.'

But what of the half who did not comment? They came from a variety of schools and backgrounds and it is difficult to believe that their pupils are more immune to racial prejudice than those mentioned above, or that such wording does not encourage prejudice, albeit subconsciously. Indeed on two occasions a teacher accepted the question without comment while another in the same school mentioned the extreme dangers of its use.

Just under half the teachers I spoke to broadly agreed with the two who said of this collection of questions: 'very establishment' and 'obviously class-biased, sex-biased and race-biased'. But the rest either did not find them so, or did not consider it relevant to their teaching of mathematics. It would have been interesting to see the teachers' reactions to questions about profits from burglary or tax evasion rather than investment. Even more revealing, had I been asking the question now, would have been the reaction to this one, suggested by Griffiths and Howson (1974).

A coal mine employs 1,000 men and loses £N per year. If it were closed down suppose 750 of the men would not expect to find another job and would have to live off social security payments. What value of N makes it cheaper to the state to keep the mine open? Is this a good way of thinking about the problem? Would you use this question? Why? Or why not?

Swetz (1978) found a problem in a Tanzanian textbook about canned peaches and 'is concerned about asking a poor man to struggle through problems of the rich'. Likewise British textbooks use questions about mortgages, investment and interest for those whose families, or who themselves, are on social security. In fact the time has perhaps already come when some people would find questions about wages offensive. Few questions ask the rich to struggle through the problems of the poor.

After further discussions with the teachers concerning attitudes, methods and topics ('If a few more people had understood what inflation meant, they might not have won the election') about a quarter of the teachers remained clearly of the view that mathematical education is, and should be politically neutral. It was interesting to see the pervasive and unanimous attitude of guilt and apology whenever a teacher felt he or she was questioning the norms of society. Four teachers expressed fear of being thought 'leftist'. Yet none was anxious about upholding the values of the right.

For some, the discussions and particularly the examples, caused a shift of position:

> 'Just looking at these questions . . . one can . . . have an influence even through mathematics which I see as being unlike any other subjects . . . You've exposed to me that it is quite easy to subconsciously . . . accidentally, inadvertently . . . put forward views which . . . you may not believe in. But . . . through a degree of thoughtlessness and ill-considered preparation you may end up putting forward social views that you disagree with'.

But no one went as far as the teacher quoted by Len Masterman (1980):

> 'For over twenty years I worked under the delusion I was teaching maths. The social pressures I put upon the kids were designed to make my maths teaching more effective. I now realise that I was really teaching social passivity and conformity, academic snobbery and the naturalness of good healthy competition, and that I was using maths as an instrument for achieving these things'.

For me, the final proof that mathematics education is by no means neutral came in answer to the question 'Has mathematics a role to play in furthering social causes and political understanding?' Two teachers, from the same institution, replied: 'Yes, definitely' and 'Oh, I shouldn't think so'.

How can anything which can elicit two such opposing but adamant replies be neutral? Politics is about conflict and there was conflict here.

References

GERDES, Paulus. (1981) 'Changing Mathematics Education in Mozambique', *Educational Studies in Mathematics*, 12.

GRIFFITHS, H. B. and HOWSON, A.G. (1974) *Mathematics, Society and Curricula*, Cambridge University Press.

MASTERMAN, Len. (1980) *Teaching about Television*, Macmillan.

SWETZ, Frank, J. (1978) *Socialist Mathematic Education*, Burgundy Press.

9
Mathematics Education Post GCSE: An Industrial Viewpoint

T. S. Wilkinson

This article first appeared in the *Bulletin of the Institute of Mathematics and its Applications* Volume 25 December 1989. It has been slightly modified by the author for this volume.

Introduction

The views which will be expressed here are of course personal and subjective: they are necessarily biased by long experience in industry in the north-east of England. However, alternative views should not be altogether unexpected. In preparing this contribution the writer felt obliged to reflect on why we teach and learn and why currently, academics, civil servants and industrialists alike should concern themselves with the teaching of mathematics.

There are of course, individual and collective benefits to be gained from education and if, as seems likely, some revision of mathematics teaching is to be undertaken, efforts should be made to increase these benefits. It will be argued that attention should focus on the collective benefit. It will also be argued that demographic pressure will oblige industry to recruit earlier and to participate in education.

Mathematics is the science which underpins all others because it provides the means by which ideas can be expressed concisely and with precision. Even so most mathematicians are aware of fallacious proofs and paradoxes. The rest of the world is not familiar with these private jokes and is invited instead to regard mathematical tools as magic wands. This policy, it will be argued, may be mistaken.

The Needs of Industry

UK Ltd and wealth creation

Wealth is created by and large by making things and selling them at a profit. Some of this wealth can then be used to solve actual and perceived social and environmental problems. Few evidently object to improvements in roads, housing, health care and disposable personal income, yet many object to profiteering and reinvestment with the objective of creating even more wealth, perhaps because few realize or admit that it is increased national wealth which permits increased national spending. (Mathematical models, appropriately qualified, should be used to stimulate more rational discussion of these issues in classrooms).

An examination of the position of UK Ltd. in the world on almost any basis reveals a relative decline in performance. True the decline is slow and discontinuous and it is not by any means a recent phenomenon but, like all deep-rooted insidious sicknesses, recognition takes time and diagnosis may be attempted in an atmosphere of panic. Self-diagnosis will in any case be difficult because few insiders have the right perspective and few outsiders have enough information or concern.

Recently the writer mingled with marine engineers and naval architects for the first time for many years to discuss the teaching of mathematics and science in schools and the needs of the engineering profession. (Beattie and Straker, 1988). Incidentally, at that meeting attention was drawn to a study in the north-east which seemed to show that GCE A Level mathematics is about two points harder than other A levels. (Fitzgibbon, 1988). At that time a shipyard closure at Sunderland was imminent (it has now closed) and the attendant publicity stimulated a certain amount of private discussion of that example of relative decline. Shipbuilding will therefore be used as an illustration but the writer is aware of other less topical examples.

The decline of shipbuilding in the UK

After hundreds of years arrangements were made in 1988 to wind up the remaining activities on the river Wear at Sunderland. Why? It has often been suggested that poor labour relations and demarcation disputes were responsible for failure. There may once have been some truth in that but not recently.

Perhaps labour was too expensive. Visits to the homes of employees did not give that impression; the writer visited several, having worked in the industry. It may have been cheaper for a time in Japan and Korea but not in Sweden.

Perhaps the materials were too expensive. At times steel was relatively expensive and even in short supply, but not recently.

Maybe the Geordies are work-shy or lack skill. It is true that training has been neglected in the last few years but shipwrights, the top men in the yards, can still be found all over the north-east working in entertainment, security and other unskilled jobs. Manual skills derived from apprenticeships and experience are commonplace in the north-east. Furthermore, several Japanese companies have established themselves in the area and Nissan has expressed particular satisfaction with the workforce and has accelerated expansion plans.

May be there is a cultural problem. Perhaps it is the exercise which the Japanese take before and during the working day or some other superficial difference to which the media regularly draw attention. Perhaps not.

Perhaps it is the weather. Even summer can be cold in the bottom of a draughty dry dock or on slipways, the traditional environment. In the late 60s, the writer attended a conference in Gothenburg where it is frequently much colder. At that time Gotaverken were building two VLCCs (supertankers) simultaneously. Supported by a small drawing office in which magnetic tapes for numerically controlled machines were produced, the ships were being built indoors, section by section. When a section had been welded to the partly completed hull, the hull was pushed out of the door to allow another section to be assembled and welded, when the last part of the hull was pushed out, the door was sealed and the dock outside was flooded. In this way two VLCCs were being built in about 25 per cent of the time then being taken by Swan Hunter to build one. All of this seemed to be accomplished without haste and with very few people. This experience and the rapid expansion of shipbuilding in the Far East convinced the writer that it is the quality and training of UK engineering staff which is inadequate.

Engineers design, develop, manufacture, commission, operate and maintain a very wide range of useful passive and active equipment. Unfortunately, university courses cover these activities unevenly, not least because few staff can claim to have had wide experience and the body of knowledge of general applicability in activities other than design and development is very limited. Furthermore, the mathematical problems which arise in manufacturing and commissioning etc., are either too simple or too difficult to teach and examine.

The needs of engineering employers

Ninety per cent or more of UK professional engineers will never be required to apply mathematics beyond the level which was formerly taught at GCE O level. The problem faced by employers and employees alike is: who will find themselves among the remaining 10 per cent?

Industry clearly requires versatility and a spread of interests and aptitudes among their staff where institutions seek to standardize training and qualifications. Specialists face problems with career progression and membership of the 10 per cent can, therefore be a disadvantage. One solution, popular in other parts of Europe, is to teach enough to professional engineers to allow them to specialize in almost anything. Unfortunately, high calibre trainees are required and long courses are inevitable.

As dramatic changes to the British system and to the status of British professional engineers and technicians are unlikely in the short term, industry must continue to recruit small numbers of specialists in engineering mathematics. Clearly, the training of the 90 per cent should be reviewed and attention should be paid to what once was known as practical mathematics pre and post-GCSE.

Employers must also be persuaded to explain their needs coherently. The transformation of manufacturing industry in the last decade or two has brought great changes in that world. It is now unfamiliar to most academics and even to older industrial staff. The machine tools which were until recently operated by 'craftsmen' have almost disappeared. By and large they have been replaced by more versatile and complicated computer controlled machines and although employers may variously believe they are operated by craftsmen or computer experts or semi-skilled labour, the world of academe would probably regard the visible operators as technicians. The truth is more complicated. Individuals no longer operate machines independently; operation is diffused among those who generate the various computer programmes and data which drive the machines.

If academic staff can find opportunities to visit the new industrial world and are able to discuss their observations with the employers who are successfully managing and promoting change, between them they will be better able to identify the skills likely to be required in future. It would be more difficult and it would perhaps be unwise to attempt to describe that world to students. The aim should rather be to discover model problems and to devise analogous problems in a world already familiar to students (or one which can be created easily), thus to exercise and develop the skills they may require.

Confidence and scepticism

The present width and depth of GCE A Level mathematics is almost certainly excessive. The material which is covered does not allow sufficient practice and consolidation so that, if there is no reinforcement at undergraduate level, the ability of moderately successful candidates to apply what was learned will diminish exponentially with time. The Higginson report (DES, 1988) was probably shelved too hastily: it was suggested that five subjects would be possible at A Level each of which would, presumably be approximately half of the width or depth of the existing A Levels. This would allow mathematical modelling, information technology and traditional communication skills to be acquired and practised. Even if many engineers received no further mathematical training they might thus gain the confidence to apply what they have learned. Employers frequently complain with just cause that new recruits are unable to solve problems and do not communicate well.

The writer's experience may be peculiar as implied by the disclaimers above, however any review of the material taught at A Level should include reconsideration of statistics.

It is of course, frequently necessary to design experiments, to make deductions from experimental observations, to test hypotheses and to find the 'best fit' to a theoretical model. However, a minimum of theory is required for these purposes. In short, if a relationship is not obvious, calculations will not help. On the other hand professional mathematicians and statisticians are characteristically sceptical and take little on trust. Others, particularly economists, politicians and engineers, would do well to copy them. Many examples of the misuse of statistics and the misinterpretation of superficial evidence are available to teachers of mathematics: they should make regular use of them. A healthy dose of scepticism would encourage the development of more appropriate mathematical models and application of the scientific method.

Investment in Training

Employers are, of course, reluctant to invest in training, particularly if it may not show a quick return. They are, quite properly, motivated by self-interest. Training benefits individuals at least as much as employers but a unique opportunity will occur in the 1990s. Demographic changes will oblige employers to put a high value on employees in general and on new recruits in particular. They will be much more inclined to recruit post-GCSE and to participate in training thereafter. There will be a mutual benefit: students will increasingly be expected to be self-financing, for example by accepting loans. A substantial proportion of post-GCSE education may thus be undertaken part-time. If so the mean level of performance at degree level will certainly improve. There is plenty of evidence that motivation and confidence increase when students gain relevant experience in the world of work and are treated as adults.

If employers, including universities and polytechnics are wise enough to recruit some of their professional engineers from continental Europe and Japan some of the gaps in UK training and experience may be exposed and filled.

Concluding Remarks

Production engineers and workshop technicians usually have to think on their feet in uncomfortable and noisy environments. In these circumstances a calculator can be a liability: their use in mathematics classrooms, though not elsewhere, should be minimized. (As a general rule, only those students who can make calculations quickly and accurately without calculators should be allowed to use them. By the same token, only those with complete command of their native language should be granted poetic licence). Of course, microcomputers with graphics capabilities can be very valuable teaching aids particularly when high quality demonstration programmes are available to illustrate physical and mathematical phenomena.

Most machine tools will soon be computer controlled but the computers need to know where the cutting edges or points are, relative to the workpiece and trial cuts are often used to establish these data accurately. The ability to make trigonometric and other geometrical calculations quickly and accurately (with the help of microcomputer or calculator) can then be very useful. The ability to interpolate, to approximate and to extrapolate confidently will ever be useful. Individuals with those abilities tend naturally to lead shop floor activities.

Whereas a diet of 'practical mathematics' may suit a substantial number of students, it may disadvantage some of those who are destined to be users and developers of higher mathematics. Many of them will find themselves among the 10 per cent of specialists identified already. To them computers will be their aides and the essential tools of their trade. Pre- and post-GCSE they should be developing mathematical models and exploring them with the help of computers and practising communication. They should be encouraged to adopt the sceptical questioning attitude of mathematicians. They may then develop an appetite for the diet of mathematics and its applications which belongs in higher education and which should arguably be delivered by qualified and able mathematicians post-18.

References

BEATTIE, J.F. and STRAKER, N. (1988) 'The Teaching of Mathematics and Science in Schools and the Needs of the Engineering Profession', NECIES, 105, Part 1.

DEPARTMENT OF EDUCATION AND SCIENCE (1988) *Advancing A Levels*, London, HMSO.

FITZGIBBON, C. T. (1988) 'Recalculating the Standard', *Times Educational Supplement*, August 26.

10
The Computer as a Cultural Influence in Mathematical Learning

Richard Noss

This article has been edited from one that first appeared in *Educational Studies in Mathematics* 19 (1988) pp. 251–268.

What is the Cultural Context of Mathematics?

Making sense of the advent of the computer into the mathematics classroom entails a cultural perspective, not least because of the ways in which children are developing the computer culture by appropriating the technology for their own ends. But what of the culture into which the computer is being introduced? Much illuminating comment in this area has come from researchers whose focus has been ostensibly with non-western cultures. For example, Gay and Cole's (1967) study of the culture and mathematics of the Kpelle of Liberia centres around the task of constructing bridges between the Kpelle's 'indigenous' mathematics and the 'new' mathematics of the school curriculum.

It is the contention of this paper that the task of bridge-building is central, not just for introducing 'new mathematics into an old culture' but equally for introducing formal mathematics into technologically developed cultures such as our own. As Mellin-Olsen convincingly shows (Mellin-Olsen, 1987), many if not most of the children in western classrooms are confronted with the mathematics of a subculture of which they are not, and perhaps have no wish to be members, where there is in Papert's phrase, no 'cultural resonance' (Papert, 1980) between their own economic and social activities and the activities in which they are invited to participate at school. While technological development may on one level appear to obscure the relationship between formal and informal mathematics (or in D'Ambrosio's (1985) terms, 'ethnomathematics'), I propose that the technology itself, specifically the computer, can be the instrument for bridging the gap between the two.

What does it mean to do Mathematics?

It is clear that most children and adults are not aware that they are engaging in mathematical activity even when they are involved in quite complex numerical or geometric activities (Wolf, 1984)[1] and that for many, the very suggestion that they participate in mathematical activity is sufficient to induce panic (Buxton, 1981). Yet mathematical ideas and mathematical ways of thinking provide powerful means of making sense of our social, economic and cultural environment. As Lancy puts it 'Grouping, categorizing, generalizing etc. represent a fundamental human need every bit as basic as the need to eat, to drink or to socialize' (Lancy, 1983 p. 64). In the sense that human beings are by definition creatures who seek to explain and control the environment, everybody is a mathematician. Much the same could be said of any systematized way of thinking about the world, for example in relation to philosophy 'It is essential to destroy the widespread prejudice that philosophy is a strange and difficult thing just because it is the specific intellectual activity of a particular category of specialists or of professional and systematic philosophers'. (Gramsci 1971, p. 323). If we substitute 'mathematics' for 'philosophy' in Gramsci's claim, we are forced to take seriously the problem of defining that which is special to mathematical as opposed to philosophical activity. I suggest that it is useful to conceive of a difference, on the one hand to avoid the danger of subsuming under the title of mathematics any activity which involves abstraction and generalization, and on the other of defining mathematics out of existence for example, as simply a way of thinking about relationships and structure. I think the difference lies in the formalism inherent in specifically mathematical activity. I do not want this to be confused with formal in the sense of a formal system; there is more than enough room in my meaning for the intuition and playing around that characterizes the process of mathematical activity but it is precisely this issue of formalism which I will argue, suggests a constructive role for the computer.

The question of what it means to do mathematics is central to understanding the possible roles that the computer might play, particularly in contexts other than its direct employment as a tool for solving a predetermined problem. This issue is the subject of some debate among those who have considered the importance of culturally embedded mathematical activity. Gerdes (1986) for example, uses the example of the regular hexagonally patterned baskets used by Mozambican fishermen to argue that there exists 'frozen' mathematics within a culture which can serve as starting points for mathematical activity within classrooms of that culture. He points out that the artisan who merely imitates the technique is not doing mathematics, in contrast to those who discovered the technique. Gerdes suggests that understanding the processes underlying the techniques is the crucial step in participating in mathematical activity. In doing so, he stops just short of defining such activity as mathematical in itself.

In discussing Gerdes' description of the mathematics embedded within the basket-weaving Keitel (1986) comments: 'It is not the point whether or not this is mathematics, in my view it is very much what Freudenthal calls pre-mathematics (with respect to cognitive levels of children in Western culture. And it is evident that such examples are excellent starting points for discovery learning, or, and that is Gerdes' concern, for embedding mathematics education in a peculiar culture environment.) The point here is that human work structures reality according to regularities which potentially are accessible to mathematical analysis.' (Keitel, 1986, p. 44).

Keitel's observation states succinctly a role that 'spontaneous' mathematics is seen to play in the learning of mathematics, namely as a starting point for more formal learning. Keitel's point is that basket weaving is not mathematics; it is what she calls the 'fore-stage' of mathematics, 'a field of problems in social reality which immanently, or at least partly, are organised in some correspondence with mathematical structures, and hence may better, or even exclusively be solved by the employment of mathematical devices'. (Keitel, 1986 p. 45). As Hoyles (1986) points out, participation in the activity itself is not mathematics unless the didactical content is such as to provoke reflection on and synthesis of the mathematical relationships embedded within the activity.

Can we consider this kind of 'intellectual material' (Mellin-Olsen, 1986), culturally-embedded mathematical activity, as more than simply starting points for mathematical learning? In the first place, there is the question of formalization. While it is perfectly possible to envisage all kinds of formalizations which can be generated by the basket-making activity, it is clear that the process of basket-making itself does not require any formalization or abstraction. And yet it is that kind of abstraction (which is essentially algebraic in character) which lies at the core of official mathematics. I will use the term mathematization to cover all the processes of mathematical formalization and abstraction, including those which are preformal in the sense of being non-written, intuitive and fragmented. Traditional attempts to use situations as the basis for subsequent mathematization have been based on attempting to develop links between the activity and the mathematical abstraction. What we have not had at our disposal was the means for learners to engage in culturally embedded activities while simultaneously mathematizing their activities. This option has been closed precisely because the technology at our disposal, books, pens, paper etc., has been inappropriate to construct mathematical environments in which mathematization can occur naturally.

The second problem is that which Gerdes refers to when he suggests that the person who discovers the technique is the one who is doing mathematics. Discovering is at best a haphazard affair and can hardly be relied on as a methodology for learning all of mathematics. Equally crucial (and related to discovery) is the learner's reflection on his or her own activities, again not unproblematic in conventional learning environments. Here again, it is possible that the available technology may be an important element. For the computer does contain the potential for focusing learners' attention on selected ideas and concepts by providing feedback in an interactive way which is not available with other technology (but not, as I shall argue below, without careful intervention by teachers).

The Relationship of Technology to Mathematics Education

At this point I want to look more closely at technology and its role in mathematics education, on the understanding that technological development is merely one form — the material form — of a more general notion of culture that consists of the sum total of deposits in the consciousness of humans. For the technology which is at the disposal of a given culture (and which dialectically has grown out of it) directly influences the kinds of mathematics which are indigenous, spontaneous or frozen into that culture.

Bishop (1979) gives many examples of the ways in which technology and culture are related to mathematical development. He cites the case of two university students in Papua New Guinea who drew a map of the campus which contained no roads; they were born in the island region where roads did not exist. Lancy in his study of Papua New Guinea (Lancy, 1983) shows that culture and schooling do have an effect on cognitive development and that despite differences in technology (between different Papuan cultures) which are reflected in the structure of language 'if you select the right domain and direct your questions to the right levels in the hierarchy, taxonomic behaviour will emerge and at a very early stage' (Lancy, 1983, p. 159).

There is ample evidence, at least from Third World countries of the distance between everyday language and activities and those of school (see for example Mitchelmore, 1983 and Berry, 1985). For the Kpelle child 'the world remains a mystery to be accepted on authority not a complex pattern of comprehensible regularities'. (Gay and Cole, 1967, p. 94) Could not much the same be said for many of the children of our schools in the west? And what difference does technology make?

The culture of a non-technological society contains a variety of contexts for the generation of mathematical abstraction but it cannot be equally rich since it is the material culture which determines the complex organization of society and the ideological and intellectual forms which accompany it. In technological cultures, practical activities have become increasingly complex and the sciences have become deeply interwoven with everyday life and paradoxically increasingly invisible.

Let us consider the culture of western mathematics, a culture in which children's schooling beyond a certain age (say 12 or 13) is based on the symbolic abstraction of algebra. To be sure our culture, like all others, contains within it every day situations in which algebra exists (albeit in a 'frozen' state) and many attempts have been made to draw links between children's experience and their mathematics. This store of situations, compared to a non-technological society, is quite rich. Could we say that algebra forms part of the indigenous mathematics of our culture? Hardly; in fact quite the reverse is true. Precisely because of the ways in which the hitherto existing technology has proved a problematic vehicle for the introduction of algebraic abstraction, there are severe cultural obstacles to developing learning environments in which children can actively engage in formalization as opposed to those which may serve 'merely' as foundations for formalism.

Meaning and Functionality

I recently gave a lecture in which I presented a class of forty pre-service students (not mathematics majors) with a number of mathematics examination questions which contained rather clear political, moral or other value judgments. The students agreed that one of these questions was both violent and racists (it is the one referring to 'Christians' throwing 'Turks' overboard in Maxwell, 1985, reproduced in this volume) and I suggested to them that they might like to recast the question in 'value-free' terms to see if such an exercise was feasible or desirable. The results overwhelmingly conformed to 'text-book' mathematics problems with reference to human beings expunged altogether and meaningless references to beads and other artefacts of the mathematical classroom replacing human beings. Just over half the replies simply replaced 'Christians' and 'Turks' with various kinds of confectionery (sweets and toffees predominating) and fruit (mostly apples and oranges). It seems that the price that is paid for removing value judgments from mathematical problems is to literally de-humanize them. In comparison a recent examination question in the UK which invited examinees to compare the total amount spent on armaments by the USA and the USSR together with the amount needed to feed the world's population, received a prolonged and unanimous attack in the national press. Small wonder then that mathematics lessons are so often 'not about anything'. (DES, 1982).

Considerable insight into the question of the dehumanization of the subject has been provided recently by considering differences in cognitive styles between men and women (Gilligan, 1982) and commented on from a mathematical perspective by Brown (1984) and more recently by Papert (1986). Mathematics, at least the mathematics of the school classroom, is typically seen as hard-edged, as a subject in which meaningless problems are posed at best about real but material objects but often about unreal and meaningless objects.[2] We need to be wary of accepting that mathematics needs to be hard-edged and dehumanized, worst of all by default; there is an interesting debate to be had on whether abstraction and dehumanization are necessarily linked (see Davis and Hersch, 1986, for one viewpoint).

I have argued that the central problem is thus in injecting meaning, and a particular personal meaning, into school mathematics (see Hoyles, 1985a). It is inconceivable that children in general will be able to utilize mathematical tools and concepts unless they feel personally involved in their use; that is unless mathematical concepts become functional tools embedded within children's own cultural environment. At the very least, we need to find objects other than sweets, toffees and fruit with which it makes sense to undertake symbolic manipulation.

Of the pedagogical and technological innovations available, it is far from obvious that the computer has a role to play. Indeed the cultural view of computers is precisely that they are deterministic, dehumanizing and cold. The implications of this cultural perspective for education and for mathematics education in particular have been far reaching and the attack on computers as a culturally destructive medium has not been muted. As Papert (1985) points out, by focusing on the machines rather than on how they fit into the ambient culture, proponents of this view are no less guilty of a 'technocentric' perspective than the hackers they attack. The criticism of computers (sic) is exemplified by Davy (1985) who asks 'What kind of culture are we developing if people have to meet its most powerful ideas through machines rather than through people? (p. 554). This takes us to the central issue. For the thrust of this paper is that the powerful ideas of a culture are always mediated through the technology which is available to that culture. Is it conceivable that our culture's knowledge and understandings could be transmitted to new generations without the employment of pens, paper, publishing technology and typewriters? How much of the science and history of human culture is frozen into the production of a single piece of paper? The question is what facets of the computational environment it is possible to exploit in order to realize a vision of the computer as a means of enriching the indigenous culture of children, rather than Davy's dehumanized version.

Computer Based Learning Environments

The key property of the computer which I want to examine here is its ability to allow its user to explore, investigate and pose problems and to offer flexible representations of situations of which at least one is on the symbolic, formal level. It is interesting that the major area in which the computer has permeated children's culture, that of computer games, is one area where, on the face of it, the player is denied access to this kind of power. But this is only half of the picture. As Turkle (1985) points out in her insightful study of the computer culture, such games do offer the player a high degree of control within a limited domain. This ability of the computer, or rather the sensitively written software, to allow users to interact in a personally powerful way is the common thread that runs through the various cultural manifestations of the computer in society. On the other hand, precisely because such games do not generally allow natural access to the symbolic, formal representation, they have little or no role in mathematics learning.

I do not want to trivialize the potential for cultural resonance which this perspective indicates. Of course there is a resonance which comes about from the fact that computers are everywhere, that they are (sometimes) fun, that they are glamorous and so on. The key point is that children see computer screens as 'theirs', as part of the predominantly adult culture which they can appropriate and use for their own ends. It may be that shooting aliens is less of an intellectually stimulating experience than writing a program but it belongs to the child, it is something which they can take from and perhaps even use to subvert, adult culture.[3]

Most of the computer culture is owned (in every sense) by official society, but just enough is appropriable by young people to feel that they can control it. One of the best examples we have of this kind of appropriation, is that of children learning to programme in Logo. Logo is a computer programming language designed for learning — there is an intentional pun here. It allows the learner extremely straightforward access both to creating interesting screen effects and to the computational/mathematical ideas which underlie them. There has been a number of longitudinal studies which have sought to analyze the power of this environment from a mathematical perspective and which have illustrated that children are able to explore and use a variety of mathematical ideas in a wide range of programming contexts. (Papert *et. al.*, 1979; Hoyles *et. al.*, 1985; Noss, 1985). These studies have confirmed Papert's claim that by learning Logo, the child is behaving as a mathematician, is essentially doing mathematics. But the question remains as to what kind of mathematics. To what extent does the mathematics of the computer culture intersect with the broader mathematical culture?

In outlining an answer to these questions, I want first to clarify what Logo is standing for in this discussion. It is intended as a placeholder for a certain kind of interaction with the computer; an interaction which allows for practical kinds of mathematical activity to take place but it is unlikely to provide such an environment uniquely. New computer-based environments are currently being designed and more established ones are being applied to create similar kinds of learning contexts. Logo, however, does provide the most extensively researched example of this kind of work and it will form the basis for what follows below.

It is helpful to distinguish three ways in which the culture of children's mathematical learning may be influenced by interacting with the computer via Logo, or, and this is the last time I shall make this qualification, with any similarly powerful computer-based tool. The first is to examine how the mathematics that children can *do* is influenced; the second is to ask what implications this may hold for what children *learn*; and finally to examine what may follow for what children may be *taught*.

What Mathematics may Children do?

I want to illustrate ways in which the computer can influence children's mathematical activity by an example. It concerns a group of seven 13-year-old Logo-experienced children working on a structured and progressive set of tasks both on and off the computer, which were designed to place the children in a situation that would allow them to 'bump up against' the ideas of ratio and proportion. The tasks involved designing a programme for generalized N-shaped figures whose complete 'solution' was certainly beyond their mathematical experience, since it would have involved the invention of trigonometry. For details of the task and the findings see Hoyles and Noss (1989). The topics of ratio and proportion have been well researched in particular from the Piagetian perspective that understanding of proportionality becomes evident only in the formal stage; later work has seemingly confirmed that children of early secondary school age (13–15 years) find extreme difficulty in thinking of a relationship between two quantities as requiring anything other than an additive operation (Hart, 1980).

Analyses of the strategies employed by the children engaged in this Logo task revealed a number of interesting characteristics. Firstly, none of the children adopted the 'additive' strategy on the computer which could have been predicted from existing research findings. Secondly, six of the seven children eventually proposed solutions which could be classified as involving a proportional strategy. Thirdly, when the same children were given a pencil-and-paper ratio test (Hart, 1980) their performance reflected the findings of Hart's study with none producing the correct answer to an item of roughly the same depth and content as the N-task which they had for the most part successfully tackled in the Logo context.

The conclusion from this exploratory study is that the computer provided the support by which children could explore and develop relationships that were just beyond their grasp with traditional pencil-and-paper technology. For example, one child recognized quite early on in the task the need to find some kind of relationship between two crucial lengths but was completely 'blocked' as to what to do about it when working on paper. Moving to the computer appeared to set him free to explore the range of possibilities, an opportunity offered at least in part by the interactive nature of the environment.

The feedback provided by the computer however, offers only a surface explanation of the results. The difficulty which children experience with multiplicative relationship is essentially a cultural one. Children's experience is largely multiplication-impoverished (witness the difficulty we have in thinking about images of multiplication for the classroom compared with addition) and contexts requiring non-trivial multiplication are not an everyday part of most children's experience.

What does the computer bring to this situation? It would be incautious to propose far-reaching conclusions for such a small-scale study. Nevertheless findings such as these are beginning to broadly converge (see, for example Hoyles, 1985b; Noss, 1986a). What evidence we have seems to indicate that it is the need for formalization, rather than merely the feedback involved, that is seminal in influencing learners' conceptions. In the example above, the computer is not teaching the child about ratio, it is enlarging the culture within which the child operates. The essence of the computer-child interaction is built on the synthesis between the child's need to formalize the relationship algebraically, or otherwise, of his or her intentions by perceiving the effect on the screen.

In proposing this explanation, I am emphasizing the opportunity afforded by the Logo environment to use symbols in a meaningful context; to pose and solve problems with symbols rather than to play with 'concrete' situations which subsequently, and often artificially require symbolization (see Pimm, 1986 for an enlightening discussion of this point in a general context). It is in this sense, that of offering a context in which mathematical formalization is a necessary part of a system to be explored, that the culture of mathematical learning may be enlarged by the computer.

What Mathematics May Children Learn?

Being involved in an activity is not a sufficient condition for learning to take place; to pose a slightly weaker version of the same statement, it does not guarantee a match between what the learner learns and what the teacher thinks they are learning. For the latter, reflection is required, as well as a conscious effort to draw the learner's attention to the 'important' relationships involved.

There is however, a significant range of learning which, in Papert's sense, is Piagetian; learning which takes place through immersion in a culture (a culture that is which may contain important vehicles for 'teaching'), the most obvious being that of the acquisition of natural language. The attempt to locate learning of this kind, generated by computer interaction, has only recently begun (see for example Lawler, 1985) and has been beset by studies which have employed poor research designs or untenable hypotheses.

So far as mathematical learning is concerned, some illumination of this problem can be given by starting from the observation that many children appear to harbour fundamental misconceptions about elementary geometrical concepts until quite an advanced age. For example, Hart's study (Hart, 1980) indicates that many children fail to appreciate that the length of an object is not changed by displacement; only 42 per cent of first-year secondary students judged that two lines, one oblique and one horizontal with their end points aligned, were such that the oblique line was longer than the horizontal one. This kind of finding is remarkable in that it implies that there is often little shared meaning between teacher and learner of common terms such as 'length'. One explanation of this robust finding may be that the term 'length' in common usage has a much less precise meaning than that which is necessary to answer the above question correctly. The *mathematical* meaning is not usually taught; it is somehow picked up (or not) from diverse settings within everyday activity. Unfortunately the two sets of meanings often do not correspond.

In an exploratory study, Noss (1987) compared the responses of 84 children who had studied Logo for one year and 92 who had no computing experience, on a set of geometrical paper-and-pencil items designed to probe their comprehension of length and angle. The children were aged between 8 and 11 years. The aim was to gain information about the kinds of primitive components of geometrical knowledge, such as length conservation, which might be mentally constructed during Logo work. Two further examples were the ability to distinguish the invariance of an angle under rotation and variation in the length of the rays which define it, and the recognition that the comparison of two lengths depends on the units of measurement employed. In all there were three categories of items for the concept of length and three for angle. The question was whether, by enriching the learning culture in which the children were involved, they would more easily be able to match their different fragmented conceptions of the idea of length and angle into a more conventional mathematical form. The findings of the study can be summarized as follows:

1. For all three angle categories there was a trend (significant in two categories) in favour of the Logo groups.
2. In two out of the three length categories in which comparison with Hart's study was possible, the Logo group's performance was almost up to the level of those in Hart's study (despite a 1–3 year age difference), in contrast to the comparison group which was somewhat lower.
3. In five out of six of the categories, the Logo girls were differentially successful in relation to the boys; in the non-Logo groups, this situation was reversed.

What kind of explanation can be suggested for these findings in terms of the cultural perspective outlined above? That the effect on the pupils' conception of angle was more marked than for the concept of length is consistent with the findings of Papert *et. al.*, (1979), who suggest that new knowledge acquired in the Logo environment has to 'compete' with existing knowledge and that the amount of time required to displace it would depend on how firmly rooted it was. Papert conjectures that knowledge about angles would be more easy to displace than knowledge about length.

This conjecture goes some way towards explaining the differential effect in favour of girls. There is evidence of a deficit of spatial abilities among some girls, most probably accounted for (though not necessarily entirely) by a socio-cultural bias against spatial experiences which tend to be encouraged among boys (Badger, 1981), but whose effect on mathematical attainment is remediable through appropriate activities (Sherman, 1980). If this is the case, then it would follow that Logo experiences are more likely to be helpful for girls to construct geometrical concepts than boys. The appropriateness of the environment to girls is, of course a key element in this chain or argument. On this point Turkle (1985) has suggested that Logo offers a programming environment which is more appropriate to girls' cognitive styles than other forms of computing.

To summarize, there is evidence that some spontaneous learning may take place in suitably designed computer-based environments and that a possible mechanism for understanding this process is to focus on the ways in which the computer influences the cultural reservoir of mathematical ideas available for children to draw on.

What Mathematics May be Taught?

The introduction of the computer into the learning environment entails more than simply a technical component (hardware + software). Firstly there is a pedagogical component which consists typically of a teacher, a curriculum, written materials and so on, means by which the interaction with the computer-based material can be structured and children encouraged to reflect on their activities (see Hoyles and Noss, 1987a, for an elaboration of these ideas). While there may be a category of mathematical knowledge which can be acquired spontaneously as suggested above, it is equally the case that learning mathematics requires a conscious awareness of mathematical structures by students and thus conscious intervention by teachers. A number of researchers have pointed out that it is perfectly possible to remain unaware of the essential mathematical ideas behind Logo programming (see for example Hillel, 1984; Leron, 1985). In addition we need to take account of the contextual component, the setting in which the problems or more generally, the learning situations are framed. It is clear that the distinction between a problem and a situation is intimately bound up with the context. For example as Lave *et. at.* (1985) have shown how for real-life problem-solving a situation becomes a problem in the course of activity in a particular setting, people and setting simultaneously create problems and solution shapes.[4]

A number of researchers has illustrated how the computer can be used to create an environment where such dialectical relationships between problem generation and solution can take place in a reasonably natural way. For example Noss (1986b) has shown how children engaged in a Logo environment switch between exploratory and problem-solving activities and Hoyles and Sutherland (1986) have illustrated how Logo provides both a rich environment for pupil-posed problems as well as a wide range of contexts for spontaneous experimental activity and collaboration.

It should be clear that the computer is being cast in a rather special relationship to the learning process, not simply as another concrete embodiment of an abstract mathematical concept. As Dorfler (1986) argues, the distinction between concrete and abstract is artificial in any case, since it presupposes that there exists an a priori union of actions and operations which is fractured in the course of learning. The key idea is that of focusing attention on the important relationships involved, a role in which the computer is rather well cast, as Weir (1987) points out, but not without the conscious intervention of educators and the careful development of an ambient learning culture.

The particular facility of the computer to focus the learner's attention and simultaneously to provide feedback seems to provide a promising framework for thinking about teaching mathematical ideas in a computer-based context. It is helpful to consider the strand of research initiated by Vygotsky, who emphasized that the key to collaboration is that it provides support to entry into the cognitive area in which the child would not be capable of solving problems unaided — the 'zone of proximal development' (Vygotsky, 1978). In a study of pupils who have been allowed to explore and use mathematical concepts encapsulated within Logo programs, Hoyles and Noss (1987b) have shown that the computer is capable of performing a similar scaffolding role[5] for developing understanding of those concepts. At the same time, we are beginning to understand the role that the teacher can play directly in the process, a role which in a computer-based environment certainly becomes no less critical but considerably more subtle (see Noss and Hoyles, 1987).

The Computer, Research and Curriculum Development

I have argued that the computer used in suitable ways can expand what it is possible to do, to learn and to teach in the mathematics classroom. This process is very much in its infancy, not least because of the limitations of the hardware which are only now gradually being overcome.

Computer-based environments are continuing a tradition within computer science which began with the artificial intelligence community in the 1960s who were more concerned with providing themselves with intellectual tools to help them define and explore situations, than with rigorous algorithm design for a solution of well-defined problems. The parallel with mathematics education is reasonably clear; it is no accident that the former community was the catalyst which provided LISP and its derivative Logo, while the latter has been responsible for languages such as Pascal (and of course BASIC). One of the processes which seems as if it might at last be happening, is that the increasing power of the hardware and software is allowing the explorers to displace the educational hegemony of the problem solvers and algorithm designers (see Noss, 1986a for a discussion of the cultural side-effect of this hegemony). This process is one in which, I suggest, mathematics educators need to play an active role.

At the same time there has been a growth of interest in researching the ways in which existing computer-based tools can be employed to create microworlds for learning about reasonably well-defined subsets of the mathematics curriculum in areas such as probability, functions and variables, the concept of limit and three-dimensional geometry (for a recent collection of papers on this research effort see Hoyles, Noss and Sutherland, 1986).

Alongside these developments, there has emerged a widespread consensus of the importance of the teacher in the learning process, a recognition that the computer may have an important role to play in influencing teacher attitudes towards mathematics and its teaching and to begin to operationalize Papert's notion of a 'computer culture' within the mathematics classroom. Research is in progress at the University of London Institute of Education[6] to explore and evaluate the use of the computer in this way and to work in conjunction with teachers on the creation of mathematical microworlds for use in the classroom.

Almost all the research and curriculum development currently taking place is necessarily locked into the framework of the existing mathematics curriculum. To be sure, those innovations which are successful will invariably have some success in eroding the traditional content of the syllabus. Nevertheless, I want to conclude with an eye to the future, by suggesting that a necessary task for mathematics education is to begin to turn towards examining how the emerging computer culture will materially alter the content of the mathematics we teach and the kind of mathematics children will be able to do and learn. To do this will require considerable human resources and a high density of machines; work which has begun in a Boston elementary school[7] (not specifically focusing on mathematics), is indicating that the social, affective and cultural issues which follow are at least as interesting as the cognitive. It is only by researching these issues, in situ, that we may begin to genuinely glimpse the potential for cultural change that the computer might bring to the learning of mathematics.

Notes

1 See also the papers in Part 2 of this volume. Editor.
2 A recent example of the extent to which this myth is culturally accepted involved a TV interview with a sociology professor who had given birth to the notion of the QALY, a Quality Adjusted Life Year, which allowed him to judge the relative 'values' of two human lives so that scarce funds could be 'scientifically' allocated to the most deserving cases. The interviewer, seeking perhaps some rationale for the idea, asked the professor whether this was a helpful way to think about the value of human life or perhaps 'only a mathematical formula'.

3 I see a parallel here with the development of 'youth culture' of the 1960s and its influence on musical taste. Think back to the way in which the musical idols of the time were denigrated by adult society for their trivialization of the official music culture and the way in which many of those idols are now seen as the representatives of the musical establishment. To be sure Elvis was not Bach, in an important sense the former did indeed trivialize our conception of music (no more so of course than existing popular music of the time); what happened was that young people claimed a part of that culture for themselves and created it themselves. In doing so they created new kinds of instruments and new musical technologies (some of which fed back to official music), and they profoundly changed and enlarged the conception of musical content itself.

4 The work of Lave *et. at.* is discussed by Evans and Harris in Part 2 of this volume. Editor.

5 The notion of scaffolding was coined by Bruner and his colleagues to describe the way in which collaboration could offer the learner just enough support to do things which he or she could not do independently.

6 With Celia Hoyles and Rosamund Sutherland and funded by the Economic and Social Research Council.

7 The 'Headlight Project' which began in 1985, is directed by Seymour Papert and his colleagues at the Massachusetts Institute of Technology.

References

BADGER, M. (1981) 'Why Aren't Girls Better at Maths? A Review of Research', *Educational Research*, 24, 11–23.

BERRY, J. (1985) 'Learning Mathematics in a Second Language: Some Cross-cultural Issues', *For the Learning of Mathematics*, 5, 2, pp. 18–23.

BISHOP, A. (1979) 'Visualising and Mathematics in a Pre-technological Culture', *Educational Studies in Mathematics*, 10, pp. 135–146.

BROWN, S. (1984) 'The Logic of Problem Generation: from Morality and Solving to De-posing and Rebellion', *For the Learning of Mathematics* 4, 1, pp. 9–20.

BUXTON, L. (1981) *Do You Panic about Maths?* London, Heinemann.

D'AMBROSIO, U. (1985) 'Ethnomathematics and its place in the History and Pedagogy of Mathematics', *For the Learning of Mathematics*, 5, 1, pp. 45–48 and reprinted in this volume.

DAVIS, P. and HERSCH, R. (1987) *Descartes Dream: the World According to Mathematics*, Brighton UK, Harvester.

DAVY, J. (1985) 'Mindstorms in the Lamplight' in Sloane, D. (Ed) *The Computer in Education: A Critical Perspective*, pp. 11–20. Columbia, Teachers' College Press.

DES (Department of Education and Science) (1982) *Mathematics from 5 to 16*, London HMSO.

DORFLER, W. (1986) 'The Cognitive Distance between Material Actions and Mathematical Operations', in *Proceedings of the Tenth International Conference for the Psychology of Mathematics Education*, London, pp. 141–146.

GAY, J. and COLE, M. (1967) *The New Mathematicians in an Old Culture: A study among the Kpelle of Liberia*, New York. Rinehart and Winston.

GERDES, P. (1986) 'On Culture, Mathematics and Curriculum Development in Mozambique, in Hoines, M.J. and Mellin-Olsen, S. (Eds.) *Mathematics and Culture* pp. 15–41 Radal, Norway, Caspar Forlag.

GILLIGAN, C. (1982) *In a Different Voice* Cambridge, Mass., Harvard University Press.

GRAMSCI, A. (1971) *Prison Notebooks*, London. Lawrence and Wishart.

HART, K. (1980) *Secondary School Children's Understanding of Mathematics: A Report of the Mathematics Component of the CSMS Programme*, University of London, Chelsea College.

HILLEL, J. (1984) 'Mathematical and Programming Concepts Acquired by Children aged 8–9 in a Restricted Logo Environment', *Proceedings of the Ninth International Conference for the Psychology of Learning Mathematics*, Holland.

HOYLES, C. (1985a) *Culture and Computers in the Mathematics Classroom* Inaugural Lecture. University of London Institute of Education.

HOYLES, C. (1985b) 'Developing a context of Logo in School Mathematics', *Journal of Mathematical Behaviour*, 4, 3, pp. 237–256.

HOYLES, C. (1986) 'Scaling a Mountain: A Study of the Use, Discrimination and Generalisation of some Mathematical Concepts in a Logo Environment', *European Journal of Psychology of Education*, 1, 2, pp. 111–126.

HOYLES, C. and NOSS, R. (1987a) 'Synthesising Mathematical Conceptions and their Formalisation through the Construction of a Logo-based School Mathematics Currciulum', *International Journal of Mathematics Education in Science and Technology*, 18, pp. 581–595.

HOYLES, C. and NOSS, R. (1987b) 'Children Working in a Structured Logo Environment: From Doing to Understanding', *Récherches en Didactique des Mathématiques*, 8, 12, pp. 131–174.

HOYLES, C. and NOSS, R. (1989) 'The computer as a Catalyst in Children's Proportion Strategies', *Journal of Mathematical Behaviour*, 7, pp. 53–57.

HOYLES, C., NOSS, R. and SUTHERLAND, R. (1986) 'Peer Interaction in Programming Environment' in *Proceedings of the Tenth International Conference for the Psychology of Learning Mathematics*, pp. 354–359. London.

HOYLES, R., SUTHERLAND, R. and EVANS, J. (1985) *The Logo Maths Project: A Preliminary Investigation of the Pupil-Centred Approach to the Learning of Logo in the Secondary Mathematics Classroom 1983–4*, University of London Institute of Education.

KEITEL, C. (1986) 'Cultural Premises and Presuppositions in Psychology of Mathematics Education', Plenary Lectures, *Proceedings of the Tenth International Conference for the Psychology of Mathematics Education*, London.

LANCY, D. (1983) *Cross Cultural Studies in Cognition and Mathematics*. New York. Academic Press.

LAVE, J., MURTAUGH, M. and de la ROCHA, O. (1985) 'The Dialectic of Arithmetic in Grocery Shopping', in Rogoff, B. and Lave, J. (Eds.) *Everyday Cognition: its Development in Social Context*, pp. 67–94. Cambridge, Mass. Harvard University Press.

LAWLER, R. (1985) *Computer Experience and Cognitive Development*, Chichester UK. Ellis Horwood.

LERON, U. (1985) 'Logo Today: Vision and Reality', *The Computing Teacher*, 12, pp. 26–32.

MAXWELL, J. (1985) 'Hidden Messages', *Mathematics Teaching*, 111. pp. 18–20 and reproduced in this volume.

MELLIN-OLSEN, S. (1986) 'Culture as a Theme for Mathematics Education', in Hoines, M.U. and Mellin-Olsen, S. (Eds.) *Mathematics and Culture* pp. 99–121, Radal Norway. Caspar Forlag.

MELLIN-OLSEN, S. (1987) *The Politics of Mathematics Education*, Dordrecht, Holland, Reidel.

MITCHELMORE, M. (1983) 'Geometry and Spatial Learning: Some Lessons from a Jamaican Experience', *For the Learning of Mathematics*, 3, 3, pp. 2–7.

NOSS, R. (1986a) 'Constructing a Conceptual Framework for Elementary Algebra through Logo Programming', *Educational Studies in Mathematics*, 17, 4, pp. 335–357.

NOSS, R. (1986b) 'What Mathematics do Children do with Logo?' *Journal of Computer Assisted Learning*. 3, pp. 2–12.

NOSS, R. (1986c) 'Is Small Really Beautiful?', *Micromath*, 2, 1, pp. 26–29.

NOSS, R. (1987) 'Children's Learning of Geometrical Concepts through Logo', *Journal for Research in Mathematics Education*, 18, pp. 343–362.

NOSS, R. and HOYLES, C. (1987) 'Structuring the Mathematical Environment: The Dialectic of Process and Content', *Proceedings of the Third International Conference of Logo and Mathematics Education*, Plenary Lecture, Montreal.

PAPERT, S. (1980) *Mindstorms: Children, Computers and Powerful Ideas*. Brighton UK, Harvester Press.

PAPERT, S. (1985) 'Computer Criticism vs Technocentric Thinking', *Proceeding of Logo 85*, Plenary Lectures pp. 53–67, Cambridge, Mass.

PAPERT, S. (1986) 'Beyond the Cognitive: the Other Face of Mathematics', Plenary Lectures, *Tenth International Conference for the Psychology of Mathematics Education*, London.

PAPERT, S., WATT, D., DiSESSA, A. and WEIR, S. (1979) *Final Report of the Brookline Logo Project, part 2*. A1 Memo Np. 545 Massachusetts Institute of Technology, Cambridge, Mass.

PIMM, D. (1986) 'Beyond Reference', *Mathematics Teaching*, 116, pp. 48–51.

SHERMAN, J. (1980) 'Mathematics, Spatial Visualisation and Related Factors, changes in Girls and Boys grades 8–11', *Journal of Educational Psychology*, 71, pp. 476–482.

TAYLOR, H. (1986) 'Experience with a Primary School Implementing an Equal Opportunity Enquiry', in Burton, L. (Ed.) *Girls into Maths Can Go*. pp. 156–162. London, Holt, Rinehart and Winston.

TURKLE, S. (1985) *The Second Self: Computers and the Human Spirit*, New York. Simon and Schuster.

VYGOTSKY, L. (1978) *Mind in Society*, Harvard. Harvard University Press.

WEIR, S. (1987) *Cultivating Minds: A Logo Casebook*, New York, Harper and Row.

WOLF, A. (1984) *Practical Mathematics at Work: Learning through YTS*. Research and Development Report No 21. Sheffield. UK. Manpower Services Commission.

11
The Contextualizing of Mathematics: Towards a Theoretical Map[1]

Paul Dowling

In the late 1950s and 60s, a major area of interest within the field of mathematics education was the context of the mathematics curriculum. It was during this period that we saw the influence of Bourbakiism in the construction of 'Modern Mathematics' or 'New Math' as an academic content unified by set theoretic ideas. The hope then — particularly in midst of panic generated by the Soviet launching of the Sputnik at the end of the McCarthy era in the US — was that a high grade, academic mathematics curriculum would produce the mathematicians needed if the West was to retrieve its position of technological leadership in the world. The result was a lot of confused teachers, pupils and parents and a song by Tom Lehrer[2]. Ultimately, the programme foundered because the mathematicians who had devised it failed to take into account the fact that most of the teachers who would have to put it into practice did not embody the necessary mathematical background to enable them to grasp the significance of Bourbakiism. Instead of constituting the basis for the unification of mathematics, set theory just became a new topic on the secondary syllabus and, in many primary schools, it was (and to some extent still is) impossible to speak about any objects — mathematical or otherwise — without prefacing one's statement with 'the set of . . . '.

Whilst 'modern Mathematics' introduced the slide rule onto the mathematics curriculum, the 1970s saw the engagement of much of the mathematics education community with electronic calculators which rapidly became very cheaply available during this period. From the late 1970s, however, interest has shifted towards the notion of the context of mathematical activity. Initially, the focus (in the UK, anyway) was on the notion of mathematics as a set of [sic] tools which could be used in a variety of contexts in working and everyday life. The mathematics curriculum was seen as responsible for providing pupils with the mathematical toolbox which they would carry with them in life outside school, in the home and in the factory.

This is what I term the 'utilitarian' approach and is represented by the research carried out on behalf of the Cockcroft Committee. More recently, a movement that I have called 'mathematical anthropology' has sought out mathematical activity in more exotic locations with the intention of challenging what are perceived as élitist views of mathematics and possibly leading the way for an emancipatory mathematics curriculum. Recent work in the US and in Brazil has used mathematics (generally arithmetic) as a focus for research in a number of contexts, using a broadly neo-Vygotskian theoretical paradigm. In this paper I want to look critically at the first two of these three approaches to the contextualizing of mathematics and, after a very brief discussion of the third, to suggest an alternative view in which contexts for mathematical activity are mapped out according to the social organization of the various sites in which it may be said to take place. My general premise in developing this theoretical map is that mathematical activity must be consistent with the social organization of its site of elaboration; such a premise is certainly consistent with the empirical failure of the 'Modern Maths' initiative.

Utilitarianism

If 27 men can do a piece of work in $11\frac{1}{3}$ hours, find how many men would be required to do the same work in 17 hours.

(Parr, 1972)

What is the context in this mathematics 'problem'? It is probably uncontentious to suggest that 'men working' is an imaginary context, and unlikely to be of much assistance in considering this sort of school maths task. Indeed it is probably true to say — as does Eugene Maier — that whilst many school mathematics 'problems' are purportedly related to life outside school

> . . . they have little in common with that life. They are school problems coated with a thin veneer of 'real world' associations. The mathematics involved in solving them is school mathematics, of little use anywhere but in school.

(Maier, 1980 and this volume)

The issue that I want to address here is precisely the ways in which we might consider the relationship between what might be described as mathematical activity, and the various contexts within which it is presumed to occur. I shall look, firstly, at the utilitarian perspective exemplified by some of the research carried out on behalf of the Cockcroft Committee and by the Cockcroft Report itself. I intend to show that this work actually illustrates both the substantive separation of school mathematics from everyday and working life, and the tendency of the utilitarian approach to define non-mathematical practices in mathematical terms.

The Cockcroft Committee was set up in 1978, in the wake of Callaghan's Ruskin College speech and the 'Great Debate', which themselves are generally understood as responses to criticisms of state education from the engineering industry and the writers of the Black Papers. The committee was instructed to pay 'particular regard to the mathematics required in further and higher education, employment and adult life generally' (Cockcroft *et. al.*, 1982).

As part of their response, the committee set up three research studies: two looking at mathematics and employment, and carried out at the Universities of Nottingham and Bath, and the third — a smaller scale study — looking at the *Uses of Mathematics by Adults in Everyday Life* was carried out by Brigid Sewell (1981). Sewell's image of school mathematics was a body of skills and knowledge which people could use in their daily practice and which are prerequisites of full participation in society, thus:

> Pertcentages play an ever increasing part in the dissemination of information, both through the news media and from central government. An understanding of the national economy assumes a sophisticated comprehension of percentages, as does much of the discussion about pay rises. For the shopper, the ability to estimate 10 per cent can be a valuable 'key' to checking other percentages — even if a precise answer seems too difficult. Since the currency became decimalized, it is a trivial matter to work out 10 per cent of a sum of money, and this can easily be used to estimate other percentages. Those who lack the skill even to calculate 10 per cent are surely handicapped when attempting to understand the affairs of society.
>
> (Sewell, B. 1981, p. 17)

The image here is of mathematical knowledge inherent in everyday practices and institutions — the news media, shopping — but which can nevertheless be abstracted and given a mathematical label — 'percentages' — so that it can be included on a school syllabus in an appropriate form. There is, furthermore, the clear implication that, if the mathematical skills and knowledge are not transmitted in school, then they will not be developed subsequently, and individuals will be 'handicapped' in their attempts to become active participants in society.

There are a number of points which can be made about this particular extract from Sewell's work. Firstly, it is certainly the case that the use of the language of 'percentages' in the media or in the retailing industry invokes the mathematical activity of those 'mathematicians' who have developed this particular branch of arithmetic[3]. However, to make use of the 'dead labour'[4] of an activity is not tantamount to an engagement with the activity; this point is made by Paulus Gerdes with respect to the mathematical activity which is 'frozen' in the production practices of Mozambican basket makers:

The artisan who imitates a known production technique is — generally — not doing mathematics. But the artisan(s) who discovered the techniques, *did* mathematics, *developed* mathematics, was (were) thinking mathematically.

(Gerdes, 1986, p. 12)

As used in the media and in shopping, it may be that the successful transmission of messages through the language of percentages frequently demands little more than the ability to order, add and subtract simple numbers and, at most, a recognition that, for example, 35 per cent can be interpreted as thiry-five out of every hundred. The ability to perform calculations with percentages is, arguably, *never* required in everyday life: most of us might be aware of the current mortgage rate, but probably have little idea of the extent of the impact on out payments of a 1 per cent increase, and I have never met anyone who has felt the need to *calculate* repayments on *any* loan — we simply trust the lender.

The point is that attempting — as Sewell does — to retrieve 'frozen mathematics' from everyday activities is *not* the same thing as highlighting *living* mathematics in those activities; one of the tasks used by Sewell in her research will serve as an example.

On the news recently it was said that the annual rate of inflation had fallen from 17.4% to 17.2%. What effect do you think this will have on prices? (If answer 'none') What do you think ought to happen if it had fallen to, say, 12%.

(Sewell, 1981, p. 33)

Clearly, the use of annual rates of inflation is dependent on the mathematical technique of calculating moving averages and, when expressed as 'inflation is currently running at . . . ' might be interpreted as using some of the concepts of differential calculus: indeed a change in the rate of inflation points towards second differentials[5]. Furthermore, a commentary on the statement 'the annual rate of inflation had fallen from 17.4% to 17.2%' would hardly be complete without a critical discussion of the particular form of weighted average used to compute these figures, and the empirical and theoretical underpinnings of the weights, not to mention the implications for various groups in society of basing economic policy on this sort of average. In practice, I suspect that most of us have an idea that inflation is endemic in our society, and that any reduction in it is a good thing. Precisely how good any particular reduction might be is almost impossible to say: a 0.2 per cent reduction might, for example, be wholly the result of a short term fall in the price of imported beef which more than offset a general increase in the prices of fresh vegetables; a situation which we might expect to leave certain groups somewhat unimpressed.

Sewell is asserting that the practices to which she refers embody mathematics as an *essence* and labels as incompetent those subjects who fail to pass what is basically a maths test. Mathematics is used both to define the essential nature of the practice and to assess the competence of the practitioner. Incompetent practitioners engage in inadequate practices and are thus negatively defined in relation to mathematics. The difficulty of adopting the alternative assumption of competence on the part of everyday practitioners is that it may well lead to the conclusion that no mathematics at all — or, at least, very little — is required in everyday life.

The study carried out at the University of Bath (1981) involved the sifting of working practices for embedded mathematics. The team devised two constructs for recording purposes: a STIM (specific task incorporating mathematics) is 'intended to be a straightforward description of a task that involves mathematics'[6] whilst a MIST (mathematics incorporated in specific tasks) is 'the abstraction of the mathematical procedures used in the STIM'.[7] The Bath researchers suggest that there is 'a vast army of people' who 'do not appear to require any formal Mathematics, not even counting nor [sic] recording numbers', nevertheless, 'all these occupations involve actions which could be described in mathematical terms'[8]. The MIST list corresponding to this army is as follows:

A set, dis-joint sets.
Mappings, one-to-one, one-to-many, many-to-one correspondences.
Symmetry, bilateral and rotational.
Rotation, reflection, translation and combinations of these[.]
Tessellating patterns.
Logical sequences (if . . . then . . .).

(Bath University, 1981, p. 26)

The terminology employed here is very much out of the 'modern maths' tradition, and might be interpreted as signifying a substantial proportion of the mathematics curriculum in primary and secondary schooling. When we look at the STIMs corresponding to the above MISTs, we find a rather more mundane description:

1. Articles are sorted into separate collections for packing or on an accept/reject basis.
2. Articles are moved into particular orientations involving moving sideways, turning over or round.
3. Articles, such as wine glasses [,] are checked for uniformity of shape.
4. Packed articles form regular patterns.
5. Assembly tasks can involve matching parts, such as connecting wires to correct terminals.
6. Tasks often have to be carried out in particular orders sometimes requiring simple decisions, but which would not often be verbalized. For example, a

creeler in a carpet factory: 'Is the spool empty? Yes! Replace with another of the same colour'. Awareness of the consequences of not following the prescribed order may be important.

(Ibid. pp. 25–6)

It should be made clear that the Bath team are not suggesting that the study of formal logic in school is necessary if carpet creelers are to be able to change spools when they're empty rather than when they're full:

> . . . in our opinion, formal study of the Mathematics associated with the items in the MIST above [sic] would be unlikely to be of much value to the operatives in the performance of their work.

(Ibid. p. 26)

Indeed, throughout the report, subjects who appear to be competent in their working practices are represented as owing little of their competence to school mathematics:

> One of the principle things to emerge from these discussions is that few people, young or old, appear to have come out of school with a very clear idea of what the purpose of studying Mathematics might be (or education generally in many cases).

(Ibid. p. 199)

We are therefore not surprised to find that:

> After talking generally with the interviewee about school and training, we then ask if he [sic] could supply us with examples of where Mathematics was useful in his work. With many people, their initial reaction was that they did not use 'Mathematics' and it was necessary to prompt them with suggestions like 'any figures that you have to write down?', 'any measuring you do?'

(Ibid. p. 8)[9]

Unlike Sewell, the Bath team are not measuring competence in terms of mathematics — indeed, they are not measuring competence at all; the report is almost a lament at the loss of utilitarianism. However, the Cockcroft Report reintroduces the utilitarian philosophy by sleight of hand (that is, without any supporting evidence!):

> However great the effort which is made, illustrations of the practical applications of mathematics within employment which are given to a group of pupils, whose members may enter many different types of job, cannot provide the immediacy of the actual job itself. Nevertheless, **it is important that the mathematical foundation which has been provided in the classroom should be such as to enable competence**

**in particular applications to develop within a reasonably short time
once the necessary employment situation is encountered**.

(Cockcroft, 1982, p. 24 emphasis in original)

The Bath research illustrates precisely the point that I have made above, and
which seems to have escaped Sewell, that is the separation of the discourse of school
mathematics (Dowling, 1986) from everyday and working life discourses. In school
mathematics, 'articles are moved into particular orientations involving moving
sideways, turning over or round' signifies 'rotation, reflection, translation and
combinations of these' and 'packed articles form regular patterns' signifies
'tessellating patterns': these mathematical notions are probably somewhat distant
from the minds of shifters and packers. The Cockcroft Report posits school
mathematics as a condition for competence which can be replaced only by —
presumably — an unreasonably lengthy period of on the job training. Certainly, the
Cockcroft Report is explicit in defining working practices as embodying
mathematics, that is, more than simply describable in mathematical terms.

The disjunction between school mathematics and everyday and working life is
illustrated in the third piece of research carried out on behalf of the Cockcroft
Committee by Robert Lindsay of The Shell Centre at Nottingham University. In an
interview with an apprentice plant operator:

> I asked Derek what sort of maths he used at work and on the [Further
> Education College] course, he told me that he reads the PH meter, adds
> weights in grammes and uses the dip stick to separate one lot of fluid into
> two equal amounts. He also uses a measuring cylinder to measure in cubic
> centimetres.
>
> Derek told me that he did not really use any other maths at work yet,
> although on the course some of the formulae were difficult. He quoted me a
> formula
>
> $$Q = \frac{KA(q1 - q2)T}{L}$$
>
> (Shell Centre, nd(a) ... p. 229)

As a result of Lindsay's early retirement and untimely death, his fieldnotes — which
are not full transcripts[10] — have not been fully analyzed, and he does not give any
information as to the prompts which he may have had to give. However, both Sewell
and the Bath team report that subjects were frequently not aware of using
mathematics at all. In view of this, it seems likely that the labelling of the use of dip
sticks and meters etc., as 'mathematics' was Lindsay's rather than Derek's idea. In
any event, Derek himself is aware of the mismatch between the emphasis on
mathematics on his college course, and the lack of it in his work. Several of Lindsay's

subjects quote the above formula — which relates to heat conduction — and evidence from the rest of the research suggests that Derek has very little chance of ever applying this particular piece of knowledge that he has so painstakingly memorized; school or, in this case, college mathematics and working life are, in Bernstein's terms (1977) strongly classified with respect to each other.

In another interview with a young technician in the Production Division of Boots Ltd:

> Duncan explains how he works out percentage yields. When an experiment is conducted and the reaction is complete, Duncan knows the molecular weight of the product. If a molecular weight of 244 produces a product whose molecular weight is 272 then 244 grammes will yield 272 grammes. But if Duncan had started with 50 grammes of the original reactants, then the theoretical yield would be 50 multiplied by 272 and divided by 244 which equals 55.73 grammes. Now if in practice we know that the actual yield from 50 grammes is 52.1 grammes, then the percentage yield is 52.1 multiplied by 100 and divided by 55.73. In this particular case the percentage yield is 93.48%.
>
> (Ibid. p. 250)

Again, the indirect speech multiplies the ambiguities of interpretation. Nevertheless, as reported, Duncan's description is purely procedural, suggesting that he has internalized an algorithm rather than mathematical knowledge relating to direct proportion. We might expect school mathematics — at least in its rhetorical form[11] — to emphasize 'understanding' rather than rote learning; the latter would be entirely effective and appropriate in the workplace; it is, after all, unnecessary for the machine operator to grasp the engineering principles of the machine that she or he operates; this applies to a surgeon using laser technology and to a pilot using information technology as much as to a factory worker employing less exotic hardware.

This separation between school mathematics and working life is also apparent in the performances of individuals within the two spheres. For example, another of Lindsay's subjects, Amanda — an invoice typist with Midlands Gas — is required, in the course of her work, to find totals of prices and VAT. When it comes to performing the same sort of activity in the context of 'mathematics' however, she appears to be completely incompetent:

> I asked Amanda about tables. She said she used to recite them every day, and the Headmaster used to come and make spot checks to see if they were known but she doesn't really remember them now. I asked her if she knew what 8 × 9 were [sic] and she said 'it's 52 isn't it, or is it 54? I asked what

method she used to get this answer and she said 'Well I know 8 × 8 are [sic] 64 and I want another 8, so that makes 52, doesn't it?'

(Shell Centre, nd(b); p. 661)

Nevertheless, the view that there is a unity to 'mathematics', however diverse may be the activities in which it is embedded is widespread among the cognoscenti of school mathematics. John, a training advisor with the CITB, is reported by Lindsay as having said that:

... every employer he spoke to said the lads don't fill in their time-sheets correctly and therefore find difficulty in seeing how their total money is arrived at. John noted however, that in the building site dart games these same people can always add and subtract and find out how many more they need to score without any apparent difficulty. They also have a reasonably good idea of how much change they should get, so John infers that the basic ability must be there and it probably hasn't been exercised enough at school.[12]

(Shell Centre, nd(c); p.585)

An interesting research question would involve attempting to ascertain the rationality behind time-sheet completion by the 'lads'. There may be very good reason for 'incorrect' completion: pre-empting accusations of attempted fraud; the empirical finding that errors tend to go in the employee's favour; etc. In any event, it is unlikely that a mathematical rationality prevails.

As I have mentioned, Lindsay did not complete his analysis of his work. However, it seems clear from his fieldnotes that he is describing working practices in terms of school mathematics and — at least in cases such as Amanda's — that he is inferring that competence in working practices is related to competence in school mathematics; he offers no non-school measure of competence in Amanda's case. The suggestion once again is that these working practices embody mathematics and that denied practitioners who are adequately versed in this essential knowledge, the practices are inadequate, they are pathologized by being negatively defined with respect to school mathematics.

Further research evidence of differentiation in terms of performance is also provided by John Spradberry in his work with London schoolchildren who had 'failed' in terms of school mathematics.

The pupils, at sixteen years of age, had failed consistently to master anything but the most elementary aspects of School Mathematics. Every year they appear in the bottom 20 per cent at examination time. Using Hayman's language, they had not developed the capacity for abstract thought. They had received and remained unhelped by considerable 'remedial' teaching and finally, they left school 'hating everyfink what goes on in maffs'. Yet in their spare time some of these same young people kept and raced pigeons. They understood the intricacies of the racing rules as a part of the folklore of pigeon-keeping; what the expert mathematician would recognise as 'mathematical knowledge' featured in, though remained undifferentiated from the rest of this knowledge. Weighing, measuring, timing, using map scales, buying, selling, interpreting timetables, devising schedules, calculating probabilities and averages and applying the 'four rules' to numbers, measurements and money were a natural part of their stock of commonsense knowledge. They were learnt as such without the rituals of the classroom. These pupils were their own teachers, learning as they went, assessing and filling the gaps in their knowledge.

(Spradberry, 1976, 237)

When Spradberry subsequently tried to draw on 'real contexts' in his mathematics teaching, he found that, after an initial period of novelty interest, the students rejected his relevant mathematics as not being *real* maths. It is, of course, only the 'expert mathematician' who is able to recognize that pigeon-keeping *is* (or at least involves) 'real maths'; Spradberry is not a utilitarian, but he is displaying a similar essentialism.

Possibly contrary to their intentions, the work of the researchers sponsored on behalf of the Cockcroft Committee supports the notion that school mathematics and everyday and working life discourses are strongly classified with respect to each other and problematizes the notion of mathematics as a tool, as espoused by the HMI: 'The aim is to encourage the effective use of mathematics as a tool in a wide range of activities within both school and adult life'. (DES 1985, p. 3) or as a means of communication which is decontextualized, as represented in the Cockcroft Report:

We believe that all these perceptions of the usefulness of mathematics arise from the fact that mathematics provides a means of communication which is powerful, concise and unambiguous. Even though many of those who consider mathematics to be useful would probably not express the reason in these terms, we believe that it is the fact that mathematics can be used as a powerful means of communication which provides the principal reason for teaching mathematics to all children.

(Cockcroft, 1982. p. 1, emphasis in original)

or as facilitating a way of knowing or understanding, as illustrated by this beautiful anecdote from Mike Cooley:

At one aircraft company they engaged a team of four mathematicians, all of PhD level, to attempt to define in a programme a method of drawing the afterburner of a large jet engine. This was an extremely complex shape, which they attempted to define by using Coon's Patch Surface Definitions. They spent some two years dealing with this problem and could not find a satisfactory solution. When however, they went to the experimental workshop of the aircraft factory, they found that a skilled sheet metal worker, together with a draughtsman had actually succeeded in drawing and making one of these. One of the mathematicians observed:'*They may have succeeded in making it but they didn't understand how they did it.*'

(Cooley, M., 1985, p. 171, my emphasis)

The mathematician's observation might almost lead one to believe that the two non-mathematicians who had produced the required item had actually done it by accident! As with school mathematics, the emphasis is on 'understanding', that is on being able to elaborate the specifically *mathematical* description of the activity: the rhetoric represents mathematics as a body of tools; the practice results in mathematics taking over and actually becoming the activity. A lawn mower is a tool, but no one ever suggests that I need to understand how it works just to cut the grass; that's someone esle's job in a modern society exhibiting a highly specialized technical division of labour[13].

The utilitarian approach to mathematics education exhibits an essentialism whereby diverse — and arguably *all* — practices are taken to signify mathematics or mathematical activity. Clearly, this is not the view adopted by Cooley's mathematicians. For them, mathematics is an activity carried out by an élite group and which confers unique forms of understanding on its practitioners. The utilitarians would democratize mathematics, but actually devalue social practices by seeking to establish them as a chain of equivalences (Laclau and Mouffe, 1985); equivalent, that is, to vulgar *non-mathematical* or *mathematics avoiding* practices, unless they articulate themselves to mathematics as a universal signified. The utilitarians will describe the practices of the shopper, the newspaper reader, the technician, the invoice typist, as inadequate unless they signify the mathematical content *which is really there*. Such signification will, of course, transform vulgar everyday and working practices into genuinely mathematical activities. Such signification of course, is contingent upon schooling.

In this section, I have argued that three pieces of research which set out to illustrate how mathematics is or can be used as a tool in everyday or working life, actually succeed in demonstrating the existence of a discontinuity in signification of the practices that they observe. The result of this discontinuity is that the utilitarian notion of mathematics as a universal set of tools can no longer be supported. It actually represents a pathologizing of everyday and working practitioners who do not see the toolbox or its contents. In the next section I want to consider an approach which effectively reverses the utilitarian perspective and looks more like a celebration of everyday and working practices in their constitution of mathematical activity.

Here again though, the signification of these practices as mathematics remains with the researchers and not the practitioners.

'Mathematical Anthropology'

> A spider conducts operations which resemble those of the weaver, and a bee would put many a human architect to shame by the construction of its honeycomb cells. But what distinguishes the worst architect from the best of bees is that the architect builds the cell in his [sic] mind before he constructs it in wax.
>
> (Marx, 1976 edn. p. 284)

Thus Marx makes his distinction between insects and arachnids on the one hand and human beings on the other: spiders and bees produce their intricate patterns simply because that's what spiders and bees do; human actors engage in planned activity. That Marx's interpretation of anthropodal psychology (or, rather, lack of it) is accurate is presumably attested to by the finding that the designs of honeycombs and webs are, essentially, invariant — that is, highly limited — within respective species. But similar repertory limitation is often found in human activities[14]. Paulus Gerdes, for example has found the maintenance of traditional patterns in Mozambican basket weaving (1986) and in Angolan sand drawing (1988a), and Patricia Marks Greenfield (1984) reports on the limited repertoire of certain Mexican weaving[15]. Western capitalist countries arguably construct the notion of material production as being associated with almost continuous innovation stimulated by market forces and informed by research and development. Nevertheless, we may assume that continuities are maintained, at least locally in time; even research is substantially routine. It is certainly possible that an individual worker might learn a process, and subsequently repeat the same process — essentially unchanged — throughout her/his working life; under such circumstances how can we maintain the distinction between the worker's productive activities and those of the bees and spiders?

Karl Marx offers a possibility in his analysis of the labour process (1976 edn). His idea is that human labour becomes objectified in its product which is a 'use-value'. To this extent there is no distinction between human and non-human producers: honeycombs and webs might also be thought of as objectified or 'dead' labour and also constitute use-values for their constructors. However, unlike spiders and bees, use-values not only emerge from production but also enter into it:

> Although a use-value emerges from the labour process, in the form of a product, other use-values, products of previous labour, enter into it as a means of production. The same use-value is both the product of a previous

process, and a means of production in a later process. Products are therefore
not only results of labour, but also its essential conditions.

(Marx, 1976 edn; p. 287)

Thus, even if a worker is only following an internalized algorithm, it is a distinctively
human activity in that that algorithm is itself the product of creative human labour.
This is the point made by Paulus Gerdes in the comment already quoted:

There exists 'hidden' or 'frozen' mathematics. The artisan who imitates a
known production technique is — generally — not doing mathematics.
But the artisan(s) who discovered the techniques, *did* mathematics, *developed*
mathematics, was (were) thinking mathematically.

(Gerdes, 1986; p. 12, emphasis in original)

Gerdes' project is, basically, anti-racist and anti-colonialist. One of the results of the
imperialist activities of European states is that certain high status cultural products —
literature, mathematics, etc. — have become defined as European creations[16], which
are 'given' to African and other non-European cultures in colonial and post-colonial
schools, most of which have been designed on a European basis. A similarly
condescending attitude is apparent in 'aid' programmes — as if Europe had not
become rich via the exploitation and, ultimately, disabling of African and other
cultures. And again in the common and academic parlance which speaks of 'Third
World' and 'developing' or 'underdeveloped' countries, and of 'primitive', 'pre-
industrial' and 'traditional' societies[17].

Gerdes argues that a considerable amount of mathematical knowledge exists,
'frozen', in African production activities. It may be called 'Pythagoras' Theorem',
but it is embodied in the design of African button weaving. Gerdes believes that
unfreezing this mathematical knowledge will lead to cultural confidence:

'Had Pythagoras not ... *we* would have discovered it'. The debate starts.
'Could our ancestors have discovered the 'Theorem of Pythagoras'?' 'Did
they?' ... 'Why don't we know it? ... 'Slavery, colonialism ...'. By
'defrosting frozen mathematical thinking' one stimulates a reflection on the
impact of colonialism, on the historical and political dimensions of
mathematics (education).'

(Gerdes, 1988b, p. 152)

Gerdes' analysis is far more thoughtful (and interesting!) than the utilitarian research
discussed above, and his intentions are critical and emancipatory rather than
functionalist. However, he remains susceptible to the charge of defining productive
activities in school mathematics terms. Pythagoras' Theorem may be used to describe
pyramid building or button weaving, but it is only in mathematical discourses that it

has the appearance of being decontextualized and ultimately becomes expressed as a symbolic generalization such as:

$$a^2 + b^2 = c^2$$

or even an iconic generalization such as Figure 11.1.

Unfortunately, another way of interpreting Gerdes' mathematical prospecting is that it is necessary to show value in African productive activities in *mathematical* terms in order for these activities to have any value at all; the rather disturbing implications of this are apparent when one notices that Gerdes makes no attempt to 'defrost' the mathematics in the activities of spiders and bees! That Mozambican basket weavers do not describe their activities in mathematical terms is the result of precisely the same conditions that lead the subjects of the Bath team similarly to reject a mathematical description, that is, in neither case are the subjects speaking in the discourse of school mathematics.

That Gerdes thinks it is important that mathematics be found in basket weaving is precisely the result of the hegemony of European rationalism, and the result is to devalue (and in all probability inadequately to describe) Mozambican and Angolan cultural practices. Cultural confidence is to be acquired by wrenching apart cultural signifiers (visible practices) and their culturally specific signifieds[18], discarding the latter, and replacing them with shiny new mathematical concepts. If school mathematics is as useless to the non-school mathematician as is suggested by some of the work discussed in the previous section, then the transplantation of signifieds might be about as useful to Mozambicans and Angolans as the decontextualizing and

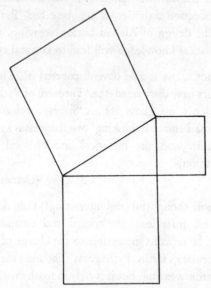

Figure 11.1 An ionic generalization of Pythagoras' Theorem

recontextualizing processes of shelling and swallowing an oyster are to the mollusc itself!

The same attempt to define everyday and productive activities as 'mathematics' seems currently to be gaining in popularity in the field of mathematics education. D'Ambrosio suggests that we need to expand the notion of mathematics to include what he refers to as *ethnomathematics*:

> Now we include as mathematics apart from the Platonic ciphering and arithmetic, measuration and relations of planetary orbits, the capabilities of classifying, ordering, inferring and modelling. This is a very broad range of human activities which, throughout history have been expropriated by the scholarly establishment, formalized and codified and incorporated into what we call academic mathematics. But which remains alive in culturally identified groups and constitute routines in their practices.
>
> (D'Ambrosio, 1985, p. 45 and this volume)

Alan Bishop asserts that mathematics is universal:

> The thesis is therefore developing that mathematics must now be understood as a kind of cultural knowledge, which all cultures generate but which need not necessarily 'look' the same from one cultural group to another. Just as all human cultures generate language, religious beliefs, rituals, food-producing techniques, etc., so it seems do all human cultures generate mathematics. Mathematics is as pan-human phenomenon. Moreover, just as *each* cultural group generates its own language, religious belief, etc., so it seems that each cultural group is capable of generating its own mathematics.
>
> (Bishop, 1988a and this volume)

Elsewhere, Bishop justifies his use of cross-cultural studies in arriving at this conclusion:

> Contrasts not only give us differences but they also make us recognise similarities, because two phenomena must be similar in some way in order for their differences to be recognised.
>
> (Bishop, 1988b, p. 21)

This rather surprising statement is presumably at the root of Bishop's belief that he has isolated a number of cultural invariants. It clearly gives the green light to unfettered anthropological imperialism: not only 'where's the mathematics in eating a pork chop?', but 'skiing is pancultural in that it *must* have similarities with religion otherwise we would not be able to tell them apart'!

To be fair, Bishop does realize that his perspective is restricted by his own cultural positioning, and qualifies his assertion of the existence of 'universals':

Perhaps a safer label would in any case be 'culturo-centric universals' i.e. universals from *our* culturo-centric position, since *we* are describing the phenomena as 'counting' etc. this then makes it plain that one can never *establish* the universality of phenomena, one is merely choosing to describe a highly extensive set of similarities in a certain way.

(Ibid. p. 55)

But the implication of such an admission of cultural relativism is that the sort of cross-cultural study that Bishop reports and has engaged in is likely to tell us more about the culture of the researcher than that of the researched. Were it to be the case that Bishop was attempting to construct an archaeology of dominant educational culture in contemporary Europe, then we might recognize that this implication had been taken seriously. However, Bishop clearly has larger plans as he suggests in outlining the second of two 'problem areas' which motivate his book:

The second problem area concerns children whose home and family does not fully resonate with that of the school and the wider society, be they in London, in Aboriginal Australia or in a Navajo reservation.

(Ibid. p. xi)

It would seem quite clear that both Bishop and D'Ambrosio, like Gerdes are defining non-European cultural activities in European terms, failing, like generations of European anthropologists before them — to problematize their own culture. Such a practice has an academically responsible history, Émile Durkheim, for example:

At the roots of all our judgements there are a certain number of essential ideas which dominate all our intellectual life; they are what philosophers since Aristotle have called the categories of the understanding: ideas of time, space [. . .], class, number, cause, substance, personlity, etc. They correspond to the most universal properties of things. They are like the solid frame which encloses all thought; this does not seem to be able to liberate itself from them without destroying itself, for it seems that we cannot think of objects that are not in time and space, which have no number, etc. Other ideas are contingent and unsteady; we can conceive of their being unknown to man, a society, or an epoch; but these others appear to be nearly inseparable from the normal working of the intellect. They are like the framework of the intelligence. Now when primitive religious beliefs are systematically analysed, the principal categories are naturally found. They are born in religion and of religion; they are a product of religious thought. This is a statement that we are going to have occasion to make many times in the course of this work.

(Durkheim, 1976 edn. p. 9)

Pierre Bourdieu (1977) argues that the very position of the anthropologist as an observer rather than a participant distorts that which he or she intends to observe:

> Knowledge does not merely depend, as an elementary relativism teaches, on the particular standpoint an observer 'situated in space and time' takes up on the object. The 'knowing subject', as the idealist tradition rightly calls him [sic], inflicts on practice a much more fundamental and pernicious alteration which, being a constituent condition of the cognitive operation, is bound to pass unnoticed: in taking up a point of view on the action, withdrawing from it in order to observe it from above and from a distance, he constitutes practical activity as an *object of observation and analysis*, a *representation*.
>
> (Bourdieu, 1977, p. 2, emphasis in original)

A return to Marx's labour process analysis might suggest an alternative approach which seeks to avoid ethnocentrism. Objectified or dead labour is precisely that — dead, and inactive in the labour process. What is important is the action of 'living labour' which

> . . . must seize on these things, awaken them from the dead, change them from merely possible into real and effective use-values.
>
> (Marx *op. cit.* p. 289)

My reading of Marx's analysis is that he ascribes a substantial amount of determinism to dead labour which appears to structure active labour. However, Marx is making another point[19], and this is not the place for a critique of *Capital*. A more promising approach might be to weaken this determinism, and understand dead and living labour as dialectically related, with the latter attempting to realize the diverse potential and overcome the apparent limitations in the former[20]. Interpreted in the context of the productive activities of cooperative societies, this might involve attempting to generalize these activities beyond their apparent boundaries, encouraging hegemonic action by non-European, non-capitalist, non-mathematical practices; certainly the sort of self-satisfied superiority illustrated by Hardy in his *A Mathematician's Apology* needs some shaking:

> I have never done anything 'useful'. No discovery of mine has made, or is likely to make, directly, for good or ill, the least difference to the amenity of the world. I have helped to train other mathematicians, but mathematicians of the same kind as myself, and their work has been, so far at any rate as I have helped them to it, as useless as my own. Judged by all practical standards, the value of my mathematical life is nil; and outside mathematics it is trivial anyhow. I have just one chance of escaping a verdict of complete triviality, that I may be judged to have created something is undeniable: the question is about its value.
>
> (G. H. Hardy, quoted in Davis and Hersh *op. cit.* p. 85–6)

Of course, Gerdes may well argue that this is precisely what he is doing. African productive activities are being adapted in the solution of pedagogic problems, and this constitutes a counter-colonialist hegemonic action. However, this line of argument will hold only so long as the resulting pedagogy is in European culture; the pedagogic problems are European and non-African. Furthermore, such pedagogy hardly constitutes a celebration of African culture: it remains, indeed, a colonization of the latter. Hegemony must begin from within the hegemonizing discourse, and not from definitions from without. The writers discussed in this section are not utilitarians, and they have produced far more sophisticated analyses than those offered by the Cockcroft researchers. Nevertheless, their analysis remains in the domain of essentialism, and what are proposed as 'conditions and strategies for emancipatory mathematics education in undeveloped [sic] countries' (Gerdes, 1985) in fact evidence the hegemony of European culture in fixing the signification of diverse cultural practices as mathematics. In the next section I want to look at some neo-Vygotskian work which places far greater emphasis on the discontinuities between sites of mathematical activity.

Cognition in Practice

Currently gaining momentum in the US and Latin America is a loose school of work which owes much of its theoretical underpinning to Soviet psychology, in particular, the seminal work of Vygotsky and its development by Leont'ev. Central to much of this work is the notion of 'activity', thus, as Michael Cole puts it:

> ... human psychology is concerned with the activity of concrete individuals, which takes place either in a collective — i.e., jointly with other people — or in a situation in which the subject deals directly with the surrounding world of objects — e.g., at the potter's wheel or the writer's desk ... if we removed human activity from the system of social relationships and social life, it would not exist ... *the human individual's activity is a system in the system of social relations. It does not exist without these relations.*
>
> (Cole, 1984, p. 151, emphasis in original)

This is to say, in particular, intellectual activity is not isolable from practical activity; rationality must be located in specific practices, and cannot be thought of as unitary and constituting an ultimate arbiter. Such an interpretation has a clear bearing on the 'lads'' time-sheets, mentioned by Lindsay's subject, and clearly challenges the notion of mathematics as a universal and decontextualized tool. School mathematics itself constitutes an activity[21], with its own goals and means, and with a distinctive rationality which cannot simply be transplanted into another activity — such as shopping.

A number of pieces of research are outlined by Jean Lave (1988), including her own work with the Adult Math Project involving supermarket shoppers. In this research, it was found that there was a sharp discontinuity in the performance of shoppers between arithmetic activities in the supermarket, and what — mathematically — constituted the same tasks in the context of a test. In particular, arithmetic performances in the shopping context — that is, whilst actually doing shopping — were remarkably accurate, scoring, on average 98 per cent, as compared with 59 per cent on the test. Furthermore, shoppers did not use arithmetic price comparison as the principle basis on which to make decisions:

> This kind of calculation occurs at the end of largely qualitative decision-making processes which smoothly reduce numerous possibilities on the shelf to single items in the cart. A snag occurs when the elimination of alternatives comes to a halt before a choice has been made. Arithmetic problem-solving is both an expression of and a medium for dealing with these stalled decision processes. It is, among other things, a move outside the qualitative characteristics of a product to its characterization in terms of a standard of value, money.
>
> (Lave *et al.*, 1984, p. 80–1)

Lave's findings are entirely consistent with those of Spradberry and Lindsay (with regard to Amanda), mentioned above. Furthermore, Lave suggests that the extent of the difference in performance between shopping and test contexts suggests that the shoppers are probably not using the same arithmetic strategies in their shopping activities as they attempted to use in the test.

The difference between Lave's position and those of the utilitarians and mathematical anthropologists is clear. The former are defining practices in industrial cultures as essentially mathematical, and pathologizing examples of such practices which fail to recognize this, the latter are likewise, defining practices in non-industrial cultures as mathematical, although without pathologizing particular examples; both, in other words, are denying the possibility of the multiple signification of the practices that they thus label. Lave, on the other hand, is drawing out differences in both signification and performance by taking something which she recognizes as arithmetic and exploring its meaning in specific contexts. Her evidence for the multiple signification of these is at once a challenge to the naïvety of utilitarianism and the reductionism of mathematical anthropology.

Furthermore, this and other work in this field (for example, published in Rogoff and Lave, 1984) goes some way towards relating the context specific nature of particular practices — whether they are apparently 'mathematical' (Scribner, 1984; Wertsch et. al., 1984; Rogoff and Gardner, 1984) or not (Greenfield, 1984; White and Siegel, 1984) — to the social structuring of the context. In the final section of this paper, I shall give an introduction to a theoretical outline for a map of possible contexts which respectively give signification to practices which — in the context of school mathematics — might signify mathematics. For the purposes of this paper I shall refer to these practices as 'mathematics', indicating that, in order to carry out some analyses, we must arrest the signifier; it is the contingent nature of the process of signification which is under consideration. What remains as programmatic is the description of how such significations are achieved, and how they may be related.

The Social Structuring of Mathematics

In the last section I shall lay out the basis for my theoretical mapping of possible sites or contexts for mathematical activity according to the social organization within the sites. I shall also give an indication of the direction of my further work in using the map as a guide for exploration.

Precisely what constitutes a context for a practice is clearly a question of breadth of focus. At the most general we might talk of such macro structures as the classical *episteme* desribed by Michel Foucault (1970), rather more specifically, we might investigate home and school practices through an analysis of spontaneous conversation as in Walkerdine (1988), further, we might consider the formation of an individual consciousness as an hierarchical society of agents (Minsky, 1987) or a parallel society of microviews (Lawler, 1981, 1985). I want to take as axiomatic that mathematical practices (as signs) are created, relayed in an educational context, and implemented and that, in a society exhibiting formal educational practices (schooling, apprenticeships), there is a field which officially mediates between creation and relaying. Furthermore, I want to suggest that these activities take place in contexts which can respectively be characterized by distinctive social structures. Following Basil Bernstein (1985) I intended to refer to the context in which mathematics is created as the 'field of production' of mathematical knowledge, that in which it is actually relayed as the 'field of reproduction', and that in which deliberate[22] mediation is carried out between the two the 'field of recontextualization'.[23] Recontextualization and reproduction are on the 'transmission' side of the pedagogic encounter[24], the acquisition side is accessible in the context of the implementation of mathematical knowledge: this is referred to as the 'field of operationalization' of mathematical knowledge.

It might be argued at this point that a teacher in a classroom situation may, in the course of her or his teaching, 'discover' — that is 'produce' — original mathematical knowledge. This is true as it stands, it is also true that the process of learning itself might be referred to as 'original production' from the point of view of the learner. However, such products can only be validated as such outside of the classroom. To draw an analogy with the production and reproduction of literary knowledge, a teacher of English might produce a poem on the board in the course of a lesson, as such this remains a pedagogic text; it does not become new literary knowledge until it is presented to publishers, referees and other producers: its validation in the classroom does not guarantee its validation in the field of production which exhibits quite different social structures. In the same way, the contexts for possible 'inspiration' for new mathematical knowledge may be diverse (and certainly — potentially, at least — include the classroom), but validation — and hence the structuring of mathematical knowledge in general — can be achieved only in the fields[25] of production. Precisely the same argument applies to the manner in which the 'product' of the active learner (e.g. in Piagetian terms; Piaget (1973)) is validated (or not) as a pedagogic text via assessment and examination procedures.

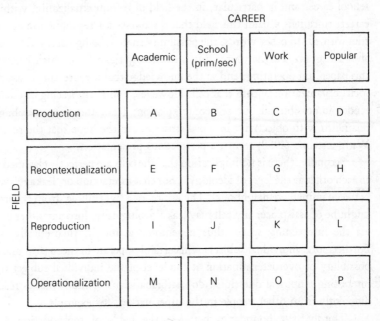

Figure 11.2

The fields defined above constitute one dimension of the social space within which mathematical practices are elaborated. However, further structure is indicated. For example, the field of reproduction within the university constitutes a pedagogic relation between students and *producers* of mathematical knowledge, whereas in the school, the relation is between students and *reproducers*. Furthermore, we have the evidence of Lave and others referred to in the previous section which suggests that mathematical strategies within consumer activities are substantially different from those reproduced within the school; we might speculate that similarly distinctive algorithms will be found in economic activities. Clearly, this knowledge has to be (or have been) produced and reproduced. Thus it might be useful to distinguish between different 'careers' within the social space. The 'academic career' refers to the general arena of higher education; the 'school career' is constituted by primary, secondary and further education; the 'work career' by economic activities within material production, professional services, etc; and the 'popular career' by consumer or domestic activities. The 'career' dimension might be positioned orthogonally with respect to the 'field' dimension, generating the *logical* possibility space illustrated in Figure 11.2. I shall end this paper by making some preliminary comments about some of the cells in this space.

A number of general, theoretical points must be made. Firstly, since I have defined as 'mathematical' those practices which we might take to signify mathematics within the context of school mathematics, I have arrested the signifier within the school career and in particular, in the field of recontextualization within the school career, since it is within this field that discourses for reproduction are collected and transformed. In other words, the possibility space is being surveyed from cell F in the above diagram. Furthermore, this is essentially the cell in which I, as the author of this paper, am operating, and so the knowledge that it represents is structured by the social organization within that cell. However, a total embracing of relativism would render impossible all but self-contemplation; I am therefore coupling ideological relativism with objectivity in social structure, in the hope that the result will be, in Wittgenstein's (1958) terms, a perspicuous presentation.

Secondly, there is no implication that the cells are perfectly classified with respect to each other at the level of ideology[26], nor that interaction or 'leakage' can occur only between immediate neighbours. The notion of a 'wormhole' from theoretical physics might be of assistance: any cell may leak, ideologically, into any other cell other than via the intervening space, thus hegemonic action is, theoretically, unrestricted. Furthermore, individual subjects will certainly operate in more than one cell, and the possibility of overdetermination at the level of the individual subject may generate interesting empirical questions: do mathematicians or mathematics teachers behave differently from other people in the supermarket, for example.

Thirdly, the boundaries between the fields of reproduction and those of operationalization correspond to those between transmission and acquisition. The field of reproduction produces transmission texts — teaching practices; the field of operationalization produces acquired texts — student practices. Transmission texts are pedagogic texts embedded within specific forms of social organization of the classroom, and which are under the control of the teacher.

Fourthly, all texts within the popular and work careers are highly specific in terms of the practices to which they relate: how can I put up these shelves (popular career); how can we produce this machine tool (work career)? Furthermore, within these careers, these practices will not signify mathematics but diverse discourses such as DIY, shopping (popular career), or toolmaking, machine setting (work career). Practices within school and academic careers, on the other hand, are unified in their signification of mathematics.

Fifthly, it should be emphasized that the space is a *logical* rather than an *empirical* possibility space, which is generated by the dimensioning of the social in terms of 'field' and 'career'. Empirically, we may well find that individual cells are empty. This general point leads on to the first specific comment which is that cell B — the field of production within the school career — is almost certainly empty. In Bernstein's conception, the recontextualizing field generates pedagogic texts by collecting (decontextualizing) texts from the field of production and re-articulating them with theories of instruction to produce instructional discourses which define pedagogic subjects ('pupils') in terms of age, 'ability', etc., in relation to principles of distribution (differentiated curriculum) and evaluation (assessment). In cell E, the texts to be recontextualized are collected from the field of production of mathematical knowledge, since it is this knowledge which is to be reproduced. In cell F, the text to be reproduced is also collected from cell A (and from cell M) — see below), insofar as the pedagogic text relates to academic mathematics. In addition, texts are also collected from work and popular careers[27] in 'vocational' and 'mundane' forms of the pedagogic text. In any event, the school career is *dedicated* to the *reproduction* of knowledge, whereas this aspect is incidental in other careers, with other fields taking the fore (production in the academic career and, possibly, operationalization in the work and popular careers). The school career, in other words, does not recontextualize and reproduce its own discourses, but those produced within other careers; it thus has no field of production.

Mobility within the respective careers is likely to be distinctive. Thus, in the popular career, subjects within the field of operationalization are constructed as future reproducers: children learning from their parents how to shop are understood as potential parents themselves. Furthermore, this is the case both within this field (children playing at being parents) and within the field of reproduction (parental expectations of grandparenthood), and this constitutes a continuity between the two fields.

In the school career, on the other hand, transmission texts within the field of reproduction (which are, themselves, recontextualizations of pedagogic texts produced within the field of recontextualization) construct a very small number of pedagogic subjects (students) as having a potential academic career, and the majority of subjects as having a potential work career; all subjects are constructed as having a potential popular career. Pedagogic subjects are thus differentiated according to their academic or work[28] destinations, and united in their popular destination. What is of particular significance is that transmission texts cannot construct a destination within the school career since i) an academic career must precede such a destination[29] and ii) pedagogic and transmission texts seek to transmit that which is produced in other careers rather than that which is produced within their own: mathematics teaching is about mathematics, not teaching. Thus, the pedagogic subject within popular and work careers is an apprentice, the career structure being vertical[30], whilst the school pedagogic subject is a proto-apprentice prior to distribution, there being no career structure within schooling[31]. The academic career structure is part vertical and part distributive, with pedagogic subjects being differentiated as having academic, school or work destinations; higher education students being at least partially autonomous in terms of domestic, consumptive and political practices, their popular destinations may already have been reached.

In this section I have made some preliminary comments on what is to be taken as a theoretical map of the possibility space for the contextualization of mathematical practices; elsewhere (in preparation) I intend to investigate in more detail the structuring within and between elements of this space. I would suggest that this map is an improvement on the constructions of a simplistic utilitarianism or of mathematical anthropology in that it focuses on the specificity of the social structuring of context rather than perpetrating hegemonic action on the part of a dominant social group.

I began by pointing at some of the recent interests in what I can now label the school recontextualizing field and considering in some detail the notion of mathematics as a set of tools — the 'utilitarian' approach. I argued that work in this paradigm actually illustrates a discontinuity of signification between the contexts in which mathematical activity is proposed to be taking place, and constitutes a pathologizing of everyday and working practitioners who fail to signify their activities in accord with the 'mathematicians'. 'Mathematical anthropology', whilst setting out to celebrate the mathematical content in diverse practices, retains an ethnocentrism which defines other cultures in its own terms. The American work of Jean Lave and others is more interesting in recognizing discontinuities between practices, but stops short at what we might expect of a sociological investigation[32]. Finally, I have introduced my theoretical map and suggested questions towards which it might point us: it may be, for example, that a focus on discontinuities between contexts is a promising strategy in considering both the failure of 'modern Maths' and the prognostication for 'investigative' mathematics which is currently in vogue.

As a recontextualizer within the school career, I want to consider how social organization determines the specificity of what I recognize as mathematics, rather than project my own recognition rules — and, ultimately, my own forms of rationality — onto other sites. Furthermore, whilst the neo-Vygotskian work recognizes context specificity, it, perhaps allows insufficient room for leakage between contexts. As with all theory, the map is to be judged on the basis of its usefulness in observing the world (and I have suggested some of the questions towards which it points us) rather than on its ultimate truth: like all maps it is other than that which it represents.

Acknowledgments

This paper would not be complete without an acknowledgment of the inspiration provided by Basil Bernstein both in terms of his own work and his criticism of mine. I must also acknowledge the academic debt owed to colleagues in the Department of Mathematics, Statistics and Computing at the Institute of Education, London University, and also to Parin Bahl who is my colleague in another context.

Notes

1 Previously published in *Collected Original Resources in Education*, Volume 13, number 2.
  The language that I shall use in developing my argument derives from the structural linguistics of Saussure (1983 edn.) and also from the writings of Foucault (1970; 1972; 1977) and Laclau and Mouffe (1985); the structure of the model is heavily influenced by the work of Basil Bernstein (1977; 1985). It is possible that some readers of this paper may be unfamiliar with some of the terminology deriving from the linguistic work and from Bernstein's more recent paper (those terms from this paper which are essential to any understanding of this text are glossed as they arise); Bernstein's earlier work is well known. Briefly, a linguistic *sign* is considered to be the unification of two elements: the *signifier* which, in language, may be thought of as the word or symbol or gesture, etc., and the *signified* which can be glossed as the concept or mental image to which the signifier relates. In this paper, I depart from a strictly linguistic use of the terms, such that the *signifier* and *signified* are both considered as *practices* which may or may not be linguistic in the usual conception of the term: thus 'adding up a shopping bill' is — in my conception — acceptable as a *signifier* (or a *signified*) which signifies (or is signified by) another practice. One result of this is that the relationship between *signifier* and *signified* is reflexive in that the terms are interchangeable: for example, 'adding up a shopping bill' (*signifier*) may signify 'mathematics' (*signified*) to a mathematics teacher; 'mathematics' (*signifier*) may signify 'adding up a shopping bill' (*signified*) to a shopper.
2 'New Math' on *That Was the Year That Was* (Reprise, cat. no. R6179).
3 It is, of course, by no means certain that such individuals and groups would have been 'mathematicians' in the academic sense.
4 I shall discuss the Marxian notion below.
5 Some years ago, Margaret Thatcher made a statement in the House to the effect that although unemployment was still rising, it was not rising as fast as before; not only does this point to the use of second differentials, but also begs the response 'and when we're all out of work, it won't be rising at all'!
6 Bath University, 1981, p. 12.
7 Ibid.
8 Ibid; p. 25.
9 The Bath research does, however, indicate that school mathematics was more or less a requirement for work in further education colleges.
10 It appears, from the general use of indirect speech, that interviews were not recorded.
11 Classroom practice may be somewhat different; Dowling, 1988.
12 There is a major methodological issue here which is not part of the current discussion, but which merits a mention. The question arises in an analysis of this sort of data as to what status to

attribute the subjects' comments. My use of them is to illustrate the discursive positioning of the subject himself. An alternative, but seriously flawed use would involve making inferences about the 'lads'; this would be to attribute the status of researcher to researched.

13 I shall return to the notion of mathematics as a tool in my discussion of the rather more competent work within a neo-Vygotskian approach to the contextualizing of mathematical activity.

14 Or, rather, asserted to be found.

15 I have yet to meet any Mexicans who will attest to this limitation!

16 Martin Bernal (1987) makes a similar point about the way in which Classical Greek culture was constructed as a European (rather than an African) creation with the development of European racism from the late eighteenth century to the present day.

17 Rather like the 'undeveloped' social practices described by Sewell as (mathematics) avoidance strategies (op. cit.).

18 Which are left unexplored in this sort of work.

19 About the possibility of disassociating the labour *process* from the *mode of production*.

20 More optimistic analyses along these lines are provided by Mike Hales (1980) and Pam Linn (1985).

21 Or, probably more appropriately, a number of activities which relate to a diversity of contexts: classroom; textbook writing; research; etc.

22 Discourse in this field concerns the production of what might be referred to as the educational or pedagogic text rather than its implementation.

23 My definition of this field is not identical to that used by Bernstein. For example, I want to distinguish between classroom practice (field of reproduction) and teacher rhetoric (field of recontextualization).

24 The use of the term 'transmission' emphasizes the fact that the teacher is attempting to communicate a 'mathematical' text and it is this which is of concern here; it is not being denied that the teacher is also an acquirer in pupil transmission.

25 I use the plural because I will postulate the existence of more than one field below.

26 Although it is being assumed that social organization — that is, objective relations — will be distinctive within each cell.

27 The particular field concerned will be contingent upon the manner of collection — see below.

28 Internally differentiated.

29 Which means that the academic destination must be provided for if the school career is to reproduce itself.

30 A 'vertical' structure is intended to indicate that there is *some* mobility in a vertical direction, not that *all* mobility is from a given cell to the next above: academic acquirers would, for example, become producers *before* becoming reproducers.

31 By this I do not intend to deny that pupils and teachers can be thought of as following career patterns in school — e.g. pupils as they progress through various age-specific stages, teachers as they attain different status points, etc. Rather, I am using the term 'career' in the very specific sense of this paper, to refer to the possible destinations within the career: no subject within the school career can have an immediate destination in a different subject position within the career — pupils cannot immediately become teachers, etc.

32 Which is not a criticism since the work in question does not pretend to offer such an investigation.

References

BERNAL, M. (1987) *Black Athena: The Afroasiatic Roots of Classical Civilization*, Volume 1, London, Free Association Books.

BERNSTEIN, B. (1977) *Class, Codes and Control*, Volume 3, second edition, London, Routledge and Kegan Paul.

BERNSTEIN, B. (1985) 'On Pedagogic Discourse', in Richards, J. (Ed) *Handbook of Theory and Research for the Sociology of Education*, New York, Greenwood Press.

BERNSTEIN, B. (1988) *On Pedagogic Discourse, Revised* in CORE, 12, 1.

BISHOP, A. (1988a) 'Mathematics Education in its Cultural Context', in *Educational Studies in Mathematics*, 19, 2.

BISHOP, A. (1988b) *Mathematical Enculturation: A Cultural Perspective on Mathematics Education*, Dordrecht, Kluwer.

BOURDIEU, P. (1977) *Outline of a Theory of Practice*, Cambridge, CUP.

CAMBRIDGE CONFERENCE ON SCHOOL MATHEMATICS, (1963) *Goals for School Mathematics*, Boston, Houghton Mifflin Co.

COCKROFT, W. *et al.*, (1982) *Mathematics Counts*, London, HMSO.

COLE, M. (1984) 'The Zone of Proximal Developments: Where Culture and Cognition Create Each Other', in Wertsch, J. V. (Ed) *Culture, Communication and Cognition, Vygotskian Perspectives*, Cambridge, CUP.

COOLEY, M. (1985) 'Drawing Up the Corporate Plan at Lucas Aerospace', in MacKenzie, D. A. and Wajcman, J. (Eds.) *The Social Shaping of Technology*, Milton Keynes, Open University Press.

D'AMBROSIO, U. (1985) 'Ethnomathematics and its Place in the History and Pedagogy of Mathematics', *For the Learning of Mathematics*, 5, 1 and this volume.

DAVIS, P. J. and HERSH, R. (1981) *The Mathematical Experience*, Brighton, Harvester.

DES, (1985) *Mathematics from 5 to 16*, London, HMSO.

DES, (1988) *Mathematics for Ages 5 to 16*, London, HMSO.

DOWLING, P. C. (1986) *Mathematics Makes the Difference: A Sociological Approach to School Mathematics*, Unpublished MA dissertation, ULIE.

DOWLING, P. C. (1988) *Division and Differentiation: A Mathematical Calculus*, Seminar Paper presented at KQC London.

DURKHEIM, E. (1976 edn.) *The Elementary Forms of Religious Life*, London, George Allen & Unwin.

FOUCAULT, M. (1970) *The Order of Things: An Archaeology of the Human Sciences*, London, Tavistock Press.

FOUCAULT, M. (1972) *The Archaeology of Knowledge*, London, Tavistock.

FOUCAULT, M. (1977) *Discipline and Punish: The Birth of the Prison*, London, Penguin Books.

GERDES, P. (1985) 'Conditions and Strategies for Emancipatory Mathematics Education in Undeveloped Countries', *For the Learning of Mathematics*, 5, 1.

GERDES, P. (1986) 'How to Recognize Hidden Geometrical Thinking: A Contribution to the Development of Anthropological Mathematics', *For the Learning of Mathematics*, 6, 2, and this volume.

GERDES, P. (1988a) 'On Possible Uses of Traditional Angolan Sand Drawings in the Mathematics Classroom', *Educational Studies in Mathematics*, 19, 1.

GERDES, P. (1988b) 'On Culture, Geometrical Thinking and Mathematics Education', *Educational Studies in Mathematics*, 19, 2.

GREENFIELD, P. M. (1984) 'A Theory of the Teacher in the Learning Activities of Everyday Life', in Rogoff, B. and Lave, J. (Eds.) *Everyday Cognition: Its Development in Social Context*, Cambridge, Mass., Harvard University Press.

HALES, M. (1980) *Living Thinkwork: Where do Labour Processes Come From?* London, CSE Books.

LACLAU, E. and MOUFFE, C. (1985) *Hegemony and Socialist Strategy: Towards a Radical Democratic Politics*, London, Verso.

LAVE, J. (1988) *Cognition in Practice*, Cambridge, CUP.

LAVE, J., MURTAUGH, M. and de la ROCHA, O. (1984) 'The Dialectic of Arithmetic in Grocery Shopping', in Rogoff, B. and Lave, J. (Eds.) *Everyday Cognition: Its Development in Social Context*, Cambridge, Mass., Harvard University Press.

LAWLER, R. W. (1981) 'The Progressive Construction of Mind', *Cognitive Science*, 5, 1.

LAWLER, R. W. (1985) *Computer Experience and Cognitive Development: A Child's Learning in a Computer Culture*, Chichester, Ellis Horwood.

MAIER, E. (1980) 'Folk Mathematics', *Mathematics Teaching*, 93 and this volume.

LINN, P. (1985) 'Microcomputers in Education: Living and Dead Labour', in Solomonides T. and Levidow, L. (Eds.) 1985, *Compulsive Technology: Computers as Culture*, London, Free Association Books.

MARX, K. (1976 edn.) *Capital*, Volume 1, Harmondsworth, Penguin.

MINSKY, M. (1987) *The Society of Mind*, London, Picador.

PARR, H. E. (1972 edn.) *School Mathematics*, Volume 1, London, Bell.

PIAGET, J. (1973) 'Comments on Mathematical Education', in Howson, A. G. (Ed.) *Developments in Mathematical Education*, Cambridge, CUP.

ROGOFF, B. and GARDNER, W. (1984) 'Adult Guidance of Cognitive Development', in Rogoff, B. and Lave, J. (Eds.) *Everyday Cognition: Its Development in Social Context*, Cambridge, Mass., Harvard University Press.

ROGOFF, B. and LAVE, J. (Eds.) (1984) *Everyday Cognition: Its Development in Social Context*, Cambridge, Mass., Harvard University Press.

SAUSSURE, F. de, (1983 edn.) *Course in General Linguistics*, London, Duckworth Press.

SCRIBNER, S. (1984) 'Studying Working Intelligence', Rogoff, B. and Lave, J. (eds.) *Everyday Cognition: Its Development in Social Context*, Cambridge, Mass., Harvard University Press.

SEWELL, B. (1981) *Uses of Mathematics by Adults in Daily Life*, ACACE.

SHELL CENTRE FOR MATHEMATICAL EDUCATION, nd(a) *Series B Report of Interviews for the Cockcroft Committee: Production*, Unpublished, University of Nottingham.

SHELL CENTRE FOR MATHEMATICAL EDUCATION, nd(b). *Series B Report of Interviews for the Cockcroft Committee: East Midlands Gas*, Unpublished, University of Nottingham.

SHELL CENTRE FOR MATHEMATICAL EDUCATION, nd(c). *Series B Report of Interviews for the Cockcroft Committee: Observations from Instructors in FE and CITB*, Unpublished, University of Nottingham.

SPRADBERRY, J. (1976) 'Conservative Pupils? Pupil Resistance to Curriculum Innovation', in 'Mathematics' in Whitty, G. and Young, M. F. D., *Explorations in the Politics of School Knowledge*, Driffield, Nafferton.

UNIVERSITY OF BATH, (1981) *Mathematics in Employment (16–18)*, University of Bath.

WALKERDINE, V. (1988) *The Mastery of Reason*, London, Routledge and Kegan Paul.

WERTSCH, J. V., MINICK, N. and ARNS, F. J. (1984) 'The Creation of Context in Joint Problem-Solving', in Rogoff, B. and Lave, L. (eds.) *Everyday Cognition: Its Development in Social Context*, Cambridge, Mass., Harvard University Press.

WHITE, S. H. and SIEGEL, A. W. (1984) 'Cognitive Development in Time and Space', in Rogoff, B. and Lave, J. (eds.) *Everyday Cognition: Its Development in Social Context*, Cambridge, Mass., Harvard University Press.

WITTGENSTEIN, L. (1958) *Philosophical Investigations*, Oxford, Blackwell.

Part 2: Introduction
Mathematics in the Workplace —
Research Views

Part 2 surveys recent work, inside and outside the United Kingdom, into the mathematics used in formal and informal workplaces.

Its first paper by Harris and Evans is a survey of the field in terms of form, content and underlying philosophy of research and it forms a fuller introduction to this section. It is followed by a paper by Harris in which she describes in detail some of the problems of analysis and interpretation of data in a large skills-based project. Analysis of open-ended questions in addition to the more straightforward counts of 'closed' questions revealed a rather more pervasive use of mathematical and numerical thinking than was sought by the particular research orientation. Implications of this aspect of workplace research are discussed.

Strässer, Barr, Evans and Wolf bring together a number of strands of analysis from their own research in a discussion of different models of learning mathematics, related to skilled performance. Their paper 'Skills versus Understanding' brings together work with a further or adult education orientation but from different perspectives, contrasting the use of algorithms in adult education to the use of modelling in mathematics teaching in schools. One contrast that is drawn between the 'skills' and 'understanding' approaches is in differences in the ability to generalize what is learned to new and different contexts. The authors present research findings on how the variation of contexts affects learning and the notion of 'context' itself is discussed.

In his article 'The Role of Number in Work and Training', Mathews draws a distinction between the knowledge and reasoning of the school subject of mathematics and 'number reasoning' as a practical need for work. By 'number' he means the sort of numerical understanding described in the Cockcroft Report as 'a feel for number'. He draws a similar distinction between 'number' and 'numeracy' the 'curriculum subject' of training and further education. Traditional training methods tackle only part of 'number skills' at work and that in a distorted and distorting manner. Mathews describes a new scheme for analyzing and accrediting workplace number competence, writing from his experience of working to effect change in traditional employer thinking on numeracy and work.

Carraher's paper contains a review of the work of David Carraher, Terezhina Carraher, Analucia Schliemann and their colleagues in Brazil and has been specially written for this volume. This unique body of research examines in detail some of the differences between the arithmetic that schooled and unschooled workers use in a range of everyday activities considering the nature and use of mathematics in workplaces and its relationship to school mathematics. After raising issues regarding mathematical knowledge it looks at several empirical studies, employing Vergnaud's theory of concepts as a framework for interpreting the results. Finally Carraher discusses some implications for educational theory and practice.

Part 2 ends with a paper by Evans and Harris that discusses some recent work of 'social practice' theorists which brings a distinctive and developmental approach to the question of how 'mathematics', as perceived by some, is used by others at work.

12
Mathematics and Workplace Research

Mary Harris and Jeff Evans

Introduction

Dowling's paper at the end of Part 2 examines some of the current thinking and research into mathematics used at work. His paper also forms a link with Part 2, which examines in more detail some of the issues relating to research into mathematics in workplaces.

There are several ways of attempting to identify and study mathematics perceived to be needed or used in these circumstances. Researchers can ask workers' employers what their workers do; they can interview workers informally or by using systematic questionnaires; they can observe, or perhaps work alongside them using techniques of systematic observation or participant observation; or they can perform a 'task analysis' or write a job description.

The choice of techniques depends to a large extent on the researchers' beliefs about what mathematics is and how it is learned. For example, people who concentrate on skills tend to perceive learning in terms of an individual's acquisition of information and particular abilities of symbol manipulation. Under such a model, teaching consists of imparting information and instruction which the student then memorizes and practices. These approaches are backed by the behaviourist traditions of learning in which the learner is seen as responding to the stimulus provided by the teacher, gradually filling his or her *empty jug* of a head or writing on his or her *tabula rasa*. Researchers with a more cognitive approach are concerned with the growth of concepts by the learner and their classification systems, structures and mechanisms. Both the behaviourists and the cognitivists tend to view the context of mathematical activity as something external to and separate from cognition. What we are calling the 'social practice theorists' on the other hand see task, performance and context as more organically related parts of a whole. They are concerned with how particular practices produce meanings and condition performance in different contexts. In the

United Kingdom research on the mathematics used in workplaces has usually come from a 'skills' stable.

Research in England and Wales

Systematic research on the mathematics used in employment in England and Wales began in the early 1980s with several un-related initiatives. From the schools' point of view the most well known is the work of the Cockcroft Committee whose brief was;

> To consider the teaching of mathematics in primary and secondary schools in England and Wales, with particular regard to the mathematics required in further and higher education, employment and adult life generally, and to make recommendations.
>
> (Cockcroft, 1982, p. ix)

The foreword, written by the Secretary of State for Education and Science and the Secretary of State for Wales began:

> Few subjects in the school curriculum are as important to the future of the nation as mathematics; and few have been the subject of more comment and criticism in recent years. This report tackles that criticism head on.

The criticism referred to came to a head in the mid seventies and was referred to by the Prime Minister of the time, James Callaghan in 'the Ruskin Speech' of October 1976 which launched the Great Debate on education. The fact that the Cockcroft Committee found that the 'overall picture was much more encouraging than the earlier complaints had led us to expect' tends to have been forgotten. 'We have found little evidence that employers find difficulty in recruiting young people whose mathematical capabilities are adequate'. (Cockcroft, para. 46).

The two main pieces of research commissioned by Cockcroft, and the effects on them and their findings of the utilitarian view of mathematics which was the brief for the Committee, are discussed by Dowling in Part 1. The work conducted at Bath University by Fitzgerald and his colleagues under Professor Bailey, included two main publications, one that came to be known as the Bath List and the other as the Bath Report. The former listed all the projects concerned with mathematics and employment that were in action across the country up to and including the time that the Cockcroft research began; these are referred to by Mathews and by Harris in this section. The existence of these projects contradicts the widespread and false assumption that the Cockcroft research was the first of its kind.[1] The Bath Report (1981) is the official report of the research conducted from Bath University and details both the research methods and results of its inquiries into:

the mathematical requirements of the employment undertaken by those pupils who leave school between the ages of 16 and 18 in England and Wales.

(Bath Report, p. 1)

The research method was interview and observation; it was felt that since the research team was not intending to 'look intensively at a well defined area' but rather 'attempting to define the area of study', then 'the field of investigation was potentially so vast that to constrain the research by a set of questions would have defeated the aim of the project' (Bath Report, p. 5). More detail of the reasons for rejecting the use of questionnaires is given in Appendix II of the Report which also draws attention to difficulties of defining samples for this type of research (see also Harris, Part 2). The findings detailed in the Report are summarized, in Chapter 3 of the Cockcroft Report and by Fitzgerald (1983a and 1983b). Summaries have their obvious uses but two negative effects of these summaries should be noted. Firstly, as a summary of a more detailed report, Chapter 3 necessarily lacks detail, emphasizes some findings over others and omits some altogether. Secondly the Bath Report does not claim that certain skills are *never* or *always* used; it is a summary of an academic view of a necessarily limited number of workplaces that does not claim to be exhaustive and it goes out of its way to emphasize how little in common some apparently similar workplaces have in their uses of mathematics. Some of its conclusions however have been overstated and over-generalized since, and this has had distorting effects on curriculum debates amongst teachers and negative effects on industry-education co-operation (see Harris, Chapter 13).

Parallel in time to the Cockcroft commissioned research (and therefore not detailed in the Report) were three research projects all within the further education sector. This research differs in perspective from the schools-oriented Cockcroft research, but since its field was also the mathematics of workplaces, its findings are relevant to mathematics in school. Aspects of this work is discussed by Strässer *et. al.*, by Mathews and by Harris in Chapters 13, 14 and 15. The three Projects were the *Young People Starting Work* Project of Sheffield University (Banks, 1981), the *London into Work Project* of the Inner London Education Authority and the work of the *Mathematics Education Group* at Brunel University. The first two projects were funded partly by the Manpower Services Commissions and used questionnaires developed in the United States and Canada and adapted for use in England; their approach, as was Cockcroft's, was prescribed by the Government's, not educators' views on education, training and mathematics. Both used research instruments from the skills-based school of industrial psychology; the research and its outcome were conceived in terms of behavioural skill, not cognitive growth, (see Harris, Chapter 13).

The research conducted at Brunel University by Ruth Rees and George Barr was of a different orientation. Its concern was with the examination performance of trainees following City and Guild's courses and the grading of levels of difficulty of the questions by means of analyses of students' responses. The work is discussed in Section 2.1 of Strässer *et. al.* in Chapter 15.

More recently, in the United Kingdom, Alison Wolf and her associates have investigated and assessed the development of arithmetic skills in particular occupations (Wolf, 1984; Wolf and Silver, 1986) and worked out practical exercises for promoting skill transfer at work (Wolf, 1989). Some of this work is referred to in Section 2.2 of Strässer *et. al.* one of whose authors is Wolf herself, and by Mathews. The work of the more recent project, *Practical Problem Solving Mathematics in the Workplace*, conducted at Sheffield City Polytechnic from 1985–7 is discussed by its Director, Drake, in Part 4 of this volume.

The Concept of Transfer

Underlying the skills approach to learning is the assumption of a process of *transfer*, the application of concepts and skills learned in one context to another context. 'Conventional academic and folk theory assumes that arithmetic is learned in school in the normative fashion in which it is taught and is then literally carried away from school to be applied at will in any situation that calls for calculation' (Lave, 1988, p. 4). Included in this assumption is the idea that transfer will occur unproblem- atically as the result of 'proper schooling' and that if it does not happen, there is something wrong with the schooling or the learner. It is this latter assumption that is behind most of the criticisms levelled at schools by employers. However, Lave (1988, Chapter 2) has demonstrated both the paucity and weakness of the research base in supporting transfer; it is an assumption, rather than a finding. Lave's view that mathematics inside and outside school are better characterized by disjunction rather than commonality lends support to the experience and research of mathematics education that emphasis on skills out of context is an ineffective method of teaching and learning. (See for example Wolf, 1984 and McIntosh, 1977). Dowling (Chapter 11) emphasizes this disjunction in his discussion of the different discourses of school and work mathematics. Degrees of acceptance of the idea of transfer are expressed by the different authors of the paper Strässer *et. al.* One view remains hopeful about the possibilities of transfer, others are more sceptical or warn against assuming that it will be very easy at all. Mathew's paper on the other hand, shows the consequences of the utilitarian idea's acceptance of transfer. Both papers draw out implications for mathematics in training and work. In spite of the evidence against it, the viability of the concept of transfer remains largely unquestioned in England and Wales, particularly so in the sectors of government that determine curriculum content. Given the overweening emphasis on skills and transfer, the existence of alternative ways of analyzing and conceptualizing activities, with just as great but very different educational implications, may come as something of a surprise.

The concept of transfer cannot cope with findings of Carraher *et al.*, of Scribner and of Lave (see Carraher, and Evans and Harris below). It may have implications for education but it also has its own internal difficulties. So far as education is concerned it contains the assumption that

> children can be taught general cognitive skills (e.g. reading, writing, mathematics, logic, critical thinking) if these 'skills' are disembedded from the routine contexts of their use. Extraction of knowledge from the particulars of experience, of activity from its context, is the condition for making knowledge available for *general* application in all settings . . . classroom tests put the principle to work; they serve as the measure of individual, 'out of context' success, for the test-maker must rely on memory alone and may not use books, classmates, or other resources for information.
>
> (Lave *op. cit.* p. 8–9)

Other Approaches

The mathematics-and-work research of Carraher, Carraher and Schliemann and their colleagues in Brazil, specially summarized by David Carraher for this volume, uses an approach that links workplace interviews with laboratory-style testing and analyzes results in cognitive psychological terms. The differences in arithmetic practice among schooled and unschooled workers discussed by Carraher and the stress of the researchers on trying to link the mathematical meanings within both practices, have implications for both theory and pedagogy.

> By studying how people come to make sense out of mathematical ideas outside school, perhaps we can learn how to promote this transition, [between symbolic representation and meaning] making meaningful representations and procedures more powerful on the one hand, and making powerful represenations and procedures psychologically meaningful [on the other].
>
> (Carraher, Part 2)

A concept of transfer is still there, but the realization of the significance of the context of the activity in conditioning mathematical meaning, raises the need for a much more detailed analysis of what is involved in describing differences in performance involving arithmetic with different activities in different contexts.

A number of studies adopting ethnographic approaches have been conducted outside the United Kingdom; these have implications for both research and education. Some of them are examined by Dowling (Chapter 11 of this volume) in his consideration of the work of what he calls the 'mathematical anthropologists', those who seek for evidence of mathematical behaviour within daily practices of mainly informal work. He sees them as arguing that various practices (for example the production of some African artefacts) embody 'frozen mathematics' and that 'defrosting' this mathematical knowledge will lead to cultural confidence; but he also interprets writers using this approach as still 'privileging' formal, abstract, Eurocentric mathematics, of judging the workplace from their point of view.[2] Although individual workers such as Harris have used participant observation techniques in attempting to analyze the mathematics done at work, there is as yet no tradition in the United Kingdom of the ethnography of workplace mathematics. The approach shows promise however. The anthropological tradition can not only show 'mathematical essences' but also the greater efficiency of less general (that is less 'powerful' in mathematicians' terms), more practical routines and strategies of thinking. Such work has a number of implications for mathematics education in school, once the narrowness of the traditional culture of mathematics and mathematics education is noted. (See Part 1 of this volume). Implications are much wider than for those of ethnic or gender 'minorities' which form the samples for such research at present.

The seminal paper on the cognitive consequences of formal and informal education by Scribner and Cole (1973) drew attention to differences in cognitive performance in formal, (including educational) settings and informal, (including working) settings. It became a stimulus for much of the research in the social anthropological tradition. The authors presented a discussion of learning which makes a number of distinctions: the dimensions of difference between the context and content of the two modes of learning include the social group basis, links between the intellectual and the emotional factors, the setting, the content of learning and the mechanism of learning. Firstly, informal learning tends to be family based and particularistic whereas the formal learning carried on in schools is the responsibility of some wider social group, which may be 'society'. In the case of non-institutional formal education, for example initiation rites in traditional societies or private tutoring among eighteenth century aristocrats, the social base may be a sub-culture. Secondly, informal learning tends to fuse the intellectual and the emotional:

> One of the most outstanding characteristics of informal learning . . . is the high affective charge that is associated with almost everything that is learned within the context. The reason for this is that the content of learning, especially in children, is often inseparable from the identity of their teacher.

> (Scribner and Cole, p. 555)

In formal learning, children have to learn to prise apart the cognitive and affective. Thirdly, the setting of informal learning is a 'natural' one, where the meaning is intrinsic to the situation, whereas formal learning is removed from the immediate context of everyday life to a special setting, deliberately organized for the specific purpose of transmission. Fourthly, when it comes to content, informal learning deals with a 'demarcated set of activities or skills with the result that the learning processes are inseparably related to the given body of material' (Scribner and Cole, *op. cit.* p. 555). The content of formal learning may comprise 'new subjects' such as mathematics, or an approach to an area of knowledge which challenges or even contradicts the 'common sense' of accepted beliefs. Informal learning on the other hand, tends to support traditional beliefs and values. Finally, the mechanisms of informal learning include imitation, identification, co-operation and above all observation: these contrast with formal learning which is acquired primarily through language; the teacher may begin with a verbal formulation of a rule and show empirical referents only later.

These distinctions have been further analyzed and developed by Greenfield and Lave (1982) and by Hoyles (1990) and they can form a useful basis for research analysis. The summary below is adapted from Hoyles (1990)

Informal Mathematics	*School Mathematics*
embedded in task	decontextualized
motivation is functional	motivation is intrinsic
objects of activity are concrete	objects of activity are abstract
processes are not explicit	processes are named and are the object of study
data is ill-defined and 'noisy'	data is well defined and presented tidily
tasks are particularistic	tasks are aimed at generalization
accuracy is defined by situation	accuracy is assumed or given
numbers are messy	numbers arranged to work out well
work is collaborative, social	work is individualistic
correctness is negotiable	answers are right or wrong
language is imprecise and responsive to setting	language is precise and carefully differentiated

In a study which is one of those considered as exemplifying the work of 'practice theorists' (see Evans and Harris, Chapter 17) Scribner (1984) went on to investigate 'informal' thinking, concerning herself with 'practical knowledge and thought for action' contrasting this view with the traditional western view of thinking from Aristotle, in which thought which is not philosophical and abstract is not to be considered as thought at all, but merely a function of artisans. By 'practical thinking' Scribner meant thinking that occurs within wider activities and which is directed towards their outcome. Scribner's phrase 'working intelligence' has two senses referring generally 'to the intellect at work in whatever contexts and activities those may be' (Scribner, p. 10) and more specifically, to the particular context of the research workplace. Her phrase 'thinking at work' with its ambiguity, is appropriate.

All research is motivated and has aims and parameters. It therefore has values and politics embedded in the assumptions from which it starts[3]. These values and politics are not always made explicit and possibly not even noticed. In Part 1 attention was drawn to the hidden curricula of mathematics and mathematics education. It is an aim of this section to draw attention to the 'hidden curriculum' of mathematics in the workplace research. Another is to raise questions concerning the aims, content and pedagogy of school mathematics curricula in the light of findings of research on mathematics in workplaces.

Notes

1 Equally misleading is the assumption that the Cockcroft research is or should be the last word on the subject.
2 The same criticism can be made of course of the Cockcroft research which was conducted by academic mathematicians judging the non-academic activities of workers.
3 Values and politics of course are implicit (at least) in the controlled availability of research funding.

References

BANKS, M. (1981) *Young People Starting Work*, Sheffield, MRC/SSRC Social and Applied Psychology Unit, Sheffield University.
BATH (1981) *Mathematics in Employment (16–18)* Final Report by Fitzgerald, A. and Rich, K. M., Bath University.
COCKCROFT, W. H. (ed) (1982) *Mathematics Counts: Report of the Committee of Inquiry into the Teaching of Mathematics in Schools under the Chairmanship of Dr W. H. Cockcroft*, London, Her Majesty's Stationery Office.
FITZGERALD, A. (1983a) 'The "Mathematics in Employment 16–18" Project: Its Findings and Implications, Part 1', *Mathematics in School*, January, 1983.
FITZGERALD, A. (1983b) 'The "Mathematics in Employment 16–18" Project: Its Findings and Implications, Part 2', *Mathematics in School*, March, 1983.
GREENFIELD, P. and LAVE, J. (1982) 'Cognitive Aspects of Informal Education', in Wagner, D. and Stevenson, H. 1982. *Cultural Perspectives on Child Development*, San Francisco and Oxford, Freeman.
HOYLES, C. (in press) 'Computer-based Learning Environments for Mathematics', in Bishop, A., Mellin-Olsen, S. and van Dormolen, J. (eds.) *Mathematical Knowledge: Its Growth through Teaching*, Dordrecht, Kluwer.
LAVE, J. (1988) *Cognition in Practice*, Cambridge University Press.

McINTOSH, A. (1977) 'When Will They Ever Learn?' *Forum*, 19, 3, summer, reprinted in Floyd, A. (ed) 1981. *Developing Mathematical Thinking*, England, Addison Wesley and Open University.

SCRIBNER, S. and COLE, M. (1973) 'Cognitive Consequences of Formal and Informal Education', *Science*, Vol. 182. pp. 553–559.

WOLF, A. (1984) *Practical Mathematics at Work. Learning throught YTS*. Research and Development Paper No. 21, Sheffield, Manpower Services Commission.

WOLF, A. and SILVER, R. (1986) *Work Based Learning: Trainee Assessment by Supervisors*, Research and Development Paper No. 33, Sheffield, Manpower Services Commission.

WOLF, A. *et. al.* (1988) *Problem Solving: Practical Exercises to Promote Skill Transfer*, Sheffield, Careers Occupational and Information Centre.

13
Looking for the Maths in Work

Mary Harris

In Chapter 3, paragraphs 103–109 of the Cockcroft Report (1982), attention was drawn to the number of projects already completed or continuing, within local education authorities and by others, in the field of mathematics and employment, particularly with respect to those students who leave school at 16. Since its publication however, much of this work has been ignored or forgotten by many people in mathematics education for whom the report and its supporting studies are often assumed to represent the first and last word on mathematics and employment. This assumption has been recently reinforced by the influence of the Cockcroft Report on the utilitarianism of the national curriculum.

Some of these seventy or so schools-industry mathematics projects provided interesting workplace examples of uses of mathematics and the reports of their activities, detailed in what came to be known as the Bath List (1980), still make relevant reading for anyone seriously interested in continuing their work or developing work of their own. Some of the projects analyzed employers' entry tests and surveyed employers' perceived needs; some surveyed the skills used in their local employment and the skills employers complain about and compared them with the skills assessed in public examinations: groups of teachers and industrialists produced core curricula with agreed minimum goals; conferences of educationists and industrialists debated the value and possibilities of core curricula and published their debates; Training Boards published lists of 'basic skills'; joint schools and industry projects published examples of industry oriented work done in schools.

One major education-industry project whose research was carried out during the same period as the research for the Cockcroft Report (and which therefore does not feature in it) was the *London into Work Project* of the Inner London Education Authority, jointly funded by the Authority, the Department of Education and Science and the Manpower Services Commission. One of its aims was:

To improve the link between the education system and the world of work by finding out about and profiling samples of jobs within the local labour markets and by applying the information thus obtained to the development of learning programmes and of coherent systems for the vocational preparation and education of young people.

(Townsend, 1982).

Its sample was 1000 young people in a range of jobs that represented the employment situation in the Inner London Education Authority area. The findings of the project were intended to have a direct effect on curricula in the Authority's further education sector and they also went to inform the development of the list of core skills to be defined later for the vocational programmes of the Manpower Services Commission. (See Mathews Chapter 14).

Sampling presented a number of problems for the research team, not least because there was no one way of notifying the existence of school-leaver jobs. Some jobs were notified to employees through the schools careers service, some through state and private employment agencies, some in the local press or radio. A reasonable picture of the employment scene in London at the same time was eventually obtained and a classification system for jobs sought. All such classifications present difficulties[1] not the least being the wide variation of what employers both describe and expect within a single employment 'type'. The scheme eventually chosen was the one used by the ILEA Careers service. An interesting aspect of this particular sample was the very wide range of jobs that were examined. London has never been a one-industry city so the sample contained a range of small firms a particularly large number of which employed fewer than fifty people. This had implications for the research findings in that a small employer tends to expect his or her employees to be more flexible in what they do than a larger and more hierarchical company in which roles are more rigidly defined. At the time the project was working, small manufacturing was moving out of central London while service industries were beginning to develop within it. The sample therefore contained a particularly wide spread of types of job.

At the time that the London into Work project was proposed, there was little positive working contact between the mathematics curricula of schools and further education within the Authority except for local initiatives and a limited number of joint courses such as the Foundation Courses provided for leavers of some special schools and the experimental, EEC-funded Bridging Courses. The ILEA Schools Mathematics Inspectorate however, were taking an active part in the development of a new limited grade GCE course and examination for school leavers called *Everyday Maths* to be moderated by the London Regional Examination Board. Its content was intended to be immediately relevant to life outside school and therefore meaningful to pupils and the curriculum was to be planned from the bottom up, starting where the students were and building on that, rather than imposing abstract mathematical demands of higher level courses in dilute form. It was an unusual venture for the time and the thinking behind it foreshadowed many of the recommendations later made by the Cockcroft Report. When they came, many of the Cockcroft recommendations fitted easily into its philosophy.

> No topic should be included unless it can be developed sufficiently for it to be applied in ways that the pupils can understand. (paragraph 451)
> The syllabus . . . should not be too large, so that there is time to cover the topics which it contains in a variety of ways and in a variety of applications.
> (paragraph 454)
> Examination papers should enable candidates to demonstrate what they do know rather than do not know, and should not undermine the confidence of those who attempt them. (paragraph 521)

A second unusual feature of the course was its concern with problem-solving, that is with a more process-oriented approach than was usual in non-academic courses at that time.

The first planning meetings about the projected school course took place during the period in which the questionnaires to be used in the London into Work project were being modified for use in London, but the two ventures were then entirely separate. The four questionnaires were entitled 'Basic Calculations, Measurement and Drawing', 'Communication Skills: Listening and Talking, Reading and Writing', 'Practical (manipulative)' and 'Problem Solving' respectively. Three of them were being adapted from Canadian industrial research questionnaires (Smith, 1975, in Townsend, 1982) and one from research in the United States (Cunningham, undated in Townsend, 1982). As part of their adaptation they were sent for criticism and approval to the School Mathematics Inspectorate, arriving at about the time that up-to-date evidence of the use of mathematics skills and examples of problem solving in employment were being sought for the development of the school's *Everyday Maths* course. The reaction of the Schools' Mathematics Inspector was to appoint an advisory teacher (Harris) with experience in research, curriculum development and materials design, to analyze the data as it came in and use it both for informing the development of the new course and in designing learning materials to support it. A project called *Maths in Work* came into being with the Advisory Teacher as its Leader — and sole incumbent.

There were immediate data-handling problems many of which were administrative but many a function of the questionnaires themselves. Obviously the scope of the latter was wider than the research of the Cockcroft projects which was concerned only with mathematics. However, the assumption that all the data on mathematics would be generated by the Basic Calculations and Problem Solving questionnaires and them alone, turned out to be unfounded.

It had always been intended that the large amount of data that would be generated by the questionnaires would be handled by computer, so each question on all questionnaires, concerning the frequency of use of each particular skill, was designed to produce a numerically codeable answer. For example, the form of the question was always 'How often in the course of your work do you (add fractions) (have discussions with your supervisor) (use a screwdriver) (file information or objects)?' Job holders could respond at one of four levels 'Very Often', 'Fairly Often' 'Hardly Ever' or 'Never', the three positive and one negative responses being coded on a four point scale. As a check on the accuracy of the figures however, a second part was included in each question in which the job holder who used a skill particularly often was asked to explain in detail how and why the particular skill was used. (See Townsend (1982) for details of the design and use of the questionnaires). Thus, although it was only ever intended that the numerical data would be used, two distinct types of data were generated, the *skill counts* and the much more open-ended verbal accounts of the *contexts* in which high-frequency skills were used. The research purposes of the further education curriculum for whom the survey was conducted would be satisfied with the figures and the computer analysis.

The aims and the resources of the schools' interests however differed considerably from those of further education. The concern of the Maths in Work Project was to be solely with mathematics; it had no reason at first to do other than analyze the figures from the Basic Calculations questionnaires as they came in and since the Project Leader had no access to computing facilities, she worked by hand. The clearest way of processing the numerical data into a readable form was to plot them in terms of the percentage of jobs in which the particular skills were used. With this end in view, the three positive scores (very often, fairly often, hardly ever) were aggregated and the first graph appeared as Figure 13.1.

The effect of the clarity of such a presentation was an immediate request from further education for graphs of the other skill areas. Eventually, when all the data were in, a complete set of graphs was produced for the final sample of 968 jobholders. Obviously the form of all the graphs was similar but variation between them immediately raised further questions; for example the Listening and Talking graph was much more 'top heavy' than the Basic Calculations graph and contained a higher proportion of positive responses (Figure 13.2).

Since the questionnaires were primarily concerned with comparing jobs on measures of skill frequency, the acts of communication listed on the Listening and Talking questionnaire were simply counted. The existence of the context part of the question however, meant that some details of the content of the communications would be available if the skill scores were high. By searching the completed questionnaires for details of contexts of these skills, it would be possible to discover the message that the acts of communication conveyed. It was here, in the Communications Skills contexts (and later in the Practical and Problem-Solving skills contexts) that a great deal of information about the use of mathematics in work was

Basic Calculations

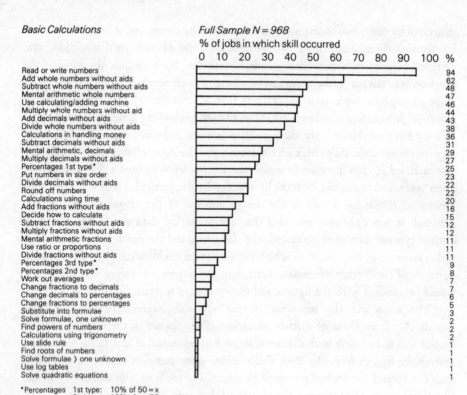

Full Sample N = 968

% of jobs in which skill occurred

Skill	%
Read or write numbers	94
Add whole numbers without aids	62
Subtract whole numbers without aids	48
Mental arithmetic whole numbers	47
Use calculating/adding machine	46
Multiply whole numbers without aid	44
Add decimals without aids	43
Divide whole numbers without aids	38
Calculations in handling money	36
Subtract decimals without aids	31
Mental arithmetic, decimals	29
Multiply decimals without aids	27
Percentages 1st type*	25
Put numbers in size order	23
Divide decimals without aids	22
Round off numbers	22
Calculations using time	20
Add fractions without aids	16
Decide how to calculate	15
Subtract fractions without aids	12
Multiply fractions without aids	12
Mental arithmetic fractions	11
Use ratio or proportions	11
Divide fractions without aids	11
Percentages 3rd type*	9
Percentages 2nd type*	8
Work out averages	7
Change fractions to decimals	7
Change decimals to percentages	6
Change fractions to percentages	5
Substitute into formulae	3
Solve formulae, one unknown	2
Find powers of numbers	2
Calculations using trigonometry	2
Use slide rule	1
Find roots of numbers	1
Solve formulae 〉 one unknown	1
Use log tables	1
Solve quadratic equations	1

*Percentages   1st type: 10% of 50 = x
2nd type: 10% of x = 50
3rd type: 10 = x% of 50

NB.   These figures represent the numbers of people who answered specific questions. The same people sometimes show that they are in fact using some of the above skills when they answer questions from the other questionnaires. The figures above should therefore be taken as minima, not absolutes.

Figure 13.1   'Looking for the Maths in Work'

found. Furthermore the information was often at variance with the skill frequency responses to the Basic Calculations questionnaires. For example, one job holder when asked about the use of percentage calculations from the Basic Calculations questionnaire denied that she used them. When asked how she dealt with irate customers under the Communication Skills questionnaire however, she explained that they were usually angry because of the prices of the suits in the outfitters where she worked and that she calmed them down by 'knocking 15% off'. Most of the hairdressers in the sample denied that they used ratio and proportion when they were asked about the mathematical skill, but when asked a practical question about mixing things they explained that they frequently mixed hair dyes and that unless these were mixed in the correct proportion (and they used the word 'proportion' quite comfortably in this 'non-mathematical' context) damage could be caused to the customers' hair.

Listening and Talking

Full Sample N = 968

Percentage of jobs in which skill occurred

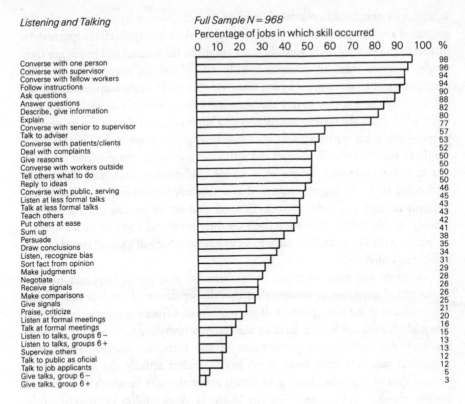

	%
Converse with one person	98
Converse with supervisor	96
Converse with fellow workers	94
Follow instructions	94
Ask questions	90
Answer questions	88
Describe, give information	82
Explain	80
Converse with senior to supervisor	77
Talk to adviser	57
Converse with patients/clients	53
Deal with complaints	52
Give reasons	50
Converse with workers outside	50
Tell others what to do	50
Reply to ideas	50
Converse with public, serving	46
Listen at less formal talks	45
Talk at less formal talks	43
Teach others	43
Put others at ease	42
Sum up	41
Persuade	38
Draw conclusions	35
Listen, recognize bias	34
Sort fact from opinion	31
Make judgments	29
Negotiate	28
Receive signals	26
Make comparisons	26
Give signals	25
Praise, criticize	21
Listen at formal meetings	20
Talk at formal meetings	16
Listen to talks, groups 6 −	15
Listen to talks, groups 6 +	13
Supervize others	13
Talk to public as oficial	12
Talk to job applicants	12
Give talks, group 6 −	5
Give talks, group 6 +	3

Figure 13.2   'Looking for the Maths in Work'

The reasons for these differences in response were not immediately obvious from the data though it appeared to have something to do with attitude[2] but it was clear that much more mathematical thinking was going on than was being revealed by answers to questions about mathematical skills as such. For the job-holders a questionnaire about basic calculations, however sympathetically administered by the specially trained researchers of the London into Work project, was still an arithmetic test of a sort. Under such circumstances many of the job holders, who would not have been in the survey at all if they had been higher attainers in school mathematics, denied their use of calculation when it was so clearly defined. However, when asked the less threatening questions about listening and talking, they often revealed a use of mathematical understanding and numerical skill they had previously denied.

Information about the use of mathematics embedded within the context data of the Communication Skills questionnaire, repeatedly illustrated differences between the origin, usage and techniques of mathematics at school and at work. In the former case, many students' mathematical activity is initiated as the result of instructions written in a text book, on a work card or on a chalkboard by the teacher in the context of a mathematics lesson. In the widespread skills-and-applications approach to mathematics teaching, numerical skills are taught in the abstract and pupils are then taught to apply them to examples usually devised for the purpose, for real life examples are often surrounded by too many variables and contain numbers that work out less well. In work however, mathematical activity arises from within practical tasks, often from the spoken instruction of a supervisor and always for an obvious purpose which has nothing to do with numbers working out well. Thus students taught to react to isolated, abstract and written commands in the specialist language and carefully controlled figures of a school mathematics class, find themselves confronted with the urgent spoken, if not shouted, instructions in a completely different context and code; 'Knock 15% off that lot so we can get rid of them quickly' or 'Have you got enough there for three seven and a quarter inch lengths?' or 'Ring up the Depot and ask them to let us have another half gross of the 43715s by first thing tomorrow'.

As more and more examples of mathematics emerged in responses to non-mathematical questions, it became clear that the separation of the four sets of data from each other and the separation of the mathematics from the context in which it occurred, determined by the form of the research instruments, could not possibly represent the mathematical performance of any particular worker. The figures themselves said very little about what any individual actually did as work. The context data on the other hand gave details about the way in which an individual reacted, thought and behaved. For the Maths in Work project in contrast to the further education emphasis on skill training, the work became not so much data analysis (How many jobs use fractions?) but situation analysis (What are the demands on this person of her work that stimulates her mathematical thinking and action?).

It is relevant here to compare more closely the research methods of the London into Work project with those of the Cockcroft researchers. The latter are discussed in detail in the Bath Report (1981) Chapter 2 and Appendix 2. Among the reasons given by the researchers for not using questionnaires was that 'there was evidence in earlier studies' (not quoted) 'that the questions may condition the responses.' (paragraph 2.5 Appendix 2). The brief of the Cockcroft researchers was to consider

> the mathematical requirements of the employment undertaken by those pupils who leave school between the ages of 16 and 18 in England and Wales.
>
> (paragraph 1.1 Chapter 1).

From this brief, they defined the mathematical requirements of the employment investigated as 'those processes or methods in use by the employees which could be viewed as mathematical. We consider that a process could be viewed as mathematical if it required thought about a mathematical concept such as shape, size, number, orientation'. The chosen research method, having rejected questionnaires, was 'interview and observation' (paragraph 2.1 Chapter 2). When it came to the interviews, the report continues (paragraph 5.4 Chapter 2)

> After talking generally with the interviewee about school and training, we then asked if he [sic] could supply us with examples of where mathematics was useful in his work. With many people, their initial reaction was that they did not use "mathematics" and it was necessary to prompt them with suggestions like, "any figures you have to write down?", "any measuring you do?".

Thus "thought about a mathematical concept such as shape, size, number, orientation" was reduced to the easily identified figures that form the public stereotype of "mathematics".' Readers of the Cockcroft report's Chapter 3 *The Mathematical Needs of Employment* frequently remark on the arithmetical nature of the way in which *mathematics* is finally interpreted. For example paragraphs 64 and 65:

> 64.   The report of the Bath study draws attention to the fact that it is possible when observing work in progress to describe certain aspects of it in mathematical terms. For instance, the mathematical concept of geometrical symmetry is present within many manufacturing processes; it is possible to describe different methods of stacking and packing in geometrical terms or in terms of sorting and classifying. However, even if the mathematical concepts involved have at some time been encountered in the classroom, the employee will probably not consciously analyse in such terms the operations which are being performed; nor, if he were to do so, would he necessarily be able to do his job any better.
> 65.   On the other hand, many jobs require the employee to make explicit use of mathematics — for instance, to measure, to calculate dimensions from the drawing, to work out costs and discounts. In these cases the job cannot be carried out without recourse to the necessary mathematics.

Here again, 'the explicit use of mathematics' is reduced by the writers to the numerical skills of measuring and calculating with a set of assumptions that could have been designed to maintain existing arithmetic syllabuses. Geometrical concepts including symmetry which may have been encountered in the classroom are not expected to have any relevance to work and when they are noticed in work are assumed not to have any relevance to school mathematics. The employee is assumed to probably not use the classroom experiences in the work situation and the writer of the Report assumes, without offering any evidence, that even if he were to do so he would not necessarily be able to do his job any better. The section quotes no examples where the concept of mathematical symmetry is crucial to the manufacturing process, such as in garment assembly, and one can speculate whether or not its authors would regard such an activity as mathematical at all given that it need require no 'explicit' measurement nor calculation. (It should be noted that the Bath Report does quote the use of symmetry for example by shoe technician apprentices. These people also used thirds, sixths, twelfths and sixteenths; see below.) The Bath Report's own definition of mathematical requirements and their inclusion of shape and orientation have been much diluted and have even disappeared from the employment section of the Cockcroft Report. If mathematics is perceived as the application of number, then questions will be asked and answers received about applications of number and that particular view of 'mathematics' will be reinforced. But if mathematics is perceived as, for example, the symbolic representation of the search for pattern and generalization in spatial and numerical phenomena, then totally different questions will be asked and a more cognitive perception of mathematics illustrated in the answers. It would appear that the assumptions that underlie unstructured interview questions are just as open to the charge of conditioning the response as those of structured questionnaires and in the case of both the Cockcroft and the London into Work research, the assumptions about what workplace mathematics at $16+$ is expected to be, are fairly clear.

The effects of the narrow arithmetical focus of both the Cockcroft Report and the skill count research projects naturally reflected back into mathematics education in a way that was both distorting of what happens at work and distorting to subsequent debate. One example is what could be called the 'fractions fracas'.

A continuing complaint of industry has been and still is that school leavers are unable to manipulate fractions. The Cockcroft Report states (paragraph 75) that:

> Although fractions are still widely used within engineering and some other craft work these are almost always fractions whose denominators are included in the sequence 2, 4, 8, . . . , 64. This sequence is visibly present on rules and other measuring instruments and equivalences are apparent.
> The need to perform operations such as $2/5 + 3/7$ does not normally arise, and manipulation of fractions of the kind which is commonly practised in the classroom is hardly ever carried out.
> (paragraph 300) . . . it is difficult to find everyday situations which require fractions to be added or multiplied.

A number of points should be made here. In fact the Bath Report does give examples of fractions in use (for example Chapter 4, 4.4) including the use of non-school algorithms for division. Secondly, if the equivalences really were apparent from the rulers, then there would have been no cause for complaint. Third, the fact that the Cockcroft researchers found it difficult to find everyday situations which require fractions to be added or multiplied does not mean that such situations are generally difficult to find. Possibly committee members were not DIY practitioners who bought timber from small general stores. In spite of its emphasis on the importance of measurement at work (paragraph 79) and in spite of its acknowledgment that industry necessarily still uses imperial units (paragraph 82) the Cockcroft Report, written by mathematicians and mathematics educators, still seems to assume that the world in which these imperial units are used is a world where numbers always appear either whole or else neatly sliced on only one sort of ruler.

This playing down of its own research findings was taken by many teachers to mean that fractions are not used at work. With evident relief, many argued fiercely against teaching them, naming the Cockcroft Report in their support. Evidence from other research including that of the London into Work project (see Figure 13.1) of the continued practical use of fractional qualities were denigrated or dismissed in the light of the 'official' findings of Cockcroft and pleas by the Maths in Work Project for serious debate about the teaching of the principles of measurement and the understanding of the behaviour of numbers smaller than one in metric and imperial systems became drowned in almost wilful misunderstanding and accusations that the plaintiff was advocating a return to the mindless school algorithms which Cockcroft, not surprisingly, had not found at work. Industry continues to demand 'fractions' of school leavers confusing their practical use with the archaic algorithms through which they test the school leavers for them, which have nothing to do either with practice or with mathematical thinking. By couching its demands in terms of discredited procedures it fails to convince teachers of the still widespread use of imperial units which, in the nature of practical activity do not conveniently come whole. To employers, the thought that numbers less than one may not be a serious topic of study in school is incomprehensible. To teachers, employers seem to be demanding no more than an archaic and anti-educational pedagogy.

Meanwhile the insistence on 'skills' by both industry and the skills-and-applications mode of thought in education, almost ensures that the use of imperial units (and imperial units are listed in the national curriculum) is taught as yet another algorithmic skill, that of conversion from metric units. The topic of 'fractions' remains a generally non-serious subject in mathematics education and employers still complain about school leavers who cannot manipulate them. The fractions fracas illustrates above any other the dangerous gaps of communication, limited views and differences in perceptions of mathematics that still exist between the discourses of education and industry.

The distorting emphasis on numerical skills over mathematical thinking in the London into Work research did nothing to alleviate the situation. As we have seen, the figures for the use of arithmetic skill were artificially low in the London into Work data and did not reflect its wider use revealed in questions not about arithmetic. Overlaps in question content between the questionnaires added to the distortion; for example a question about using calculators was answered negatively by some jobholders who, when asked a practical question about using keyboards, described how they used electronic tills which calculated money for them.

All in all, the skill figures alone represented very little of what went on, in mathematical terms. They could in fact be said to represent no more than a measure of the measuring instrument itself. The mathematics that is revealed in the London into Work survey occurs 'between the lines' of the non-mathematical questionnaires and the majority of it was never analyzed because no funding was provided for the purpose. The evidence was there but 'unavailable' because it was uncountable and prejudged as unimportant. The skills that were counted were of course the countable ones, devised to be so, the rest was ignored and the figures of the London into Work survey went, with others, to inform the development of what was to become the core skill lists for such programmes as YTS and CPVE. It is interesting to note at this point that the kind of numerical skills being researched both by the skill-based research projects such as London into Work and by Cockcroft are almost identical to those named in nineteenth century curriculum lists for artisans, the arithmetic grades of the Lancastrian schools of the early nineteenth century for example, worked through tens and units, addition, subtraction, multiplication and division, fractions and reduction (Howson, 1982). Modern arithmetic skill lists are also recognizably similar to recommended curricula of the Elizabethan age.[3] (See the chapter on Robert Recorde in Howson, 1982).

It is the lists of assessable skills that people remember as 'mathematics', from Robert Recorde to today; the questions not the processes of achieving the answers in such a way as to invite further questions. In a climate that accepts that mathematics-for-labourers is minimal, numerical skills and little more, every new curriculum initiative for such a defined group generates or revives lists of skills all remarkably similar to those of previous projects and previous ages. The status quo is maintained and the popular and industrial image of workforce mathematics is fossilized in a state that is disowned by mathematicians and rejected by most in mathematics education.

Within a climate of increasing emphasis on skills, the work of the Maths in Work Project Leader became particularly difficult. She had the job of publishing the data in a form teachers could use and she also had the job of designing and producing teaching material for mathematics based on the data. The wealth of the context data and the relative meaninglessness of the figures on their own made her determined not to pursue the currently accepted form of learning material which was illustrations of arithmetic skills identified at work with accompanying practice examples and with the evident aim of remaining within the conventions of traditional arithmetic syllabuses.

Meanwhile the data had to be published and because it was so large it had to be subdivided in some way that could give a flavour of the whole. Although work was continuing on the figures by the statisticians employed on the London into Work project in efforts to group the skills into job families, this work was not available to the Maths in Work project which continued to work single-handed. In fact it was some school students who asked for a grouping. On a visit to a class of low attaining boys one day, the students became interested in the graphs (of which Figs 13.1 and

13.2 are examples) and began asking questions about what they needed to know to get certain jobs. Most of them wanted to be motor mechanics so a promise was made to return to the survey and find out what motor mechanics needed. The report, *Seven Motor Mechanics* contained all the information available from the survey, both skills count and context data with no editing or interpretation. A different way of plotting the figures proportionally was devised whereby areas of squared paper were differentially shaded to represent the answers 'Very Often', 'Fairly Often', 'Hardly Ever' and 'Never'. One thing emerged immediately; there was a very low use of arithmetic skills and a relatively high use of problem solving skills. The evidence contrasted with employer perceptions of mathematics in work. Most employers' entrance tests assess arithmetic — which was not used — and ignore problem solving — which is perceived as a management skill. Further reports, *Twenty-four Hairdressers* and *Forty-four Warehouse Workers* were published and they too showed a higher use of problem solving than arithmetic and mathematical thinking revealed by responses to the questions of the Communication and Practical Skills sections.

The reports were used as was intended, to inform the development of the Everyday Mathematics course and they also informed the development of the BBC Schools Television series 'Maths Counts' but only this very small number of job contexts (75) was published. The main meat of the data however was in the contexts, still on the raw questionnaires and extractable only by hand and it would take the life time of one person to extract it all. Altogether the details of only 200 jobs out of the grand total of 968 were extracted because no money was ever made available for the purpose. It took about an hour on average to transcribe each job onto cards and the work had to fit in with the rest of the Project's work which included design and production of learning materials, organization and teaching of courses, work in school, work with teachers outside school, administration and correspondence. A vast body of expensively collected data that could have informed both industry and education as to how mathematical thinking and behaviour is generated in workplaces was never used because the possibility of its existence was ruled out by the pre-determined approach adopted by the research. There is no doubt that in due course it will be thrown away unanalyzed and out of date, an expensive testimony to a limited view of mathematics in work.

The message learned during the formal research phase of the project was that powerful forces continue to regard mathematics in work as a set of minimal arithmetic skills and that such a view is impoverishing and basically meaningless unless the context is specified. The practical lesson however was more positive. The significance of the unofficial finding that it was the context of the activity, not the questions about specific arithmetic skills that revealed the mathematics, became the basis for the learning materials developed by the Project and discussed by Harris in the Postscript to this volume.

Notes

1 See the Bath Report for a justification of the scheme used by the Cockcroft researchers.
2 For a more sophisticated description of the phenomenon see Dowling in Chapter 11 and Harris and Evans in Chapter 12 of this volume and Lave, J. (1988) *Cognition in Practice*. Cambridge University Press.
3 Readers may like to amuse themselves by comparing the mathematics course recommended by the Hadow report page 218 (1927) with that of the foundation list of the Cockcroft Report page 135 (1981) and the Revised Code of Regulations of 1862 with the National Curriculum of 1989.

References

BATH LIST (1980) *A survey of Mathematics Projects involving Education and Employment*, University of Bath, May 1980 and supplement.

BATH REPORT (1982) *Mathematics in Employment (16–18)*, for the Cockcroft Committee, Bath University.

COCKCROFT, W. (1982) *Mathematics Counts, Report of the Committee of Inquiry into the Teaching of Mathematics in Schools*, under the Chairmanship of Dr W. H. Cockcroft, HMSO.

CUNNINGHAM, T. W. (undated) *The Occupational Analysis Inventory*, Centre for Occupational Education, North Carolina State University.

HOWSON, G. (1982) *A History of Mathematics Education in England*, Cambridge, Cambridge University Press.

SMITH, A. (1975) *Generic Skills: Research and Development*. Advanced Development Division, Canadian Department of Manpower and Immigration, Ottawa, Ontario.

TOWNSEND, C. (1982) *Skills Needed for Young People's Jobs, Vol. 1*, IMS Report No. 40, Institute of Manpower Studies.

WOLF, A. (1984) *Practical Mathematics at Work*, Manpower Services Commission.

14
The Role of Number in Work and Training

David Mathews

Introduction

The subject of mathematics is usually treated as a dimension of a person's achievement which transcends interests or occupations. English (in the UK) is treated in the same way. No such emphasis is given to music or craft or design subjects or history except on the local basis of particular schools or particular families. Through the national curriculum there is an attempt to evaluate science to this status, but it is mathematics and English which are treated as basic, fundamental, essential.

Mathematics is a powerful tool in the world of work. From research to administration it serves functions of modelling, calibrating, recording, communicating and many more. But it is its impact in the work of people who are not ostensibly mathematicians that has been the cause of the most popular heated debates over the last decade. From the point of view of people in work there are two reasons why the relationship between mathematics and work is important. First, there is the question of the immediate interest of the learners and to what extent their needs for preparation for starting work or continuing in work or advancing their careers are being met. Second, there is the question of the enrichment of learning of mathematics itself and the cognitive growth of the learners. The two interests may be characterized respectively as illuminating work through mathematics and illuminating mathematics through work.

In both work and training however, there is a tendency for some uses of number either to go unrecognized or to be confined to the formal processes of mathematics and its application. The usual treatment of number has followed an agenda determined almost entirely by people's own experience of the mathematics curriculum, particularly that of their secondary education. The apparent exhaustiveness of mathematics has contributed to this treatment, if not totally determined it. Excluding issues of attainment, two problems arise, firstly, some aspects of number usage which are not popularly associated with the subject of mathematics tend to go unobserved, unsupported in learning and unaccredited and secondly, mathematics itself is treated in isolation from its applications.

Identification of Mathematics in Work

There is a number of ways in which the limited understanding of the place of number in work may be revealed. These include:

- Studies in the field that can be described as 'education-industry liaison'.
- Employers' and others' prescriptions for training.
- Job, skills and task analyses carried out for the parochial interests of employment, including selection procedures.
- Examination of what trainers actually *do*.
- Examination of the importance given within employment to the concept of *numeracy* (see below).

One summary of the common features of all these examples is that training and other processes at work appear to have become rather too obeisant to education in elevating theory above practice, to the extent of corrupting the view of practice to fit the theory. Hence number-rich tasks are characterized in terms of the formal mathematics to which they approximate, rather than what they are in their own right.

Education and Industry Links

The Cockcroft Report (1982) gave considerable weight to the mathematical needs of employment. Through research studies which were specially commissioned and through the submissions of a variety of bodies, the Cockcroft Committee of enquiry identified a wide range of mathematics techniques which were both needed in employment and taught in schools. The jobs examined were exclusively those entered directly by school leavers at ages 16 or 17. The report also indicated that there were various activities in work which were capable of being described in mathematics terms but in which the strategy of carrying out the activity was not recognizably mathematical. A further category of skill was that in which the worker was so used to the activity that he or she no longer thought of it as mathematics. Other studies have been less insightful. For example, an influential booklet on mathematics and engineering (EITB, 1980) quite rightly identified the need for marking out and the use of formulae and calculations but it did not look at the more subtle judgments of length or time or position which are quantitative in character, though not immediately recognizable as the formal operations of standard arithmetic. Further studies served as communication channels for employers' views — sometimes moderated by teachers, sometimes aggravated by them. Very often a list of the perceived most important mathematics topics of 'basic skills' would be the central outcome. The tone of many such reports was not so much conservative as regressive, seeking a return to arithmetic often accompanied by demands for a return to correct spelling and grammar. The *London into Work* project (Townsend, 1982 and see Harris in Part 2 of this volume) and work at Sheffield University (Banks, 1981) using the Job Components Inventory were among a number of more dispassionate investigations of the needs of school leavers. But even these reports remain firmly rooted in the use of mathematics topics as a way of describing work activity involving basic calculations, dimensions, shape and so on. Occasionally in all these studies there would be some aspect of judgment or perception identified under calculations, mathematics or perhaps problem solving which would break the barrier of formal arithmetic operations, but these were few and far between and quite unsystematic. The problem with such investigations whether by interested or disinterested parties, is the nature of the questions asked and the assumptions of those people of whom they are asked: both questioners and questioned have frames of reference or perceptual sets which are heavily dominated by the schools' mathematics curriculum, both taught topics and assessment and which tend to colour both the questions and the responses.

Prescriptions for Training

The flurry of abuse and reportage across the education-employment boundary which characterized certain parts of the Great Education Debate has somewhat abated. What has followed has been a rather greater recognition from employers and the vocational education and training system that they cannot blame all their troubles on the secondary education system. Trainers may be less convinced of where the 'fault' lies, and may be less inclined to assert that there is fault to be attributed. And as employers come more and more to recognize the potential role of older workers, they come to see that continuing tirades against schools would appear absurd. They certainly need a more positive strategy to adapt their own practices to deal with whatever phenomena they detect.

The CBI in its recent examination of training (1989) has proposed the idea of nationally agreed common learning objectives (CLOs). Among the areas in which CLOs would be defined, it is argued, is 'Applications of Numeracy'. The interpretation of this is given as follows:

> Understand, interpret and use effectively numerical information whether in written or printed form; identify which form of numerical communication

is most effective (maps, flowcharts, models) for a given situation depending upon intended recipients; appreciate those situations where problems cannot be tackled using numerically based solutions.

This is a view of number as communication, a language for a quantitative, technological world. It starts from a body of knowledge, in this case numerical rather than mathematical. It then looks at where there is need or opportunity for the numeracy needs to be applied. The contrast with lists of computations and the types of fraction an engineering craftsman should be able to add is refreshing, but it is still a desperately partial view.

Jobs, Skills and Task Analysis

To obtain a clearer view of the role of number in competent performance it is necessary to start with the job or jobs in question rather than with a list of mathematics topics. There are many processes of analysis used in employment which do just this. Analysis of competence or work activity however, is not a matter simply of applying a metaphorical magnifying glass to work. There is no absolute underlying structure which can be discovered merely by looking more and more closely at what someone does or how they do it. Observation and formal analysis are both influenced by commitments or intended uses of the outcomes of the analysis and by assumptions or beliefs about the underlying nature of skill or competence expressed in the type of construct which the analysis identifies.

Some analyses take a similar view of number in work to that already described, their frameworks (for example Smith, 1978?) drawing almost exclusively on curriculum mathematics topics. Others (for example McCormick, 1969) adopt a perceptual and cognitive framework, looking at mental or information processes which come from a learning psychology rather than an education/training commitment. Such analyses reveal that much of what involves magnitude or specifically number in work is as much to do with estimation and classification as it is with arithmetic operations. In such frameworks arithmetic operations have their place but are embedded in wider classes of activity, such as combining and analyzing information, transcribing, compiling and co-ordinating. Aspects of magnitude and dimension can also be identified in terms of spatial and diagrammatical skills. As MacFarlane-Smith (1964) observed, perception of form and of the relationship between forms without the intervention of a system of symbols or language is as valid an aspect of mathematics as those parts which depend on the symbolic coding we call number. Yet further analyses illustrate a perception of the skills of work and of many tasks in life, which are quite different from a 'topic' description of mathematics. Fleishman and Quaintance (1984) identify mathematical reasoning and number facility as two separate abilities from a list including information ordering, visualization and inductive reasoning, defining them respectively as;

> *Mathematical Reasoning* The ability to reason abstractly using quantitative concepts and symbols. It encompasses reasoning through mathematical problems in order to determine appropriate operations which can be performed to solve them. It includes the understanding or structuring of mathematical problems. The actual manipulation of number is not included in the ability.
>
> *Number Facility* This is the ability to manipulate numbers in numerical operations; for example add, subtract, multiply, divide, integrate, differentiate, etc. The ability involves both the speed and accuracy of computation.

Regardless of whether Fleishman and Quaintance's or anyone else's analysis is valid, the examination of skilled behaviour which such work entails, clearly breaks the boundaries of the most commonly experienced mathematics topics at secondary level.

What Do Trainers Actually Do?

An important factor on the way in which work skills are interpreted is the influence of members of staff of organizations who are responsible for selection and training. Often they have very little power in the organization: their budgets are comparatively small and are extremely vulnerable to pressures on investment and cash flow. Nevertheless, because they interpret employers' perceptions and the results of any skills analyses they may have conducted or commissioned, they hold the key to number in competent performance. It would be possible to describe number in work as suffering from the Dead Hand of Training.

Training, or more broadly learning, ranges from the casual, *ad hoc* and in the work-place to the formal, off-the-job and systematic. There are now developments which aim to make better use of the on-going workplace experience of workers (see for example Levy, 1987) seeking to judge the effectiveness of planned learning experiences by their outcomes. For many trainers and workers however, the measure of quality in learning is still the degree to which it is provided off-the-job in a dedicated teaching environment and this input dominance is reflected in attitudes to skill and competence. Some training approaches even go to the trouble of taking people away from the workplace and then creating simulations of workplace activity as their principal mode of training. We might be grateful for this approach in the case of airline pilots but it can seem perverse in the case of a salesman or an engineer. The use of off-the-job training to enable a more general reflection on what happens in the workplace is however to be encouraged, particularly when a cross-curriculum approach is used for identifying various uses of number (see for example BTEC, 1986). Off-the-job studies of underpinning theory as a means of referencing technical information, also has an almost unarguable case but it does not follow that off-the-job is always best and there have been unfortunate effects particularly in the case of number and mathematics. Wherever number is found in an explicit form in an activity, a calculation or an interpretation of statistics for example, and where that activity is part of identified training needs, there is a tendency to abstract it and teach it formally off-the-job. This teaching may be concurrent with a worker's time in the job, or it may precede it in full-time education or training but because it has been abstracted it often does not reflect practice. Decorators for example do not judge how many rolls of wallpaper they need by measuring the room in the classroom manner unless the paper is hand-blocked and costs £150 a roll. Most people at work do not judge volume by multiplying length by breadth by height. Once the decision is taken to train a skill off-the-job, the strongest link with its application is broken and the skill is treated as something that can exist in free-floating form.

This is what happens to simple arithmetic or other mathematics that is noticed, but many aspects of numerate behaviour may be ignored both off-the-job and in the workplace because they are more subtle or because they are perceived as having no equivalent in formal curriculum mathematics (as in many of the judgments and estimations) or because the formal mathematics to which they relate is perceived as being too advanced for the particular class of learner. Again, the tendency to formalize and isolate is often reinforced by trainers' own personal histories. Many specialist trainers and line managers responsible for training revert not only to subject matter but to teaching styles which they themselves experienced as learners and this affects the training of young people in particular. All too often trainers have never been challenged to identify ways in which people learn: their concern is more often to produce a good looking classroom presentation. But it is through assessment that perhaps the greatest offence is perpetrated. In both formative and summative, assessment of the identifiable (as school sums) and the identified (with red ticks or crosses) results in the predominance of the easily assessed over the worth knowing. This leads to what can be described as an 'accreditation gap', the difference between the achievement with which someone is credited and that which they really have achieved.

Numeracy

Many discussions of number in work concentrate on young people and on apparent problems of inadequate achievement, identified in the issue of 'numeracy'. The term was coined as a parallel to the term 'literacy'. Like literacy it is socially defined and therefore subject to change over time. Precise definitions of numeracy are less important than the interpretation of the term. It tends to be seen in terms of low level, computational or interpretative skills and is based on one person's conforming to someone else's expectation of certain types of skill and knowledge. The expectation may include both work and non-work aspects. Numeracy's greatest limitation as a concept is that, at any time, it is popularly taken to be an absolute; a person is either numerate or not. As a lower limit of social acceptability, numeracy is subject to all the conservative or traditional influences that bear on the definition of thresholds. Like literacy, poverty or conformity to social norms, numeracy is based on yesterday's perceptions rather than today's. The real demands of work, by contrast, are not so neatly ordered. Focusing only on numeracy ignores the problem of many 'numerate' people who fail to deal adequately with the number aspects of their work (and non-work) activities. Many managers would be affronted to be thought anything but numerate, but might themselves not be sufficiently skilled in number for the statistical and mathematical models which they encounter in their own jobs.

Both numeracy and its partner-in-arms 'basic skills' are often defined by lists of largely computational 'skills'; mathematical operations defined out of context. Alternatively numeracy, though not basic skills, may be defined as a highly generalized quality of perceiving things numerical or using number as a language, as in the example from CBI above. It is not even enough to say that there should be more to numeracy than computational skill. The problem lies in taking mathematics as taught in schools as a reference, rather than the needs and experiences of people in and outside work. By taking the wrong reference point, we may be missing half the problem, or there may be less of a problem than is imagined. Moreover, by taking the needs of work, we can free ourselves from concentrating so much on 'low achievement'.

New Perceptions

Another construct which has suffered from restrictive definitions is *intelligence*. The joke about intelligence being what intelligence tests test is well known. As in the case of numeracy, there is not one operational definition but many and like numeracy again they may all be top-down definitions. However much they may be based on statistical investigation, the degree of measured intelligence has almost always been the degree of conformity to a hypothetical construct. In recent years however, there has been a loosening of the ties with statistical, predominantly norm-referenced models and a greater interest in finding out what it is that people are using by way of cognitive quality when they do things. No investigation is free of assumptions and investigators' perceptual frameworks but the naturalistic approaches being adopted do at least break some of the barriers imposed by an overweening scientific method.

The work of Sternberg and Wagner (1986) and others point a way to capturing *practical intelligence* and competence in daily activity. A similar position of starting with the effective performer of tasks applies to the study of 'folk mathematics', a series of phenomena largely ignored on employers' premises but certainly seeming to carry lessons for trainers. How many trainers have observed workers of all ages demonstrate huge facility in scoring at darts while leaving unquestioned their approach to the teaching of number?

A Framework for Dealing with Number Work

Persuading those with influence in the world of work to change their view of skills of number will be no easy matter. Neither is it easy to inform educators about how number is used in work. What can help is to use a framework for looking at work which helps to discriminate between its different aspects so that it need not be thought of as some amorphous whole with skills attached as separate entities. Traditional analyses and common perceptions of jobs have tended to focus on discrete tasks rather than the job as a whole and a good deal of the application of number lies outside narrow tasks. The elements of such a framework being used by those who wish to change the approach to learning and accreditation within work are offered here. In it the following are perceived as key functions in human resources, development, vocational education and training:

Selection
Identification of training need
Formative assessment and the provision of feedback
Design of instruction
Learning processes
Summative assessment and accreditation.

These functions tend to be dominated at present and so far as number is concerned, by views of mathematics similar to those expressed in employers' complaints or definitions of numeracy described above. This constricted view of number is one part of the existing, generally unimaginative view of competence and achievement. The Job Competence Model (Mansfield and Mathews, 1985) is now being used to counter this view. It is a simple typology, highlighting critical aspects of successful performance of a job which traditional qualifications have tended to ignore. Four components of job competence are distinguished:

Tasks
Task Management
Contingency Management
Dealing with the role or job environment

Task Skills are those skills used routinely in activities which are well-defined and have a conclusion and an outcome which is usually discernible.

Task Management Skills are the skills exercised in combining the different tasks of a job, scheduling, dealing with the variability in demands of different tasks, dealing with responsibilities which intrude on routine tasks.

Contingency Management Skills are used to deal with events which are liable to occur, but are unpredictable or novel. Contingencies include things going wrong and plans and expectations being changed by external factors.

Skills of Dealing with Role/Job Environment are to do with mediating between tasks to be carried out and the natural and artificial constraints placed on the role or job. They include the skills of working with other workers and with people from outside the work organization. They are equally to do with the physical environment of the occupation and with the demands or standards which are imposed — time, safety, cost and so on.

Core Skills

Within the four components of job competence one can look at the nature of work or performance in a variety of ways. The framework from which the Job Competence Model was derived however, was that of the MSC Core Skills. These were defined as part of the development of the Youth Training Scheme in the early 1980s. Covering number and its application, communication, problem solving and practical skills, the core skills are selected broadly, to support functions of design, learning and accreditation within the vocational education and training curriculum. Although the skills were identified in the context of YTS, they were not intended to be exclusive to it but were also aimed at supporting the development of a technologically competent workforce and workers and learners of all ages and levels. They are intended to illuminate the competence of managers and technologists quite as much as that of jobs entered by young people straight from school. The Core Skills are formally defined as *those skills which are common in a wide range of tasks and which are essential for competence in those tasks* (MSC, 1964). They were assumed to have a generic quality and each represents a class of sub-skills. The generic and class qualities are carefully selected and this is particularly important in the case of number. The number skills are not a selection of mathematical operations as a minimum or central or basic set in the way that is found, for example, in definitions of numeracy. Instead the number core skills represented proficiency in actions or judgments which entail application of number in the sense of magnitude, quantity, dimensionality and shape. So far as is possible, the skills are described in terms of the *outcome* of the actions or judgments.

Core skills of number and its application

1   Operating with numbers
1.1   Count items singly or in batches
1.2   Work out numerical information
1.3   Check and correct numerical information
1.4   Compare numerical information from different sources
1.5   Work out the cost of goods and services

2   Interpreting numerical and related information
2.1   Interpret numerical data or symbols in written or printed form
2.2   Interpret diagrams and pictorial representation
2.3   Interpret scales, dials and digital readouts
2.4   Identify items by interpreting number, colour, letter codes or symbols
2.5   Locate places by interpreting number, colour or letter systems.

3   Estimating
3.1   Estimate quantity of observed items or materials
3.2   Estimate quantities required for a process
3.3   Estimate portions or shares
3.4   Estimate dimensions of an observed object or structure
3.5   Estimate weight, volume or other properties
3.6   Estimate the time needed for an activity
3.7   Estimate the time an activity has been going on
3.8   Estimate rate of use of items or materials
3.9   Estimate the cost of goods and services
3.10 Estimate and compare shapes or angles
3.11 Estimate the size of gaps or holes and the fit of items
3.12 Estimate required sizes of gaps or holes and the fit of items
3.13 Estimate size or shape for the purpose of sorting
3.14 Estimate settings for tools, equipment, machinery

4   Measuring and marking out
4.1   Measure the dimensions of an object or structure
4.2   Measure out required dimensions and shape
4.3   Measure weight, volume or other properties
4.4   Measure out required weight or volume
4.5   Measure the time a process or activity takes

5.   Recognizing cost or value
5.1   Compare the cost of different goods and services

5.2   Compare the relative cost and benefits of buying or using goods and services

5.3   Recognize the value of items in order to take appropriate care of them.

The absence of arithmetic operations on the list followed from the recognition that they formed only a part of what is involved in dealing with number. Surrounding even the simplest arithmetic operations were two processes; *mathematizing* or formulating the action or judgment in mathematical or number terms and *interpreting* or attributing meaning to the result of the operation, particularly through appropriate action.

Other aspects of dealing with number would not commonly be described as operations at all. Many kinds of judgment, under the generic title of estimating represent for many people the most common encounters with magnitude and dimension. Ironically number or mathematics interest within the 103 core skills is not confined to the number area. In certain contexts core skills within communication and problem solving in particular entail mathematical thinking. For example:

Provide information in writing and by means of tables and diagrams
Plan the order of activities
Plan the arrangement of items
Plan how to present information
Monitor the availability of stocks or materials
Adjust heating, lighting, ventilation

Core skills are intended to be used as an instrument of what is called *work based learning*. This is learning characterized by its being directed to the work role, drawing on learning in the workplace and elsewhere (Levy, *op. cit.*). The Core Skills indicate potential for transfer of skill from one context to another. In this they differ profoundly from say, numeracy syllabuses or other mathematics syllabuses in which skills are meant to be acquired in some general sense in order to be applied universally. The core skills give some structure for reflecting on activity and indicate what is common between one activity and another, thus enhancing the prospect of transfer or the development of concepts. They are primarily an instrument for analysis, by the learner or worker as much as anyone, rather than a specification of a syllabus or learning context. Their relationship with context is particularly important in this respect. A worker cannot have or achieve a core skill in some absolute, free-ranging sense. He or she achieves a version of it, a member of that particular class of skill within a context. There are more likely to be connections between members of a single class, that is different instances of the core skill, than between classes but this does not mean that there are general skills assumed to be applicable in any context.

New Standards and National Vocational Qualifications

Under the aegis of the National Council for Vocational Qualifications and with the financial support and policy imperative of the Government, the whole range of vocational qualifications in England, Wales and Northern Ireland is currently being restructured. The key aspect of this process is the development of new forms of occupational competence standards on which the qualifications are based. The form they take is defined precisely: they are distinct from other frameworks used to describe or assess individuals and they are based on work performance not learning or training performance. The standards thus reflect role expectations, the expected outcomes of workers' activity within the job, rather than some quality or prior qualification which entrants are expected to bring to the job.

Occupational standards are specifications. They consist of criteria against which an individual's performance in the tasks and activities of work may be compared and judged as competent or not competent and the criteria are set by the operational requirements of employing organizations. Standards do not describe people; they are not personal attributes; they describe the performance which is expected of people, what their performance ideally should be. It is National Vocational Qualifications which are concerned with people; they provide the statements that an individual has achieved certain standards.

The ability, potential or personal effectiveness of a job holder or candidate remains important, but is distinguished from performance in the job which the standards specify. Whatever it is that makes a good manager or secretary may be a proper subject for selection or training but it is standards that define what a good manager or secretary is expected to do; the activities through which someone learns to do a job may be important but they do not determine the standards. The knowledge and skills which may form the content of training may be essential for learning to do the job but they are not themselves the job and provide no direct benchmark for performance in the job.

What has happened is that the construct basis of standards and qualifications has been changed. That basis is now much more oriented towards the operational requirements of organizations and the economy. While prejudices and assumptions about number skills are not going to change overnight, job competence standards may create an opportunity to investigate and recognize skills of number as they are used, rather than as they are defined by a formal discipline. However a warning is necessary. In the development of standards for managers, the greatest difficulty encountered has been in persuading them to accept the notion that they should take responsibility across the full range of what is usually felt to be management functions. In addition to development of staff for which they are responsible the thing that managers try most often to dodge is finance. And contributing to statistical analyses. And mathematical modelling.

References

BANKS, M. H. (1981) *Young People Starting Work*, Social and Applied Psychology Unit, Sheffield University.

BTEC (1986) *Common Skills and Core Themes*, London, Business and Technician Education Council.

CBI (1989) *Towards a Skills Revolution — a Youth Charter*, London, Confederation of British Industry.

COCKCROFT, W. H. (ed.) (1982) *Mathematics Counts*, London, HMSO.

EITB (1980) *Mathematics and Engineering — an Illustrated Guide to Basic Skills*, Watford, Engineering Industry Training Board.

FLEISHMAN, E. A. and QUAINTANCE, M. K. (1984) *Taxonomies of Human Performance*, New York, Academic Press.

KNOX, C. (1977) *Numeracy and School Leavers — A Survey of Employers' Needs*, Sheffield City Polytechnic.

LEVY, M. (1987) *The Core Skills Project and Work Based Learning*, Sheffield, Manpower Services Commission.

MCCORMICK, E. J. (1968) *Position Analysis Questionnaire*, Lafayette, Purdue University.

MACFARLANE-SMITH, I. L. (1964) *Spatial Ability*, University of London Press.

MANSFIELD, B. and MATHEWS, D. (1985) *Job Competence: A Description for use in Vocational Education and Training*, Blagdon, Further Education Staff College.

MSC (1984) *Core Skills in YTS: Youth Training Scheme Manual*, Sheffield Manpower Services Commission.

SMITH, A. D. W. (1978? undated but distributed in 1978) *Generic Skills — Keys to Performance*, used by Townsend op. cit.

STERNBERG, R. J. and WAGNER, R.K. (Eds) (1986) *Practical Intelligence*, Cambridge, University Press.

TOWNSEND, C. (1982) *Skills Needed for Young People's Jobs*, Sussex, Institute of Manpower Studies.

15
Skills Versus Understanding

Rudolf Strässer, George Barr, Jeffrey Evans and Alison Wolf

This article first appeared in *Zentralblatt für didaktik der Mathematik*, **21**, 6, 1989, pp. 197–202.

1. Introductory Examples: Description of the Problem

In a textbook for future electricity craft technicians, one may find the following task:

> In a torch, you have batteries of 1.5V. With a resistance of 15Ω for the torch (bulb etc.) a current of 0.3A flows through the bulb. How many batteries do you need?

With Ohm's Law, the solution is easy:

potential difference = resistance × current
= $15\Omega \cdot 0.3A$
= $4.5V$
$4.5V : 1.5V = 3$

You need three batteries.

An analog task for commercial trainees may be:

> Calculate the annual interest to pay to a bank that gave a loan of £15,000 at an interest rate of 3% pa.

Here a different 'rule' would need to be used.
Interest = principal · rate
= $£15,000 \cdot 0.03$
= $£450$

A look at the tasks and the solution shows a specific flow of activities: with a rule or formula identified, numbers are inserted (the question of dimension of numbers may be suppressed) and a new number (the 'solution') is calculated. This is rather a schematic procedure, apt to habitualization and very helpful in producing an answer to the question raised — provided the rule, the formula is known. One can forget about mathematics and physical laws or business practices and take this algorithm, this procedure as a *tool*. Apart from their orders of magnitude, the numbers given in the tasks look the same. Differences between the tasks seem to be restricted to the words from outside mathematics — the context.

The task can also be treated in a different way: both situations are modelled by means of proportionality (with potential difference and interest being multi-linearly dependent on resistance/current and capital/interest rate). The dimensions in the formula show the relations of numbers and variables implied in the situation. *Concepts* (as multi-proportionality, current, resistance and interest rate) are tied together in the formulae. Different mathematical/conceptual *means* could be used to solve the problems. Apart from the algebraic solution given in the example, one could use a table as a numerical solution or a graph (as a geometrical solution) to represent the proportionality. The variety of representations may be linked by computer technology, offering additional ways to cognitive mastery of the situation.

So we can distinguish two ways to cope with the tasks: the modelling of the problem, eventually using a variety of conceptual and representational means (the 'understanding' solution) as opposed to using algorithms to produce numbers relevant to the situation (the 'skill' solution). If we state this in more psychological terminology, we could use a distinction offered by Skemp (1976), who provided us with the terms 'instrumental understanding' (rules without reasons) as opposed to 'relational understanding' (meaningful undertsanding).

In adult education, courses oriented towards vocational training and employment often seem to take instrumental understanding alone as their goal. They start from the question 'what mathematics does e.g., the typical operator, need to be competent in his or her job?' The course planners analyze workplace activities to determine the basic functions of a job. In England for example, it is current practice for training to be built around bundles of 'competencies' derived from such analyses and expressed in terms of criteria and tests which may involve mathematical skills and content. However, the unstated implication is often that the only mastery needed is of the particular skills required for the job. Understanding of the rules behind the skills may be considered an extra, a bonus.

In contrast to this approach, the Cockcroft Report (1981) for example brought forward the idea that mathematics is important to all people because it 'provides a means of communication which is powerful, concise and unambiguous' (p. 1). Following this argument, all members of society need to understand mathematics. Adults in particular sometimes fight for the possibility of acquiring this relational understanding. As a consequence of this struggle for understanding, adult education may have to offer the possibility of teaching this level of understanding and not restrict itself to mere procedures and algorithms.

But how are we to distinguish training, in the sense of acquiring particular skills, from the teaching of relational understanding? One way is in terms of people's ability to generalize the use of concepts and skills to different, maybe new concepts. One can look at how successfully they apply concepts and skills to unfamiliar problems. This is the type of question currently discussed by policy-makers under the heading of 'transferable skills' because they assume that narrowly defined skills are likely to become obsolete in the near future because of continuing technological progress. In this particular discussion, general capabilities and flexibility, together with an understanding of the workplace processes which reach beyond one's immediate occupational responsibilities are seen as necessary to the future work-force. The problem then remains of defining what are the boundaries of 'familiar' contexts and how to separate situations from one another. Instead of detailing these terminological questions, we present some findings related to these questions.

2. Questions and Findings on Skills' Development and Understanding

2.1 *Levels of understanding*

For this part of the paper, we concentrate on mathematics as the topic to be taught and learned and more specifically on the computational aspects, while completely ignoring modelling aspects. With this restriction, the hierarchical structure of mathematics immediately comes to mind which in turn implies that there may be different levels for a given skill corresponding to individuals' progress and development. This conceptual framework informed research at Brunel University designed to provide a coherent structure for the mathematical content of the curriculum. (For a brief description see Rees and Strässer, 1988, p. 7). To rationalize the elementary mathematics curriculum of the City and Guilds of London Institute, a framework of mathematics levels of increasing complexity was developed around which contexts could be designed which would be appropriate to specific courses. The research studies revealed that both mathematical content and the context may create difficulties: in the course design the inter-relationship between these two is therefore important and may be illustrated by the following examples.

If we come back to the torch task cited above, we can illustrate the levels of mathematics involved.

1. Calculation of R
   V:R (potential difference:current = resistance, constant at a given temperature).
   V = 12V, I = 6A, R = ? *Level I Maths.*

2. Calculation of R – more difficult.
   V = 12V, I = 0.6A, R = ? *Level II Maths.*
   because of division by a number between 0 and 1
3. Calculation of I
   V = 12V, I = ?, R = 7Ω *Level III Maths.*
   because it is calculation with natural numbers, but I is 'on the bottom' i.e. the inverse.
4. Problems involving direct proportionality of V and I belong to levels II and III.
5. A mathematical mesh analysis of DC circuits brings in more complicated situations, see diagram below.

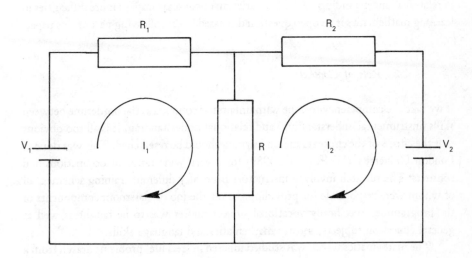

This situation may be modelled by
$$V = I_1 \cdot (R_1 + R) + I_2 R$$
$$V_2 = I_1 R + I_2 \cdot (R_2 + R)$$
which is *level III* or *level IV* maths depending on the numbers to be inserted.
If this is written as a matrix, like
$$\begin{matrix} V_1 \\ V_2 \end{matrix} \begin{pmatrix} R_1 + R & R \\ R & R_2 + R \end{pmatrix} \cdot \begin{pmatrix} I_1 \\ I_2 \end{pmatrix}$$
we come to higher levels. Complex numbers and matrixes would be used for AC circuits.

The above view is only one way of looking at levels, namely defining levels by use of numerical and mathematical characteristics and not analyzing the problem of the transposition of the formula implicit in the passage from level I to level III. The view presented is near to that taken by many training programmes in that it breaks the task to pieces and relies on a mathematical structure of the numbers and operations involved. As a sort of complement, it would be challenging to draw on a psychological hierarchy of levels in order to structure the development of skills and/or understanding.

The loan task outlined above provides a simple illustration of how the context of a mathematical task can vary. The mathematical concepts and skills in the loan task and the electricity task may look the same, but the apparent similarity needs analysis, see below. Understanding the relationship between the symbols may be the major difficulty in these tasks; an additional difficulty is the switch from one context to the other. Making this switch successfully involves being aware that the same mathematical operation or structure is involved in both tasks. This switch from one context to a different one may be called 'transfer' and constitutes a major achievement in relational understanding. Whether variation of contexts implies more difficulties in learning mathematics is an open question discussed in the following part of this paper.

2.2 Role of Contexts

If we take the competency to cope with unfamiliar contexts as the borderline between skills (instrumental understanding) and relational understanding, it is all too obvious that the effects of the contexts met in learning should be researched. This was done at London University (Wolf, *et. al.*, 1989) for a very wide range of occupations and contexts. The research involved instructors from fifty different training schemes, all of whom were responsible for providing the off-the-job or classroom components of the programme. Specifically vocational subject matter was to be taught as well as general education subjects, mostly mathematics and language skills.

The mathematics which was studied inhered in 'real life' problems drawn from a wide range of occupations. However, the problems were rarely addressed directly in vocational or adult education courses or even recognized as having common and recurring features. Similarly the mathematical operations involved were neglected for the most part in school and adult mathematics courses. The three main categories studied were:

1. budgeting in situations of uncertainty, where various general decision rules must be selected and used as more or less appropriate,
2. allocation of time slots. This set of operations occurs in a wide range of contexts for example hotel bookings, timetabling of appointments and allocation of airline seats. The tasks used for teaching and data-gathering during the research programme also included the associated area of designing and drawing up tables or plans to display information.

3. Stock control and specifically the identification of trends in demand or sales from numerical data and corresponding decisions about stock holding and ordering. (Although particular ways of filling in stock sheets may be taught as part of highly specific vocational training it is rare in the UK for people to be given general instruction and practice in the underlying principles of stock control. The identification of trends from data is demanded increasingly often by adults.)

The training schemes involved in the study were split into three groups. In one group, instructors gave students a number of teaching exercises which were all set within the same context, one which was familiar to the learners from their own working lives. In the second, instructors used teaching exercises where the problems were structurally the same, both in relation to each other and in relation to those used by the first group, but set in a wide variety of different and largely unfamiliar contexts. The third served as a control group and had no special teaching exercises or sessions. The three groups were uniform in the pedagogical decision to use whole tasks for teaching and learning as opposed to an approach which concentrates on small sections and sub-routines, supposedly hierarchically arranged. In classroom terms this meant giving more time to quite large and complex assignments involving skills other than mathematical ones, and less to the traditional small and self contained problems of mathematics teaching.

The results of the research consistently supported the use of varied contexts to encourage the development of generalizable skills, taken as a measure of relational understanding. The groups which received varied exercises consistently improved at a faster rate than did either of the other two groups, even though intervention time was very limited and their instructors received no special training. The effect was especially apparent in their performance on problems set in an unfamiliar context.

The pattern was the same across all three of the different problem-solving areas which were studied. However, interactions were apparent between the effects of this varied approach and the difficulty of the items. On the items which were extremely hard, the rate of progress was far lower than on those of moderate difficulty (items were defined as 'very difficult' when, on the first occasion of testing, a third or less of the sample obtained the correct answer). Equally, the trainees who made the least progress were those who had no prior experience in 'real life' of the problems they were studying. This is of course quite consistent with the way in which varied contexts affect learning.

This research further encourages the use of varied presentations and contexts from actual classroom evidence. Variation of contexts (as well as the whole task approach) tends to encourage the development of general understanding in a way which concentrating on repeated routine applications of algorithms does not and cannot. Stating this, it is important to remember that teaching in this way means that students often will not and indeed should not expect to get everything 'right'. If problems are too difficult (i.e., too unfamiliar) there may be too little on which to hang learning and therefore rather little progress. In general, the challenge of integrating into one's mental schemata something which is difficult actually encourages learning and understanding. Supporting this interpretation is the fact that trainees who did not see the immediate relevance of the tasks they completed were hardly less likely than their peers to find them interesting or motivating.

2.3 Context and performance revisited

In the discussion so far, it is clear that the term 'context' has been used in different senses. In part 1, the context is specified by the wording. In 2.2 on the other hand, the researchers 'simulated workplace practice, vocabulary, paperwork etc. as faithfully as possible', though they could not observe the trainers carrying out tasks at the workplace (Wolf, *et. al.*, 1988 p. 1). It is clear that school, off-the-job training and working on the job provide different sorts of contexts for learning and for doing mathematics — as do non-work everyday contexts such as supermarket shopping (Lave, 1988).

Researchers working within the framework of activity theory or practice theory (see Lave, 1988 and Walkerdine, 1988 for somewhat different views) argue that practices (such as learning maths at school, training for work at college and working as a clerk) are pervasive in shaping or 'constructing' the learner's (or practitioner's) view of the task and the context. Thus, in order to understand and describe the context, we need to understand the practice(s) which is/are 'called up' for the person concerned.

Practices can be characterized as having the following aspects:

- a set of ideas through which the world is seen and meaning is given,
- a language or 'code',
- goals, values, 'affect',
- material and institutional resources available (e.g., textbooks, training manuals, equipment, professional association),
- enduring social relations and 'positions' of power among the persons involved (e.g., teacher-student or supervisor-clerk).

Much work has been done using observation of people engaged in particular work or everyday practice (for a summary of this work see Lave, *op. cit.*) or involved in teaching and learning mathematics at school (for example Walkerdine, 1988). Somewhat different methods have been used with the aim of specifying and describing, in as detailed a way as possible, a *range* of different contexts of use of mathematics. This approach used semi-structured interviews at a UK polytechnic with social science students, many of whom were mature students (that is they had done paid work and/or brought up a family before enroling). The main aims of the interviews were to assess which practice was called up for the students by a particular task and to appreciate the students' performance within this practice (for details of the tasks and the methodology see Evans, 1988).

The analysis of the interview transcripts is continuing and critically assessing the following ideas;

1. We can make a reasonable judgment about what activity or practice a respondent has called up by using specific indicators such as the particular terminology used (for example using 'trend' rather than 'gradient' might indicate that the student is drawing on his or her practice as clerk rather than on college mathematics).
2. The practice which has been called up will condition not only the 'correctness' of the performance but also the language and reasoning used with the problem.
3. More confused, less 'correct' performance may be observed when school/college maths is called up, not only because of memory failure, 'misconceptions' and so on but also because of differences, unfamiliarity and the emotional charge that is part of a practice. Put another way, familiarity is affective as well as cognitive.
4. Non-school practices and the numerate aspects do not necessarily always bring up positive feelings. Therefore, the learning of school maths is not necessarily helped by drawing on practical mathematics examples.
5. Nor is the 'transfer of skills' or concepts from school maths to practical maths — or from non-school practices to school or college maths — at all straightforward because of differences in language, resources, social relations and emotional associations between different practices. (Lave, 1988 and Walkerdine, 1988).

2.4 Teaching understanding

A discussion of which approach to take when teaching, whether aimed at instrumental or at relational understanding, should not forget that the choice is determined by constraints which cannot be overcome by a simple change in the teaching practice. (See Rees and Strässer, 1988, p. 10f. For research on teachers' views on this alternative see Bromme and Strässer, 1988). One should also note that mathematics in the following part of the paper can be understood in a broader sense than that in which it was used in parts 2.1 to 2.3

Some of the constraints stem from the role mathematics plays in everyday and workplace situations: at least for the vast majority of situations mathematics is so integrated into the context that it does not appear as a separate part of the activity, as a separate domain of knowledge. This is especially true for the large number of vocations which are technically not very developed (as for instance hair-dressing) and/or are organized along Tayloristic principles (broken down into small components with the intention of making the work more efficient and more easy to control, for example the work on an assembly line). In vocations like these, mathematics — if used at all — is turned into algorithms, habitualized in customs and professional practices, performed with the help of blank forms, tables, formulae, spreadsheets and/or computers with (sometimes) appropriate software. In this type of situation, it is difficult for the practitioner (and sometimes for the researchers) to identify the use of mathematics. Moreover, in situations like these, teaching or learning mathematics can only be legitimized by stressing its potential for new situations and its cultural and educational value (but rarely with its explanatory value for the vocational situation at hand).

The story is different in small, especially self-employed enterprises, in working situations using scientific, advanced terminology and/or in an organization of work oriented to 'job enrichment', 'job enlargement' and reduction of job hierarchies (in larger companies). Sometimes, these job situations require greater technological and/or scientific, understanding and self-control and may bring in higher mathematical requirements. In situations like these, mathematical knowledge — though addressed as a specific part of the knowledge necessary to cope with the situation — is often only used as a helpful tool. Taken as such, it will be cut off from the disciplinary links to mathematical concepts (with their descriptive power). It will also be separated from mathematical procedures (with the operative possibilities linked with the procedures and algorithms; for a deeper analysis the 'complementarity' of the descriptive and operative power of mathematics, see for example Arbeitsgruppe Mathematiklehrerausbildung, 1982, p. 202f). Consequently, vocational teachers, even when training for working situations like these, tend to treat mathematics as a tool which will unquestionably help people to cope with future work situations. Even in 'advanced' job situations, it will be difficult to separate mathematics from the context.

When teaching mathematics to adults, teachers have to face additional constraints due to the teaching and the classroom-situation. For mathematics teaching in general, Chevallard (1985) has analyzed the 'didactical transposition' which turns scholarly knowledge into knowledge taught in classrooms. Preparing knowledge for teaching purposes presupposes that the knowledge is public and normally leads to a certain compartmentalization ('désyncretisation') and depersonalization in order to make knowledge teachable in schools. If these constraints are met, knowledge can be transformed into knowledge-to-be-taught. It is then given the form of a text with a beginning and linear order (Chevallard, *op. cit.*, p. 57–63).

If we compare this to the way vocational and non-vocational mathematics is used, we have to take into account that use of mathematics implies the modelling of a problem, of an extra-mathematical situation in terms of mathematical concepts, the solution of the mathematical problem and a re-interpretation of the mathematical solution in terms of the situation one has modelled. It is obvious that this circular structure of the application process will impede its easy teachability. Compartmentalization may be one reason for the difficulties faced when teaching the use of mathematics, because model building and re-interpretation demand a change of compartment. A simplistic way out is to relinquish one of these two domains, for example to omit mathematics as a separate subject in adult education. Sections 2.2 and 2.3 above contribute to a more sophisticated handling of this dilemma.

As was mentioned above, for developing knowledge to be taught, a piece of knowledge has to be public. (Chevellard, 1985, Chapter 5). This necessity of being public in fact works in favour of teaching mathematics in adult education. Mathematics (as part of science) tends to be public knowledge, a fact that promotes the teaching of mathematics as opposed to the training of a vocational/professional situation. In contrast to mathematics, vocational knowledge may be protected from being public as a consequence of traditional practice (see for example the medieval guilds) or for economical constraints (for example patents or licences). For teaching this knowledge, these factors may lead to a certain neglect of non-scientific, specifically vocational knowledge and may work in favour of mathematical knowledge, which has the added advantage of science, namely being socially highly valued.

3. Conclusion

If the dichotomy of this paper's title is to be commented on in one sentence, one could be tempted to say that employers only want skills while adult learners tend to seek understanding. However, this perspective obviously is not adequate for two reasons. First, even employers know that sometimes understanding leads to better performance on the job and job satisfaction. Policy makers and educationalists use this fact to legitimize training and education which aims at overcoming the restrictions of mere skill development and instrumental understanding. Second, adult learners as well as children are not always interested in the more complex, more difficult relational understanding required to transfer knowledge from one context to others.

There are also objective difficulties for learning and teaching of understanding in the broader sense, namely restrictions due to the organization of the human mind which tends to store knowledge only as relevant to the context it is learned in, (the constructivist view or the importance of subjective domains of experience — see Bauersfeld's (1985) 'subjektive Erfahrungsbereiche'). Additional restrictions come from the way teaching is organized, especially from the widespread compartmentalization and linearity of the teaching process. Every learning and teaching programme has to decide which type of knowledge or understanding it aims at, what potentials it has and what restrictions it has to cope with. There is no easy solution to these contradictory constraints.

References

ARBEITSGRUPPE MATHEMATIKLEHRERAUSBILDUNG (1981) *Perspektiven für die Ausbilding des Mathematiklehrers*, Köln, Aulnis.

BARR, G. and REES, R. (1985) *Developing Numeracy Skills*, London, Longman.

BAUERSFELD, H. (1985) 'Ergebnisse und Probleme von Mikroanalysen mathematicshen Unterrichts', in Dörfler, W. and Fischer, R. (Hrsg) *Emprische Untersuchungen zum Lehren und Lernen von Mathematik*, Wien — Stuttgart, Teubner.

BROMME, R. and STRÄSSER, R. (1988) *Mathematik im Beruf: Die Beziehungen verscheidener Typen des Wissens in Demken von Berfusschullehrern*, Occ. Paper No. 101 of IDM April, Bielefeld.

CHEVALLARD, Y. (1985) *La Transposition Didactique*, Grenoble, Pensées Sauvages.

COCKCROFT REPORT 1982 Committee of Inquiry into the Teaching of Mathematics in Schools, *Mathematics Counts — Report of the Committee*, London, Her Majesty's Stationery Office.

EVANS, J. (1988) 'Contexts and Peformances in Numerical Activity among Adults', in Borbas, A. (Ed) *Psychology of Mathematics Education Conference XII*, Hungary 20–25th July, 1988, Proceedings, Veszprem, OOK Printing House pp. 296–303.

FRENCH, P. and BARR, G. (1983) '364 Adds up to Confidence', *Times Educational Supplement*, March, 1983.

LAVE, J. (1988) *Cognition in Practice*, Cambridge, Cambridge University Press.

REES, R. and STRÄSSER, R. (1988) *Teaching to Use Mathematics for Work and Cultural Development*, Survey Lecture to Action Group 7 at the 6th International Conference on Mathematics Education (Icme-6) Bielefeld: Occasional Paper No 13 of IDM, December 1988.

SKEMP, R. R. (1976) 'Relational and Instrumental Understanding', *Mathematics Teaching*, 77 pp. 20–26.

STRÄSSER, R. (1987) 'Mathematics Training for Work' in Morris, R. (Ed) *Studies in Mathematics Education*, vol. 6. Paris, UNESCO, pp. 99–109.

WALKERDINE, V. (1988) *The Mastery of Reason: Cognitive Development and the Production of Rationality*, London, Routledge and Kegan Paul.

WOLF, A. (1989) *Learning in Context: Patterns of Skill Transfer*, Training Agency, Department of Employment, Research and Development Monograph.

WOLF, A., KELSON, M., FOTHERINGHAME, J., GREY, A. and SILVER, R. (1988) *Problem-solving: Exercises to Promote Skill Transfer*, COIC, Sheffield.

16
Mathematics Learned In and Out of School: A Selective Review of Studies from Brazil.

David Carraher

On a shady street bordering a city square in Recife, Brazil, residents from the neighbourhood are doing their weekly grocery shopping among rows of fruit and vegetable stands. The following exchange occurs between a customer and a 10 year old vendor:

> **Customer**: I'll take two coconuts (at Cr$ 40.00 each. Pays with a Cr$ 500.00 bill). What do I get back?
> **Child**: (Before reaching for customer's change) 80, 90, 100, 420.

In a subsequent interview given at his own home the child, who has had three years of schooling, is asked to determine the sum of four hundred and twenty plus eighty. He picks up a pencil and writes the number 42 with 8 underneath and claims that the answer is 130. (cf. Carraher, Carraher, and Schliemann, 1985. Dialogue originally in Portuguese).

The Issues

The existence of mathematics out of school

There can be no doubt about whether people learn mathematics out of school. Firstly, it has been clearly established that children's understanding of number begins before they have been to school (Piaget, 1952; Gelman and Gallistel, 1978; Hughes, 1986). Secondly, unschooled adolescents and adults routinely perform calculations at work, in naturally occurring situations (Scribner, 1984; Carraher, Carraher and Schliemann, 1983, 1985; Carraher, Schliemann and Carraher 1988 and in press 1990; Lave, Murtaugh and de la Rocha, 1984). Lastly, it can be shown that people who have been

169

to school occasionally represent, solve or think about mathematical problems in daily life so differently from how they are taught in school that is is difficult to deny that something important has been learned or discovered outside of school. For example, sugar farmers in Brazil's north-east (Abreu and Carraher, 1989) employ (1) a system of measures different from the metric system introduced in schools and (2) procedures for determining land area which actually go against what they are taught in school. We believe it can be reasonably argued that schooled people develop a substantial part of their mathematical knowledge out of school. The present analysis, together with previous analyses by others (Resnick, 1987; Saxe, 1988a; Schliemann and Acioly, 1989), bears upon this issue.

Key issues

The main question, however, is not whether people can or do learn mathematics out of school, but rather what is the nature of mathematical knowledge learned out of school and how is it similar to and different from mathematics learned in school. What are the principal characteristics of this knowledge? How is it organized? How is it acquired? To what extent is it taught by others? To what extent is it discovered or constructed by the individual? How is it deployed in actual settings or situations? What sort of problems is it particularly suited for? How extensive, flexible, unified, general, explicit, precise, abstract, and powerful is it? How is it related to cognitive development in general? How is it looked upon by the people who use it? What sense do they make of it? What symbolic representations, internal or external, are associated with the knowledge and what role do these representations play in understanding or solving mathematical problems? What modifications are introduced in representations and procedures as people confront new situations?

Some cautions

These are indeed weighty questions which cannot be solved alone nor quickly. Nor perhaps with the logical precision with which certain well defined areas of knowledge, such as mathematics itself, can be approached.

(a) False dichotomies as necessary evils

We may have to work with certain dichotomies, the classes of which are, at best, caricatures of reality. The 'school knowledge' versus 'out of school knowledge' distinction is one such example. First, we should recognize that out-of-school knowledge is not a clear type, that is, a sufficiently unified and coherent class to merit the referent 'it', as we have done above. Similarly, while the studies we will be referring to suggest that there are indeed differences between maths knowledge in and out of school, there remains the question of conceptual or empirical overlap. People tend to employ many of the same representational tools in and out of school, such as linguistic expressions, the actual words for numbers and numerical operations such as 'taking away', 'increasing', 'times', and so forth. For the great majority of mankind, it is likely that practices, representations and procedures used in school will have some bearing on how mathematical problems are handled outside of school. It also seems reasonable to suppose that one's experience in working with mathematics outside of school will play an important role in how one understands mathematics taught in school.

The content of mathematical knowledge engendered in and out of schools is not fixed but will vary depending upon the socio-historical moment and educational concerns of the epoch and of the particular institutions involved. What was originally 'extra-curricular' may become curricular and vice-versa. After the introduction of the Hindu-Arabic place value system in thirteenth century Europe, for example, there was a resistance for more than two hundred years against incorporating the 'new maths' into the official educational curricula (Menninger, 1969). Schools continued to use Roman numeral notation. Pupils were apparently taught to perform calculations, not by written algorithms but rather through the use of other representations such as counting tables, a counterpart to the abacus on which stones or counters were located in columns to indicate numerical values. During this period, on the other hand, an apprentice accountant in a Venetian trading company would be likely to learn on the job written notation and procedures more similar to those associated with present day primary maths curricula than to those of medieval curricula. Thus some mathematical knowledge previously developed principally out of school (on the job or through special extra-curricular tutoring from Rechnenmeisters) later came to be a standard part of the mathematics curriculum.

(b) Mathematical knowledge is both personal and collective

Any attempt to describe or account for the mathematical knowledge of people sooner or later encounters the following problem: mathematics refers to a corpus of knowledge and techniques collectively invented, developed and passed on within society over centuries, as well as to what individual people actually know. Mathematical knowledge is both a historical creation of social groups and personal construction (Carraher and Carraher, 1988). This dual nature of mathematical knowledge will pose challenges which cannot be reasonably handled by a psychological analysis since psychology, particularly individual psychology, can not fully account for the nature of mathematical knowledge which individuals develop. It further raises important epistemological issues. How is the mathematical knowledge of individuals shaped or influenced by culture on the one hand (in particular, by the way in which mathematics is represented, treated, used or taught in society) and by psychological development on the other? What are the relationships between collective and personal representations of mathematics? How is collective mathematical knowledge appropriated by individuals in social activities? Although we know that, historically, mathematics developed in human activities such as farming, commerce, astronomy and navigation,[1] we still know precious little about how engagement in present day activities is related to the development of a person's mathematical knowledge.

With these caveats in mind, let us contend with describing mathematical knowledge that develops informally outside of school. How does this compare to mathematical knowledge developed in and engendered by schools?

Informally Learned Mathematics: A Review of Three Studies

We would like to focus our attention here upon informally learned mathematics, that is mathematics learned due to little or no formal explanation and instruction, in the course of social activities such as work. Informally learned mathematics can be thought of as related to Scribner's (1984) 'practical thinking . . . thinking that can be explicitly defined as embedded in larger purposive activities and that functions to carry out the goals of those activities' (p. 2). This will be contrasted to mathematical knowledge which arises in formal contexts which Saxe has described as those contexts specifically designed to further learning (in which) a more knowledgeable individual arranges materials and communications in a way specifically designed to facilitate the learning of particular concepts that are potentially applicable to a wide variety of activity . . . (Saxe, 1988, p. 2)

School failure

Several years ago colleagues (Carraher and Schliemann, 1988) at the Federal University of Pernambuco carried out an investigation into the mathematical knowledge of children from lower and middle-class social economic strata in Recife, a capital in Brazil's impoverished north-east. There were good reasons for wishing to have a careful look at children's mathematical abilities and understanding. For one thing, the results could offer a profile of the children's strengths and weaknesses and perhaps lead to a more skilful diagnosis and handling of learning difficulties. Class related school failure was commonplace throughout the country. At the time of the study, the national average for failure and dropping out was 38 per cent and only 29 per cent of the students who entered first grade would enter fourth grade three years later (Cunha, 1981). Many of these pupils, most from lower socio-economic classes, would fail in mathematics. Quite a few people, including academics, considered such failure rates to be understandable and normal, considering the conditions of poverty in which millions of Brazilians had been forced to live. Malnourishment was commonly taken to be a main cause of school failure of the poor.

By the end of the year, 32 per cent of the lower-class pupils who took part in the study failed mathematics, as compared to 2 per cent of the middle-class youths. These failure rates were thus rather typical. However, they were very disquieting, when juxtaposed with the cognitive assessments, the principal finding of which being that there were no substantial differences between the middle- and lower-class children in mathematical skills or understanding after one year of schooling in maths. Such results raised disturbing questions about the neutrality of schools in assessing the cognitive competences of children. They also suggested that we look more carefully at what lower-class children *did* know.

Street maths and school maths

According to one study of the local economy in Recife, 48 per cent of the families interviewed were found to derive part of their income from the 'informal sector' of the economy (Cavalcanti, 1978), which includes a host of semi-skilled jobs such as shoe repairing, sales and occasional handyman work in plumbing, electrical maintainance and installation, as well as bricklaying. It is common to see children accompanying their parents in these occupations, particularly in commerce, and as they become familiar with the basics of selling, they will assume increasing responsiblity in serving customers, determining sums of purchases, receiving payment and calculating change. These children receive an informal apprenticeship, so to speak, about quantities and their inter-relations.

In the wake of the School Failure study, my colleagues and I wondered how children who worked as vendors in the informal sector would fare in arithmetical tasks in their everyday work settings. Under the guise of customers making purchases at an open air market in Recife, we interviewed youths who worked as street vendors. Since we recorded the verbal transactions it was possible to look carefully later at how the problems were solved. An example follows;

Customer; How much is one coconut?
M: 35.
Customer; I'd like ten. How much is that?

M: (Pause) Three will be 105; with three more, that will be 210.
(Pause) I need four more. That is . . . 315 . . . I think it is 350.
(Carraher, Carraher, and Schliemann, 1985)

What is going on here? *We* viewed the problem posed as one of multiplication: precisely 35 × 10. But the child did not multiply in the normal sense of the word, nor drop the zero, as we do for multiplying by ten. Rather, he solved the problem mentally through a procedure involving repeated additions which could be formalized as:

$$(3 \times 35) + (3 \times 35) + (3 \times 35) + 35 = 350$$

This procedure was based on the following items of knowledge:

3 × 35	= 105 (apparently already known by the child)
105 + 105	= 210
210 + 105	= 315
315 + 35	= 350
3 + 3 + 3 + 1	= 10

The formal description does not really do justice to the child's knowledge. First, the child's knowledge was more co-ordinated than the above list suggests: we knew that there was some sort of monitoring or keeping track of the subtotals as he worked towards the goal of ten coconuts. Second, the formal description certainly does not represent how the child represented the problem to himself. Nonetheless, it seemed to capture some relevant structural properties of the solution procedure.

After we had interviewed five street vendors on a variety of problems involving real or possible purchases, we requested and obtained permission to interview the children in their homes. We explained that we were interested in studying how people understand mathematics. Problems were devised using the same numbers as in the street setting or numbers slightly altered, increased or decreased by a factor of ten, for example. The problems given in this more formal setting were of two types; verbal problems with explicit reference to real world objects and quantities, and computation problems such as 'How much is 35 time 10?'.

Two findings emerged. First, there were substantial differences in the success rates across the two settings. Whereas the vendors solved correctly 98 per cent of the total 63 problems given in the street setting, they solved 74 per cent of the verbal problems and only 34 per cent of the straight computation problems.[2] One of the implications we drew was that it could not be assumed, as educators often do, that applied problems would be more difficult than pure computations. They appeared in fact to be easier. And yet the relative success in the streets could not simply be attributed to 'concrete reasoning' since the presence of real objects in no sense reduced the mathematical demands of the problem. The second finding has already been mentioned above; namely, that the vendors used procedures which were qualitatively different from those taught in school.

We believed that we had come across an important phenomenon peculiar to people who worked outside of school in particular types of professions. A subsequent study (Carraher, Carraher and Schliemann, 1987) showed us that things were not quite so simple as that.

Written and oral mathematics

A follow-up study (Carraher, Carraher, and Schliemann, 1987) of sixteen third-graders *not engaged in employment* provided a closer look at out-of-school mathematical problem-solving strategies under experimentally controlled conditions. Three situations were used. In the *simulated sale or shop condition* the interviewer posed questions analogous to those of the market place, in which the child was to imagine himself selling or buying items located in front of him. In the *word problem condition* children received problems similar to those in the previous study. An example is:

> Mark went to see a movie. He spent 30 cruzeiros for the bus and 175 for the movie ticket. How much did he spend altogether?

The *computation problem condition* used problems without reference to objects, such as 'How much is two hundred and fify-two minus fifty-seven?' All problems were presented orally. The children could solve the problems as they wished. Paper and pencil were provided in case the child wished to use them. The numbers used were carefully controlled so that no differences between results in different conditions could be attributed to the particular values used. The order of conditions was also counterbalanced across subjects.

Again, there were differences between the success rates in the three conditions. The pupils' scores, out of a possible ten items, averages 5.7, 5.6, and 3.8 in the three

Table 16.1: Percentage of Responses by Procedure and Operation

Operation Used	Procedure	
	Oral	Written
Addition	75%	68%
Subtraction	62%	17%
Multiplication	80%	43%
Division	50%	4%

(adapted from Carraher, Carraher and Schliemann, 1987)

respective conditions. As Table 16.1 shows, these success rates varied depending upon the operation used but when the pupils opted to solve the problems in their heads, by what we termed 'oral maths', as opposed to 'written maths' algorithms taught in school, they were more likely to be successful (although the difference is not pronounced in the case of addition).

As Table 16.2 shows, there was a very strong relationship between the problem situation and the procedure adopted by the children. In the simulated sale condition in more than four of every five cases oral procedures were adopted over written ones. In the word problems, pupils were somewhat more likely to opt for oral rather than written procedures. Finally, in the computation problems, only about one answer in ten was reached through oral means. Hence children tended to elect school type procedures in situations associated with schools (Where else does one perform mathematical operations on numbers devoid of referents?). Problems associated with extra-school situations tended to be solved through extra-school methods.

Table 16.2:   Percentage of Cases in Which Children Chose Oral Rather Than Written Procedures (by Operation and Situation)

		Situation	
Procedure	Shop	Word Problem	Computation
Addition	88%	50%	10%
Subtraction	80%	63%	10%
Multiplication	88%	69%	12%
Division	89%	71%	29%

(adapted from Carraher, Carraher, and Schliemann, 1987)

Let us now look more closely at the nature of the oral mathematics strategies. Two general classes of procedures (which we called 'heuristics') were identified. *Decomposition* referred to strategies, used normally in addition and subtraction problems, which involved breaking up a quantity into smaller quantities and carrying out the solution in a piecemeal fashion. The following is one such example.

Experimenter: What is two hundred fifty-two minus fifty-seven?

Eva: Take away fify-two, that's two hundred, and five to take away, that's one hundred and ninety-five.

The child has decomposed 57 into 52 + 5. She then subtracts 52 from 252, obtaining 200. Finally she removes 5 from 200 to obtain 195.

Repeated groupings were characteristic of oral solutions of multiplication and division problems. A multiplication is performed through successive additions and the operation is terminated when the proper number of times has been processed.

Division is handled through repeated subtractions. Stopping occurs when the quantity to be divided has been fully distributed. The following examples are typical:

José Geraldo: Situation: store. Computation problem given: 15×50. '*50, 100, 150, 200, 250*' (The child was keeping track of the number of 50s in one hand. When that hand was finished, he went on.) '*250, 550, 600, 650, 700, 750*' (The child started by adding chunks of 50 five times to 250. He doubled this chunk, getting ten 50s and then went back to adding individual 50s).

Francisco: Situation: word problems. Computation: $75 \div 5$ 'If you give ten marbles to each' [child], that's fifty. There are twenty-five left over. To distribute to five boys, twenty-five, that's hard.

Experimenter: (That's a hard one)
'That's five more for each. Fifteen each'. [The problem was solved by successively subtracting the convenient quantities distributed while keeping track of the increasing share that each child received; 10 marbles were given to 5 children, which accounted for 50 marbles, and the remaining 25 were distributed among the 5 children — 5 to each, totaling 15 for each child.]

(adapted from Carraher, Carraher, and Schliemann, 1989)

We conclude: (1) mathematical problem solving strategies are related to the problem situation; and (2) oral mathematics is not peculiar to vendors or to employed people.

A Framework for Discussing Mathematical Concepts:
Vergnaud's Model: concepts, invariants, representations and situations

Vergnaud (1983, 1989) has developed a theoretical model of concepts which can serve to bring order to the results discussed so far. His model is based upon the idea that concepts always involve three aspects: invariants, representations, and situations. *Invariants* refer to the properties or relations associated with the concept. Density, weight, time, colour, velocity, and so forth are examples from the scientific realm. In mathematics, number, equivalence, distributivity, proportionality, symmetry are some examples. It might seem that what Vergnaud means by invariant is the concept itself, but this interpretation would be incorrect. Concepts are necessarily represented. Invariants are intimately linked to *symbols* or *symbolic representations*[3] Invariants are expressed in a number of ways: through natural language, through written language, through algebra, diagrams, tables and so forth. Invariants are thus embodied in or expressed through symbolic representation. To give an example, a proportional relationship[4] between two physical dimensions could be represented in diverse ways, for example, by the explanation that, 'for every increase of A units in the first dimension there will be an increase of B units in the other dimension'. A graph expresses proportionality through the linear property of a curve, which captures visually the fact that increments in X are proportional to increments in Y. The linearity of the curve expresses the invariance in the proportional relationship between the two dimensions. Another way of expressing proportionality is through the expression $A:B = C:D$. None of the representations *is* proportionality, but each does express, instantiate, or capture proportionality in its own distinct manner. As Vergnaud (1983) points out, there will be advantages or disadvantages to certain representations depending upon what relations to other areas of knowledge one wishes to explore.

Finally, concepts are always tied to *situations* which imbue them with meaning. It is not perfectly clear what Vergnaud means by *situation*, but it appears to be broad enough to include both conventional social situations (commerce, math classes, sports events, test situations) and real or imagined problem situations- 'times problems', 'problems in which both things go up together', etc. (Carraher, Carraher, and Schliemann, 1987).

For most of us, there will persist the temptation to treat concepts as equivalent to Vergnaud's invariants or (for those of us raised in an operationalistic psychological tradition) as some necessarily inseparable alloy of invariant and symbolic representation. This failure to separate invariant from representation is seen in the pedagogical practice of treating the concept of multiplication as equivalent to the ability to do column multiplication, or subtraction, for instance, as the ability to subtract according to present-day methods of subtraction.

However, if we are to take Vergnaud's model seriously, we should not confuse the invariants, the underlying properties and relations, with their embodiments in symbolic representations. Further, we must treat as different concepts those in which the invariants are the same but the representations or situations are different.

A situation involves many invariants or properties, as do representations. Presumably a concept is richer when the same invariant is coherently tied to multiple representations and situations. Through such relations concepts are related to other concepts. An example is provided by the relations between the concepts of fraction, ratio, proportionality, linear equation, rational number, division, multiplication, and so forth. It is due to this overlapping nature of concepts that Vergnaud introduces the notion of *conceptual field*. One such field is that of 'multiplicative structures'. Another related but nonetheless separable field is 'additive structures'.

The epistemological and pedagogical implications of Vergnaud's model are likely to be substantial. However, we shall restrict ourselves here to the implications for the research we have been reviewing.

Since invariants are distinct from symbolic representations which express them people may recognize, know about, or handle invariants in multiple ways. We will argue that this is a reasonable way to interpret the above cited studies.

The basic invariants of addition and subtraction

Some of the basic properties of arithmetic are identity, inverse, commutativity, and associativity. In mathematics, additive identity is the principle that a number is equal to itself; that is, formally, A = A where A is any number. In real world situations the A's would correspond to quantities[5] and identity would take on a broader meaning. Piagetian studies of number conservation are relevant here; they show that by 8 years of age most children recognize identity for discrete and continuous quantities. The identity of other physical quantities — length, area, and volume — may take much longer to develop. Identity is closely tied to the idea of invariance. The fundamental idea of invariance is that some property such as number is construed or treated as 'invariant' even though other properties — the intervals between objects, for example — may have been altered.

For addition, an identity operator is a symbol which when added to any number does not change the number's value. The idea is formally expressed as:

$$A + O = A$$

In real world situations the idea is that adding nothing to some quantity will leave the value unchanged. While this may seem trivial, the meaning and manipulation of zeros commonly pose a challenge to many children in carrying out written addition and subtraction algorithms[6]. Indeed, roughly half of the 'buggy algorithms' identified by Resnick and Omanson (1987) involve errors in handling zeros. By contrast, the oral mathematics procedures we described above typically involve rounded intermediary values which, in written form, would include zeros, as in Eva's decomposition of 252–57 cited above.

In mathematics, every number has an additive inverse which when added to the number results in zero:

$$A + (-A) = 0$$

Inversion seems to correspond roughly to the idea of subtraction as the inverse of addition. Again, this property may seem to be self-evident, but it is not.

Addition, but not subtraction, is commutative in that the order of the terms does not effect the total value. Thus,

$$A + B = B + A$$

Studies related to this come from the study of young children's counting. The essential idea of association is usually expressed as:

$$(A + B) + C = A + (B + C) = A + B + C$$

and explained as follows: 'it doesn't matter which order you sum a set of numbers, the result will be the same'. If we simplify things a bit we will see the idea that

$$(A + B) = A + B$$

or even

$$A + B = (A + B) = D$$

Taken as a procedural statement, the above would mean that two quantities, when summed, yield a third quantity: if you sum A and B you get D. It can also be thought of as a declarative statement expressing the logical relations among the quantities; namely, that a given quantity[7], let us say, D, can be thought to be equivalent to two smaller quantities [A and B] taken together.

Additive composition can be viewed as the understanding that numbers can be thought of as compositions of other, smaller numbers (Resnick, 1986). Let us broaden this slightly to include the composition of quantities.

$$A = (B + C) = B + C$$

It is this understanding which allows children to mentally enact what we have termed 'decompositions', that is, break up or partition a quantity into several smaller quantities. We thus use decomposition in the sense of a Piagetian operation: an internalized mental representation of action.

This is meant to extend, so that the component quantities themselves can be broken down. A formal description such as that given below captures to some extent the idea:

$$
\begin{aligned}
&\text{Example} \\
A = \{B + C\} \qquad\qquad & 39 = \{20 + 19\} \\
= \{(D + E) + C\} \qquad\qquad & = \{(15 + 5) + 19\} \\
= \{D + (E + C)\} \qquad\qquad & = \{15 + (5 + 19)\} \\
= \{D + F\} \qquad\qquad & = \{15 + 24\}
\end{aligned}
$$

When subtraction is added to the model, equivalences such as those below can be taken into account.

$$
\begin{aligned}
& & \textit{provided that} \\
(M - T) = \{M - (U + V)\} \qquad\qquad & T = U + V \\
= \{(N + P) - (U + V)\} \qquad\qquad & M = N + P \\
= \{N + P - U - V\} \\
= \{(N - U) + (P - V)\} \\
= (X + Y) \\
= Z \qquad\qquad & Z = X + Y
\end{aligned}
$$

Do people ever do such manipulations? In a word, yes. Moreover, they do so frequently, and in ways which have not been specifically taught to them. Resnick provides a clear case in her analysis of a first grader's adding 'one hundred and fifty-two to forty-nine':

> **P**: 'I would have the two 110's, which equals 200. Then I would have 50 and the 40 which equals 90. So I have 290. Then plus the 9 from 49, and the 2 from 52 equals 11. And then I add the 90 plus the 11 . . . equals 102.
> **I**:102?
> **P**: 101. So I put the 200 and the 101 which equals 301'. (Resnick, 1986)

In analyzing a third grader's protocol, T. Carraher (1988) lays out the steps in the reasoning in the problem, 'two hundred fifty two minus fifty seven.

> You take the two hundred. Minus fifty, one hundred and fifty. Minus seven, one hundred and forty-three. Plus the fifty you left aside, the fifty-two, one hundred and ninety-three, one hundred ninety-five.

Steps used:

$$252 - 57 = (200 + 52) - (50 + 7)$$
$$= (200 - 50) - (52) - (7)$$
$$= (150 - 7) + (52)$$
$$= 143 + (50 + 2)$$
$$= (143 + 50) + 2$$
$$= (193) + 2$$
$$= 195$$

(adapted from T. Carraher, 1988, p; 10)

If we compare the above decomposition with that of Eva for the same problem, it is easily seen that the same solution can be reached in slightly different ways, based upon different ways of breaking up the quantities[8]:

$$(252 - 57) = (200 + 52) - (52 + 5)$$
$$= 200 - (52 - 52) - 5$$
$$= 200 - 5$$
$$= 195$$

The above analysis becomes especially interesting when compared with that of the school taught mathematics routines for addition and subtraction. This can be seen in Carraher's (1988) analysis of the school algorithm for subtraction, in which we have included a few extra steps:

$$252 - 57 = (250 + 2) - (50 + 7)$$
$$= (240 + 10 + 2) - (50 - 7)$$
$$= (240 + 12) + (50 - 7)$$
$$= (12 - 7) + (240) - (50)$$
$$= 5 + (240 - 50)$$
$$= 5 + \{(200 + 40) - 50\}$$
$$= 5 + \{(100 + 100 + 40) - 50\}$$
$$= 5 + \{(100 + 140) - 50\}$$
$$= 5 + \{(140 - 50) + 100\}$$
$$= 5 + 90 + 100$$
$$= 195$$

(freely adapted from Carraher, 1988, p. 10)

The line of analysis has led my colleagues (Carraher and Schliemann, 1988) to conclude, rightly so I think, that school taught written mathematics routines are based upon the same properties as those of oral mathematics. We have seen that both written and oral mathematics are based upon the additive composition of quantities and furthermore employ the associative, commutative, identity, and (to some extent, the) inverse properties of arithmetic.

There are also differences between oral and written maths. The written algorithms for addition and subtraction always entail (1) expressing quantities according to a place value notation system in which each digit corresponds to a multiple of a power of ten; (2) working along such representations from right to left and (3) carrying and borrowing (when necessary). In schools, written subtraction is always handled as a stepwise series of computations beginning with units and proceeding upwards to multiples of increasingly larger powers of ten. Borrowing occurs when a subtraction cannot be carried out at the ongoing level of analysis, when one must subtract, for example, as in the above problem, 7 units from 2 units[9]. The underlying logic of the borrowing procedure requires a redefinition of the working minuend so that the subtraction for a particular column can be correctly performed. This redefinition is based upon the breaking up and regrouping of quantities.

There are reasons for believing that these features do not apply to the mental representations made by the people whose oral maths we have been describing. As we mentioned earlier, place value notation and column-organized algorithms are historically recent cultural inventions which superseded earlier forms of notation and calculation. We know for certain that people of earlier epochs were able to do calculations without doing them as schools teach students to do them nowadays (Menninger, 1969; Carraher, Carraher and Schliemann, 1984). Further, it has been shown (Carraher, 1985) that young children (5 to 7 yrs) can understand the meaning of number values such as 'thirteen', 'twenty-six', 'thirty-five' and 'forty-two' and express them through a token money system, without understanding the same values when expressed through Hindu-Arabic notation. In the same study, Carraher found that unschooled adults who could not correctly write numbers larger than nine not only understood values such as 'eight hundred and forty-three', but could carry out computations such as 'one hundred and thirty-four plus sixty-seven', representing the values in a token money decimal system.

We now have a clearer understanding of the similarities and differences between oral and written maths. The children who could do maths out of school while showing great difficulty with schooled approaches did indeed demonstrate a knowledge of the fundamental invariants of addition and subtraction. The difficulties they showed when attempting school taught methods related to the deployment and meaning of the particular procedures and representations adopted by schools, rather than to the fundamental arithmetical logic of addition and subtraction. In other words, they have concepts of addition and subtraction which share the invariants of school advocated concepts but differ in the symbolic representations.

Mathematical Knowledge, Work Experience and Schooling

The above studies have shown that important mathematical concepts appear to develop outside of school without specific instruction. The concepts and procedures considered would appear to arise through an individual's social interactions in everyday activities such as commerce and the production of goods. How far will self-discovery take people in the development of mathematical knowledge? Let us look at three more studies from Brazil in order to clarify this issue as well as the role of the school in the process of learning.

Construction foremen: proportions

Construction foremen in Brazil are responsible for making decisions and providing supervision regarding the myriad activities involved in constructing a building, whether it be a one family residential dwelling or a multi-story suite of offices. Legally speaking, they work under the supervision and guidance of a chief engineer and an architect, who would in principle be held accountable were there to be a serious defect in the building's construction or were a serious accident to occur on the job, such as the building's collapsing. In practice, foremen are often given a larger range of duties and responsibilities than envisioned in the legislation. In the construction sector, one often encounters the expression '*alugou a placa*' to describe cases in which the engineer (or architect) has essentially 'rented out his sign' to the builder, that is, provided a construction sign with his name on it as the supervizing engineer in order to satisfy building ordinances. In such cases the engineer only occasionally sets foot on the building site. This leaves the construction foreman with many decisions to be made of many natures. He must supervise the men, decide what building materials will be needed in the short run so orders can be placed, make decisions about the precise location of walls, stairways, doors, and pillars. He will have architectural blueprints in his possession and refer to them repeatedly to guide him regarding the construction needs. Although a foreman may tend to work on projects of the same nature, the construction of homes for example, each project will have substantial differences from the last. Middle- and upper-class houses for example, tend to be 'tailor made' by the architect for the future owner. This means that construction is far from a routine affair.

A foreman is commonly of humble origins and has only a rudimentary education. Some will have gone to school for a few years, possibly to secondary school; some will have never attended school at all. As a rule, he will have risen to his present position after several years of work on construction sites employed in a range of activities. He is a man with practical experience. In some cases he will know how to construct a building from ground up, having personally worked at everything from digging foundations, raising walls, to plumbing, installing electrical conduits and laying kitchen floor tiles.

Carraher (1986) interviewed seventeen Brazilian construction foremen with between zero and twelve years of schooling. Her main emphasis was upon the development of the concept of proportionality, which is in the area of multiplicative structures. She created a task in which the foreman was to determine the real length of a wall shown scaled down in a blueprint. Normally, the blueprint should identify the real-life lengths or contain a scale ratio such as 1:100 which would allow them to determine the real-life measures. In her task, only three pieces of information were available for inferring the missing length: the blueprint length of the target wall and the real and blueprint lengths of another wall represented in the same blueprint.

Approximately one-third of the foremen attempted to solve the problem by trying out scales which they already worked with on the job. This would work out in two of the four problems given, but on novel scales, they would be stymied. However, approximately two-thirds of the foremen attempted to 'discover the relation' from the givens. There emerged among these formen the apparently self-invented procedure of what Carraher termed 'rated addition'. Rated addition involves maintaining invariant the ratio of units on one scale to the units on the other, based upon the two measures from the identified wall. For example, one illiterate foreman, with twelve years of experience on the job, attempted to find the real life measure of a wall drawn to 8cm knowing that another wall marked 5cm on the blueprint was in reality 2 metres long.

> **J. M.:** 'On paper it is 5 centimeters. The wall is to be 2 meters. Now, one thing I have to explain to you. This is not a scale that we usually work with'.

E: 'That's right'.

J. M.: 'This one we'll have to divide. We will take 5 centimeters here and here 2 meters . . . This one is hard. One meter is worth 2 1/2 centimeters'. (This is the simplified ratio). 'Two meters, 5 centimeters' (marking off the centimeters on the measuring stick and counting the corresponding meters). 'Three meters, 7 1/2 (cm), 3 meters, but there are 5 millimeters more' (to get from 7 1/2 to 8 cm). (He proceeds, finally determining that 5mm on the blueprint must correspond to 20cm in reality).

The foremen study shows that advanced concepts such as proportionality can indeed be learned on the job. Further, it is clear that self-invented methods can reach into the realm of multiplicative structures. Finally, schooling was not at all related to success in solving the scale problems.

Bookies, permutations and probability

Schliemann and Acioly (1988) provide another look at the role of work in the understanding of multiplicative relationships. Their investigation concerns the mathematical understanding of bookies who work in a numbers betting game referred to as the Animal Lottery (*Jogo do Bicho*). The game can be described as follows. At the end of each workday in Brazil five four digit numbers are drawn by the game's organizers. Players who place bets on the actual numbers chosen or make 'class bets' under which the drawn numbers fall, win a certain number of times the amount of money they bet. As in most games of chance, the amount won will be approximately proportional to the odds: the greater the odds against the bet, the greater the potential winnings.

We will not go into all of the details nor discuss the curious terminology and attitudes regarding the bets. Suffice it to list here the several types of bets players can make: (1) direct bets on chosen numbers, such as 9623; (2) group bets on 'animals' each animal being associated with numbers which end in four consecutive two digit terminations, (a bet on the ostrich would encompass numbers ending in 05, 06, 07, or 08 since these are its associated numbers); (3) bets upon numbers ending in particular two- or three-digit terminations, such as –897; (4) bets on so-called 'inverted' or permuted digits, such as permutations generated by 35210; (5) and permuted bets with repetitions of the same digit allowed.

When a player comes to make a bet, the bookie must determine how much should be paid and register the bet on a paper slip, a copy of which is returned to the player. The cost of the bet to be paid is essentially the 'total number of numbers bet on directly or indirectly' times the amount wagered on each possible winning number. If the player wishes to bet on more than one of the 5 numbers to be drawn, that must be taken into account also.

So, for example, if a player wishes to place a 5 cruzado bet that the number '7864' will be the first number chosen, he must pay 5 cruzados. If he wishes to bet on *any permutation* of the digits 7, 8, 6, and 4, he would have to pay twenty-four times as much, or 120 cruzados, since there are exactly twenty-four permutations of four distinct digits taken four digits at a time (4 times 3 times 2 times 1). If he wants his bet to cover each of the five selected numbers, he pays five times that, or 600 cruzados.

Needless to say, there is a considerable amount of probability theory implicit in the game. The simple matter of determining how many numbers a person is betting on is no simple matter even for those of us who have had secondary school probability. Of the twenty bookies in their study, only four had between nine and eleven years of schooling and hence only these might have actually studied probability theory. Of the rest, five had between five and eight years of schooling, seven had between one and four years and four had never been to school.

It happens that the bookies do not actually have to work out permutations and none of them did so in the present study. Rather, they either answered from previous experience or looked up the number of permutations in tables. In some bets, such as determining the 'inverted' bets among seven digits for thousands (four digit numbers) and hundreds (three digit endings of four digit numbers), the bookie would have to sum the permutations for the thousands with the permutations for the hundreds, calculating the amount due on the basis of this total.

The tables merit consideration here. Bookies are provided with two-dimensional tables in which they can locate the number of permutations for a bet by looking up the number of digits being bet on and running over to the column which corresponds to the type of bet being made. By reading from such tables a simple 'inverted' [i.e., permutated] bet on four digits requires multiplication by twenty-four, as we noted earlier. The same sort of bet on eight digits requires multiplying the 'per bet' amount by 1680. So although the betting involves some very complex invariants, the tables take over the burden of calculating the number of relevant permutations (and combinations) and effectively shield the bookie from the underlying mathematics.

In the sample of bookies studied, most of the bets were of a routine nature. Of the 1,701 bets observed in the natural setting, 1,253 or three out of every four bets were on one number and so required no calculation whatsoever. Only twenty-one bets (4.7 per cent of the total) were bets on permutations or arrangements of the digits in a given number.

While the authors provide considerable data on how calculations were actually carried out, two ideas stand out. Firstly, bookies almost never commit mistakes in their calculation (although it should be added that they usually work with multiplications of round numbers); secondly, they display oral mathematics solutions observed in the studies mentioned earlier. We will move to some of the results which look at their understanding of probability and the role of schooling. All of the bookies were interviewed regarding their understanding of the nature of probability concepts

Table 16.3:   Interview Questions on the Probabilistic Aspects of the Game.

1. *If, each day, you bet on a certain (two-digit number), how many days do you think, on the average, you would have to bet to win?* (does the person recognize the possibility of finding order in a game of chance?)
2. *Is it possible that the same thousand can be drawn four times in the same week?* [extremely low likelihood]
3. *If someone bets on the same number for ten years, does he earn more than if he did not?* (profit of the games organizers requires that the odds in the long run be on their side)
4. *Are there any numbers which are more likely to be drawn than others?* (equal likelihood for each number)
5. *Which kind of bet is more profitable for the bookie?* (a sophisticated concept of the difference between the bet:earning ratio and the 'favourable outcomes': 'total outcomes' ratio.)
6. *Which are the bets people most like to bet on? Why?*
7. *Why is it that in the triple and double of groups you mulitply the value of the bet by ten and not by five, since you have in fact five prizes.* (underlying probability)

(from Schliemann and Acioly, 1989)

as they applied to the game. Table 16.3: includes the main questions included. After each question, we have summarized the reasoning that went into the inclusion of each question.

The authors classified the responses from the interviews into the two categories: empirical-arbitrary answers and logical answers. They describe the empirical answers 'as those that defined successful bets as a matter of luck'. Logical answers were those that 'related the probability of occurence of a certain number to all the others that were possible'. Examples follow, first of a bookie with an empirical view of the game, followed by another with a 'logical' view.

Jose, Schooling: one year.

E: *If each day you bet on a certain ten (two digit number) how many days do you think, on the average, you would have to bet to win?*

J: *That is hard to know because it is always a matter of luck. Because sometimes a ten is drawn and then it takes a year for it to appear again. It all depends on one's luck. I have a player who bets on the same hundred everyday. Once it appeared twice in the same week.*

E: *Are there any numbers that are more likely to be drawn* (than the rest)?

J: *Yes. It's the horse* (numbers ending in 41, 42, 43, and 44) *and the alligator* (i.e., numbers ending in 57, 58, 59, 60).

E: *Why?*

J: *Because it appears every week.*

E: *Why is that?*

J: *I know because I am used to it...*

Here is now a protocol which reflects a logical mode of answering:

Lusimario. five years of schooling.

E: *If each day you bet on a certain ten* [i.e., two digit ending], *'how many days do you think, on the average, you would have to bet to win?'*

L: *I don't know how to compute it. I don't even know if one can compute it as an exact average. But if I bet on one number the bookie is betting on 99. There are plenty of chances that that number won't appear. But there is one chance in 99 (sic). I don't know, but you have to relate it to that.*

E: *Do you think that the same thousand* [i.e., particular four digit number) *can be drawn four times in the same week?*

L: *It's very difficult. I don't know if there is a way to compute it. I know that, if it happens, the probability is very small because it competes with all the other 9999* (numbers). *Although each one has the same probability of being drawn. But here it is one against 9999*

E: *If someone bets during ten years, does he earn more than if he does not?*

L: *I think that it's more likely that one will lose. I think that if one doesn't bet one may earn more because one plays against the bookie, who is the one who earns more in the end. When one bets on a group* (four two-digit numbers of 100 total two-digit endings), *one bets on one number against 25, when one bets on the tens (two-digit endings), one bets on one number against 100: on the hundreds (three-digit endings), one against 1000. I don't know how to explain it but I think that the player always loses more than he earns, . . .*
       (from Schliemann and Acioly, 1989)

Lusimario seems to view number selection as random, with all numbers having an equal likelihood of being drawn. This is different from the notion of luck as a whimsical or mysterious force according to which 'anything could happen'. Second, his answers appear to reflect an understanding that classes of outcomes vary in likelihood of occurrence.

As Table 16.4 shows, there was an interesting relationship between amount of schooling and the tendency to understand the probability notions associated with the game: the greater the schooling, the greater the likelihood of giving 'logical' answers.

There were other intriguing findings in the bookie study. Although the lottery work does not require that bookies understand about permutations, bookies appeared to know that a player had to pay depending upon the number of ways he might win.

Table 16.4: Percentage of Logical Answers among Bookies by their Level of Schooling

Level of Schooling	% logical answers
—	
None	11
1 to 4 yrs.	16
5 to 8 yrs.	37
9 to 11 yrs	75

(from Schliemann and Acioly, 1989)

The authors provided the bookies with two tasks of combinatorial logic in order to ascertain just what they do understand.

The Piagetian task for combinatorial logic (subjects must find all permutations of three and four colours) was given to all bookies. There are, of course, six solutions for three colours and twenty-four solutions for four colours. The classification of responses is fairly straightforward. Generally speaking, subjects at stage I do not find all the solutions for the four colour task. Even if they do find all for the three colour task (level IB) they remain doubtful as to whether the cases exhaust all possibilities. Level II subjects know they have all the cases for three colours, but they either do not find all the cases for four colours (level IIA) or they find all the four colour cases with considerable difficulty and experimentation (IIB). Level III subjects find all the four colour cases from the start.

Results showed a significant relationship to schooling: the higher the level of schooling, the better the performance. No unschooled subjects found all of the four colour permutations and no subjects with nine to eleven years of schooling failed to find all of the four colour permutations.

Two other permutations tasks were introduced. The first involved determining the number of distinct football shirts which could be fashioned from three colours. In a modified version letters [c,a,s,a] were used instead of colours. Since the experimenter did not mention the conceptual relationship to the lottery game, it was possible to make note of whether the subjects saw the task as being related to the lottery task. If the bookie did not spontaneously voice an opinion on the matter, he was directly asked whether he thought the task at hand and the lottery game were related.

Again, there was a relationship to schooling, 58 per cent of the responses by the unschooled bookies indicated that the tasks were not like or related to the game. This fell to 38 per cent among the bookies with one to four years of schooling and 0 per cent for those with five or more years of schooling. There may have been various reasons why some bookies did not see the tasks as related to the lottery. One unschooled subject is cited as having considered himself wholly incapable of solving the task with letters since he did not know how to read. The dialogue is instructive:

B: 'This one (with the letters) is even worse because I don't know how to read'.

E: 'But you don't have to read. I want you to find out how may different ways there are to change the places of the letters'.

B: 'This I can't do'.

E: 'What if you tried to like you do in the game'.

B: 'This one is too complicated because to read is more difficult than to deal with numbers. I know how to do a few computations but I can't read at all. I don't even know how to write my name'.

(from Schliemann and Acioly, 1989)

Candy sellers; schooling and abstraction from the situation

Saxe's investigation (1988) of youths who sell sweets in the streets of Recife, Brazil, provides another look at some of the relations among mathematical knowledge, schooling, and work experience. The candy sellers he studied must be distinguished from the youths studied earlier by Carraher, Carraher, and Schliemann (1984) in that, with the exception of the 5 to 7 year olds, they were totally responsible for purchasing the sweets at wholesale prices, determining the prices for their customers, and selling the sweets in the streets of Recife. We are thus talking about a one-person enterprise rather than a family operation. Since the annual inflation was high at the time of the study, roughly 250 per cent, the vendors would constantly be required to revise their selling prices.

The vendors varied in schooling from under 2 years to 7 years, and in age from 5 to 15 years. There were also two comparison groups of children not engaged in selling. One group was from 'the same commercial environment as the sellers' and the other from a rural setting.

Several tasks were devised. An *Orthography* task required the respondent to read and compare the values for multidigit numbers between 146 and 5000, which was 'within the range that they addressed in their practice'. In the *Bill Identification* task they were to identify the value of twelve samples of (a) paper currency, (b) paper currency with the Hindu-Arabic numerals occluded, and (c) the same numerals removed from the bills. In a *Currency Comparison* task they were to compare 14 pairs of currency units, identifying the more valuable of each pair and stating the relative worth (telling how many of the smaller denomination were equal to one of the larger denomination). In the *Bill Arithmetic* task children had to show they could count piles of mixed-denomination currency and perform computations such as giving the change of a Cr$7600 purchase for which Cr$10000 was paid. A *Ratio Comparison* task required the youths to determine which pricing ratio would be more profitable, for example, selling 1 lollipop for Cr$200 or 3 for Cr$500. [Actual currency and sweets were arranged accordingly in front of the child.]

The resuls can be summarized as follows:

1 *Orthography.* Among the sellers, understanding of Hindu-Arabic notation appears to be very strongly influenced by schooling[10]

2 *Bill Identification* was uniformly high and unrelated to population origin and schooling. Identification of values from black and white photocopies of the Hindu-Arabic numerals of paper currency varied, however, with level of schooling.

3 *Currency computation* was uniformly high and unrelated to population origin, age and schooling.

4 *Bill arithmetic* was strongly related to experience in dealing with commerce (approximately 70 per cent for the sellers against 25 per cent for the rural group with the urban non-sellers falling somewhere in between).

5 *Proficiency in ratio comparisons* was also strongly related to experience in commerce. Roughly 70 per cent of the responses by sellers were correct, as compared to 50 per cent for the urban non-sellers and slightly less than 30 per cent for the rural counterparts. It was also very strongly related to age (roughly 15 per cent, 60 per cent and 85 per cent for the 5 to 7 yr olds, 8 to 11 yr olds and 12 to 15 yr olds respectively). Yet performance here was essentially unrelated to extent of schooling. (These results suggest that the population groups were not well matched on age.)

Saxe's data also show that 'sellers with greater levels of schooling were more likely to use paper and pencil solution strategies involving standard algorithms than sellers with less schooling experience' (p. 16). Saxe suggests that the school strategies are more than simply different ways of doing maths, but that they engender a way of thinking about the mathematical aspects of the sales which is not so encouraged by the oral maths. He compares two sellers of the same age level but with considerable differences in schooling to illustrate his point.

In discussing how he determines his prices in order to guarantee a profit, the seller with a first grade education approaches the issue by hypothetically considering successive sales. The profit from each sale (of three candy bars for Cr$1000) sums to yield the box profit:

Marcos: 'I count like this' (illustrating a count of the [candy] bars in groups of three by a value of Cr$1,000). 'These two (groups of 3) bring Cr$2,000; these two (groups bring) Cr$2,000, these two (groups of 3) Cr$4,000, these 2 Cr$6,000 . . . these two Cr$10,000. I count like this 'cause I'm going to sell the chocolate at 3 for Cr$1,000, and this way the full box will bring Cr$10,000'.

E: 'How much will you profit after selling the full box?'

Marcos: 'Since the box cost me Cr$8,000 and I'll sell the full box for Cr$10,000, my profit will be Cr$2,000'.

E: 'Do you think your profit will be good?'

Marcos: 'It's not going to be very good. But if I sell two for Cr$1,000, it's going to be hard to sell, and if I sold 4 for Cr$1,000, I'd lose too much'.

(Saxe, 1988)

Saxe contrasts this protocol to Luciano, who has completed 5th grade. Luciano spontaneously produces several written calculations in which the 'unit wholesale price' appears as well as the 'unit retail price'. By subtracting them, he obtains the unit profit, that is how much is earned on each candy bar, even though he sells them four at a time. Saxe's comments on the differences are instructive:

> The first grader's solution has a direct mapping on the actual operations of exchange, each count of three by Cr$100 represents a transaction. In contrast, the [relatively] schooled child conceptualizes the mathematical relations with reference to an intermediate value that is not a part of any aspect of the acutal transaction itself — the wholesale price per unit — a value that is accessible by the use of the standard division or repeated addition of a multiplication algorithm . . . Such a solution strategy is both distant from the seller's anticipated transactions and it is also powerful — the solution strategy is one that can be used across many types of problems irrespective of any particular setting convention. (p. 11)

Farmers: areas, and lengths

Farmers who commercialize their production are immersed in a context in which mathematical thinking is inescapable. As Abreu (1988) notes, sugar farmers in Brazil's North-East must calculate the areas of plots of land being worked on in order to purchase correct amounts of fertilizer and to determine a worker's pay based upon the amount of clearing and tilling that has been done. Sugar mills pay farmers sums based on the weight of the sugar produced, among other things. Farmers in turn pay harvesters and truck loaders normally on the basis of estimates of the weight of the sugar cut and loaded. Each of the thirty-two farmers studied by Abreu kept written numerical records. This may seem natural but it must be emphasized that eleven of the farmers had never been to school and, with the exception of their own record keeping, did not know how to read. We can thus regard the record-keeping as a response to the demands of the situation. At the end of the week, workers are paid, and the farmer sums the daily production of each worker to determine the worker's production during the week, upon which the pay is based. Having the daily production in writing not only relieves the farmer of the burden of having to store it in memory; it also provides a degree of assurance to the worker (who may have actually watched the farmer enter the values in the book) that the values will be preserved till the end of the week.

The area of land is determined through a system of measures and techniques which are not taught in schools. Instead of using the metric system, farmers work with *bracas, cubos* and *contas*. The braca refers both to the wooden rod used in pacing out distances and to the unit length of the rod, 2.20 metres.

Measurement of the area of a quadrilateral is taken by the farmer walking from one corner of the plot to a neighbouring corner, touching the rod front tip on the ground in front of him and revolving the rod around that point as the farmer proceeds forward and passes the point. The other tip touches the ground and the first is lifted off tracing an arc around the new center as the farmer continues walking. During this procedure the farmer has been counting the rod-lengths. When he reaches the second corner his count will be equal to the length of the side in braca units. He continues to measure the other three sides in the same manner.

With the linear measures determined, he then calculates the area, which is found by summing the opposite sides, halving each sum, and then multiplying the two resultant numbers. This corresponds to the formula:

Area = { (Side 1 + Its Opposite)/2} times { (Side 3 + its Opposite)/2}.

This formula happens to be correct for rectangles but incorrect for other four sided figures; nonetheless it provides a reasonable approximation for most practical purposes[11]. (cf. Abreu and Carraher, 1988). Triangles are treated as a special case of a quadrilateral with one side equal to zero. In explaining how they determine the area of a triangle, farmers will commonly draw a triangle and write the lengths next to the corresponding sides, with a zero at one corner to indicate the side of 0 bracas.

Answers are given in *contas* (a unit equivalent to a 10 braca by 1 braca rectangular plot) and each conta consists of 100 cubos (or cubes) equivalent to a 1 by 1 braca area. The farmers seem to truly understand the relationships among the measures, as the following explanation shows[12]:

Interviewer: *What's a braca?*

J: *A meter is 100 cm and a braca . . . bracas accepted by agronomists are two meters long but for people from the countryside the braca is two meters and twenty . . . It's a stick like that one standing up over there* (he points to it). *Well, we measure for example a conta ten by ten* (bracas), *so it's not necessary to cube because ten by ten is 100. 100 cubes is a conta and this here* (the figure) . . . *nobody ever made this for you, m'am!? Look here, ten bracas, four sides of ten, here there are 100 cubes inside. Now the cubes are these things* [pointing to a square].

E: *This is a cube?*

J: *Here inside the square you'll find 100 squares like this.*

E: *Each one is a cube?*

J: *It's a cube, it's one braca like this (runs finger around the perimeter of a cube.)*

            (cf. Abreu, 1988)

The farmers can readily convert between metric units and their own system, not just in area determination but also in fertilization dosages, where they can go from bags/conta to tonnes/hectare. Approximations are used. By using whole number measures of area, areas can normally be expressed while staying within the natural number system.

Technically speaking, the unwritten formulae used by farmers are incorrect except for the special case when a quadrilateral is rectangular. Nonetheless, the system is meaningful and somewhat accurate for quadrilaterals, provided they do not depart much from rectangularity.

Grando (1988) studied fourteen farmers from the south of Brazil and compared their mathematical problem solving with that of students from the same region. Some of the data from her study are analyzed by Carraher (1988). One of the intriguing findings from this analysis was in the differences in how problems were approached. In a problem which involved contents relevant to farming, such as distributing the location of plants in a given area, farmers tended to provide answers which, when incorrect were generally sensible, that is within a reasonable range for the information at hand. For example, in determining how may pieces of wire 1.5 m in length could be cut from a 7-meter roll, farmers answers 'fell between 4 and 7 pieces with 93% giving the correct answer' (Carraher, 1988). Among the fifthgraders, fully 40 per cent of the answers were 'absurd' or nonsensical (greater than 7) in that they would allow for there being more pieces afterwards than there were meters of wire to begin with, 15 per cent of the seventh grade answers were also absurd. The students consistently displayed a much greater variety of strategies than the farmers. But as Carraher notes, in solving problems they were far more likely to lose the meaning or sense of the problem. Carraher argues that these differences indeed reflect different approaches to problem solving: 'while oral mathematics generates computation strategies on the basis of semantic relations, written mathematics generates solutions on the basis of rules for exchanging values from one column to another'.

Overview and Conclusion

In reviewing the above studies of mathematics at work it has become clear that there is no unitary 'work mathematics' nor a unitary school mathematics. Some professions place varied mathematical demands upon workers. Foremen for example, must understand concepts such as volume and area and be able to work with proportional relationships which are challenging even to students with several years of formal instruction. Other jobs, such as that of bookies, may require constant use of mathematcs, but in what could be described as routine tasks. As we noted with the bookies, props such as tables may be devised to allow a worker to solve a problem without his or her having to develop an extensive knowledge of all of the invariants relevant to the solution.

Work provides challenges and opportunities for people to develop mathematical knowlege which, as Carraher (1988) has argued, is meaningful. Mathematical knowledge developed at work appears to be meaningful in the sense that the representations used have clear referents. As a result, people are able to monitor their reasoning, checking the appropriateness of their conclusions. Self-invented methods may be necessarily meaningful: perhaps only if people truly understand a problem can they discover on their own a method which will work. But even where a community passes down procedures for solving a problem, as in the case of the farmers, there appears to be an attempt, successful in this case, at understanding the approach. So the major virtue of self-invented or intuitive mathematics is meaningfulness; its major liability consists in the limited conditions to which this knowledge may be useful or relevant.

The role of schooling in the development of mathematical knowledge is not a simple one. If the 'street maths' and 'written and oral maths' studies were to be taken in isolation, it might appear that the mathematics taught in schools is irrelevant or even detrimental to solving maths problems in real life (even though this was not concluded by the authors). However, several observations from other studies suggest that this would be an unwarranted conclusion. The error rates associated with written maths approaches seem to drop considerably when one looks across a wide range of schooling. It could be argued that the 'stronger' students are more inclined to continue in schools in Brazil. However, it seems more reasonable to suppose that students take several years to become comfortable with place value notation and with column-oriented algorithms for calculating, in which one operates not directly upon quantities but rather on algorithms, the relations of which to the original quantities are not direct. One could thus argue that the mathematics taught in schools constitutes a long-term investment, the payoff of which can only be felt in the long run. But what sort of payoff?

The studies reviewed offer some hints. The investigation of bookies suggests that through schooling, they begin to note similarities and treat as similar situations which, on the basis of contents involved (colours, numbers, or letters), would appear to be different. It has been suggested that the candy sellers with relatively high levels of schooling begin to distance themselves from the characteristics of the sales situations in which they work and begin to analyze their sales in terms of more abstract notions, such as 'profit per unit purchased', a derived quantity never directly encountered in their work. The work with the farmers from the south of Brazil suggests that schools may promote the generation of multiple solutions which are removed from the immediate situation. We also saw in this case that the cost for leaving the immediate situation can be high; students showed many inappropriate strategies for representing problems. Are these 'teething problems' that will go away or do they reflect a failure of schools?

Resnick (1986) characterizes well the paradox which schools find themselves up against in the emphasis upon formalization in maths:

> On the one hand, the expression, $A + B$, takes its meaning from the situations to which it refers. On the other hand, it derives its mathematical power from divorcing itself from those situations. (p. 30)

She then elaborates on this paradox in discussing the 'dual role of formal mathematical language as both signifier and signified':

> To become truly proficient at mathematics one must be able, eventually, to reason with and on the formal symbols themselves. Part of the power of algebra, for example, is that once an appropriate set of equations is written to express the quantities and relationships in a situation, it is possible to work through extensive transformations on the equations without having to think about the reference situations for the intermediate expressions that are generated. In this sense the 'meaning' of algebra is encompassed within the formal system; a meaningul expression is one that is legal within the formal system, and the application of the correct transformation rules insures that all expressions that are generated will be legal. But every algebraic expression can also be interpreted in terms of the situations (quantities, relationships, etc.) to which it refers, and every transformation rule can be justified in terms of the ways in which quantities behave under certain kinds of transformations. These references provide an alternative meaning for algebra expressions and transformation rules. (p .37).

In science, power refers to a relationship between work and time. Powerful tools and methods are those which accomplish a lot of work in relatively little time. The power of formalizations lies in their helping us accomplish mathematical tasks such as determining answers and expressing relations succinctly and efficiently. But tools, whether they be hammers, chisels, computers, algorithms or systems for representing numbers and relations, are only powerful if we know how to use them and adapt them to the problem at hand.

Some conceptual tools, particularly those which represent the collective efforts of dedicated thinkers over several centuries, would not be discovered or invented by students on their own, even with considerable prodding and encouragement. (The place-value notational system and Newton's laws of motion are I believe, two cases in point.) Fortunately, children do not have to make such inventions[13]. Educators have had bestowed upon themselves the role of introducing students to many of these tools so that in the course of their relatively brief academic careers students might learn how to use them and, presumably also how and why they work as they do. Ironically, this often means that students will start using tools before they understand what they mean. A case in point is the learning about number:

the child already has at hand words and symbols for the numbers. He first learns [the names of] these external symbols for number and only later masters the meaning of them. (Alexsandrov, 1962, p.11)

It is likely that mathematics instruction will continue to require that students use symbolic representations before they have developed a full understanding of their potential meanings. Until and unless students understand the relations underlying such tools, their mathematical knowledge will remain more procedural than conceptual and consequently be far less powerful and useful than it might otherwise be. By studying how people come to make sense out of mathematical ideas outside of school, perhaps we can learn how to promote this transition, making meaningful representations and procedures more powerful, on the one hand, and making powerful representations and procedures psychologically meaningful.

There remain fascinating issues to study, among which lies the question of theorems in action. Theorems in action are:

mathematical relationships that are taken into account by [people] when they choose an operation or a sequence of operations to solve a problem. These relationships usually are not expressed verbally . . . so theorems-in-action are not theorems in the conventional sense because most of them are not explicit. They underlie [people's] behavior, and their scope of validity is usually smaller than the scope of theorems.
(Vergnaud, 1988, p. 144)

While the present studies support the idea of theorems in action, it is still unclear how we should best account for people's practical theorems. Vergnaud appears to believe that theorems-in-action entail more than problem-solving in a manner which does not violate the invariances. Any 'cook book' approach, that is any non-understood but correct procedure will properly transform quantities to produce a correct answer. But a blind, rule-following approach does not seem to fall within the purview of theorems-in-action. Rather, Vergnaud seems to think that a theorem-in-action will entail a recognition that, despite tranformations (enacted, in this case, upon the quantities) something is understood to remain unchanged along several crucial points in the process of solving the problem. It may well be the case that some people will consider the initial situation (252–57) as equal to the solution (195, if properly carried out), without regarding the steps along the way as involving a preservation of quantities. For others, the initial state may be thought of as one quantity (252) which is then somehow transformed to obtain a second quantity which is different from the first. For these people, following the procedure correctly guarantees that the relevant properties are successfully handled even though there is no recognition of invariants by the person. This possibility requires us to distinguish between knowledge based upon a recognition of invariants and that which is not, that is between learning about invariants and learning procedures. This appears to be what Hatano (1982) has in mind in his contrast between 'conceptual knowledge' and 'procedural knowledge'. We probably need to develop yet subtler ways for describing different sorts and degrees of implicit knowledge. As Vergnaud himself well recognizes, invariants are not either known or not known, but involve a scope of validity (Vergnaud, 1989). How can we determine the scope of intuitive knowledge? The issue merits further thought and study.

Notes

1 Agronometric measurement seems to have played a decisive role in early mathematics (viz. de la Fuye, 1915, Newman, 1988/1956). Hessen (1932) has argued that even much later mathematical developments thought to be the product of isolated geniuses could be much more closely tied to current day events and technology than previously thought. He argues, for example, that the Cartesian coordinate system may have been a natural development from the use of navigational maps, with longitude and latitude widely used during the Age of Discovery. Likewise, he argues that Newton's physics may be more related to the seventeenth century concern with warfare ballistics than with the mere ponderings over fallen apples

2 A Friedman non-parametric analysis of variance was significant at the 0.05 level. The difference between the street condition and the computational problems was significant.

3 It could be argued that a very general concept will be so independent of particular representation that it could be treated, for all practical purposes, as representation-free and even situation-free. The concepts of 'object', 'relation', 'idea' and 'structure' might be used as examples. A limited concept would be one restricted to a small number of situations or representations. At times the limits of a concept may constitute a pedagogical challenge as when, for example, a student provides a correct definition of Archimedes' law of floating bodies ('when an object is placed in water, buoyancy is the force of the water upon the object, a force equal to the weight of the water displaced by the object') without being able to relate the explanation to the actual behaviour of objects in water.

4 We will restrict ourselves, as Vergnaud does, to those linear relationships which can be described by first-order equations in which the y-intercept b is equal to zero. In such cases the x values will be directly proportional to the y values, so will the increments along x and y. When the b intercept is not zero, only the increments are proportional.

5 The symbol '3', or numeral '3', refers to a number, but 3 meters, 3 cats, 3 km per hour and so on are quantities, which Freudenthal has referred to as 'concrete numbers' (Freudenthal, 1973, p.207).

6 It is generally held that the invention of zero was a major mathematical achievement in that it permitted the creation of a comprehensive place value notational system such as we have today. (Roman numerals are not a complete place value system although some meaning is derived from the relative placement of symbols.)

7 Association should be intuitively most clear for whole-number quantities larger than one. Associativity will hold for any quantity but in order to appreciate this for zero we must alter the concept of number to go beyond naturals. Witness the problem. $0 = A - A$ by inversion but this is not of the form $D = A + B$. If we treat minus A as $+(-A)$ then the quantity $-A$ is no longer a natural number and the 'minus' is no longer a binary, but rather a unary, operation, since it relates to only one quantity. We may be able to get around this problem formally but it is indeed hard to do so according to the intuitive representation of association as the breaking up of a quantity into two other quantities. [Indeed, how could one break up nothing into two quantities?] The reader has certainly managed to deal with these issues in the course of his or her mathematics education. However, this problem was by no means trivial historically nor may we suppose that it will be trivial to children. As we said, the introduction of zero is more complicated than it may seem at first.

8 Evidently there is a considerable degree of interpretation involved in describing the steps in different people's thinking and we may be mistaken if we take the analysis too literally. For instance, we must be cautious about whether Eva really represents something like '(52–52)' at all. In a literal sense, of course, she does not lay out the problem to herself in these terms. Her imaging certainly does not employ parentheses. Indeed she may not think of the quantities or numbers in a visual sense at all, as the ensuing discussion of place value is intended to show.

9 One could of course register negative digits for the columns in which the subtrahend digit is greater in value than the minuend digit. However, this would only postpone the operation of borrowing. In the end, in order to achieve a conventional correct place value notation for the answer, one would have to convert negative digits to positive digits.

10 But since age is not partialed out from schooling, and because age and schooling level will likely be positively correlated, it is difficult to know whether the relationship is due mainly to the effect of age. This is confirmed by the fact that age itself was positively related to items correct with the 3 population groups collapsed.

11 Some of the earliest archeological records available indicate that the Babylonians also calculated the area of quadrilaterals by multiplying the means of opposite sides (de la Fuye, 1915). The error of such an approach varies. A quadrilateral 8 by 8 by 10 by 14 bracas will yield exactly 98 cubos; according to the farmers it would yield 99 cubos, a 1 per cent overestimation. If they were to break the area into a right triangle and a square, they would, however, calculate 64 cubos for the square and 30 for the triangle, giving a total of 94, a 4 per cent underestimation. A 14 by 10 braca parallelogram with two 60 degree vertices has 112 cubos. But if the farmers were to calculate its area they would obtain 140 cubos, which is a 25 per cent overestimation.

12 *Cubos* allow farmers to represent areas of land smaller than one *conta* without having to enter the domain of rational numbers where formerly valid conventions for representing and operating on numbers no longer hold. Thus 3.47 contas is described as 3 contas and 47 cubos, that is as the sum of two quantities expressed as natural numbers.

13 There is of course a sense in which even the appropriation of knowledge entails discovery. But that is not the presently intended sense.

References

ALEKSANDROV, A. D. (1962). '*A general view of mathematics*', in Aleksandrov A. D., Kolmogorov. A. N. Lavrent'ev, M. A (Eds.) *Mathematics; its content, methods, and meaning*. Cambridge, USA: M. I. T. Press, pp. 1–64.

ABREU, G. and CARRAHER, D. W. (1989). 'The Mathematics of Brazilian Sugar Cane Farmers', in Keitel, C. Damerow, P. Bishop, A. and P. Gerdes, (eds.) *Mathematics, Education, and Society*. Paris: UNESCO.

ABREU, G. (1988). *O uso da matematica na agricultura: o caso dos produtores de cana de acucar*. Unpublished Masters Thesis. Department of Psychology, Universidade Federal de Pernambuco.

CARRAHER, D. W. CARRAHER, T. N. and SCHLIEMANN, A. D. (1984). 'Having a feel for calculations', in Damerow, P., Dunkley, M. E., Nebres, B. F. and Werry, B. (eds) *Mathematics for All*, Science and Technology Document Series No. 20. pp. 87–89. Paris. UNESCO.

CARRAHER, T. N. (1985) 'The decimal system; understanding and notation', in Streefland, L. (Ed.) *Proceedings of the Ninth International Conference for the Psychology of Mathematical Education*. V (1) pp. 228–303.

CARRAHER, T. N. (1986) 'From drawings to buildings; working with mathematical scales', *International Journal of Behavioral Development*, 9 pp. 527–544.

CARRAHER, T. N. (1988) *Street Mathematics and School Mathematics*. Address to the 12th PME Congress, Veszprém, Hungary, July, in *Proceedings of the 12th International Congress of the Psychology of Mathematics Education*, pp. 1–23.

CARRAHER T. N. and CARRAHER D. W. (1988) 'Mathematics as personal and social activity', *European Journal of Educational Research*.

CARRAHER, T. N.; CARRAHER, D. W. and SCHLIEMANN, A. D. (1985). 'Mathematics in the streets and in the schools', *British Journal of Developmental Psychology*, 3, pp. 21–29.

CARRAHER, T. N., CARRAHER, D. W. and SCHLIEMANN, A. D. (1987). 'Written and oral mathematics', *Journal for Research in Mathematics Education*, 18 (2) pp. 83–97.

CARRAHER, T. N., CARRAHER, D. W. and SCHLIEMANN, A. D. (1988). *Na vida, dez: na escola, zero; os contextos culturais da aprendizagem da matematica*. Cadernos de Pesquisa, 42, pp. 79–86.

CARRAHER, T. N., and SCHLIEMANN, A. D. (1983) *Fracasso escolar: uma questao social*, Cadernos de Pesquisa (Sao Paulo), 45 pp. 3–19.

CARRAHER, T. N., SCHLIEMANN, A. D. and CARRAHER, D. (1988) 'Mathematical Concepts in Everyday Life', in Saxe, G. B and Gearhart, M. (Eds.) *Children's Mathematics. New Directions for Child Development*, 41 pp. 71–87. San Francisco; Jossey-Bass.

CARRAHER, T. N., SCHLIEMANN, A. D. and D. CARRAHER (1990 in press). *Street Maths and School Maths*, Cambridge; Cambridge University Press.

CARRAHER, T. N. and SCHLIEMANN, A. D. (1988). 'Culture, arithmetic and mathematical models, *Cultural Dynamics 1* (2) pp. 180–194.

CAVALCANTI, C. (1978) *Viabilidade do Setor Informal. A Demanda dos Pequenos Servicos no Grande Recife*, Recife, Brazil: Instituto Joaquim Nabuco de Pesquisas Sociais.

CUNHA, L. A. (1981). *Educacao e Desenvolvimento Social no Brasil*. Rio de Janeiro, Livraria Francisco Alves.

FREUDENTHAL, H. (1973) *Mathematics as an Educational Task*, Dordrecht, Holland, Reidel.

DE LA FUYE, A. (1915) 'Measures agraires et formule d'arpentage a l'epoche presargonique', *Revue d'assyriologie et d'Archeologie Orientale*, 1915, 5 pp. 117–146.

GELMAN, R. and GALLISTEL, C. R. (1978) *The child's understanding of number*, Cambridge, Harvard University Press.

GRANDO, N. I. (1988) *A matemática na agricultura e na escola*. Unpublished M. A. thesis, Universidade Federal de Pernambuco, Brazil, 1988.

HATANO, G. (1982) 'Cognitive consequences of practice in culture-specific procedural skills', *Quarterly Newsletter of the Laboratory of Comparative Human Cognition*, 4 (1) pp. 15–18.

HESSEN, B. (1984) 'As raizes siasis e economicas do 'Principia' de Newton. Revista Brasiliera do Eosino de Eisica, 1984, 6 (1) pp. 37–55. (English version; 'The Social and Economic roots of Newton's Principia', in Bukharin, N.I. *et. al.*, 1971 *Science at the Crossroads*, pp. 147–212, Frank Cass and Co.

HUGHES, M. (1986). *Children and Number: difficulties in learning mathematics*, Oxford; Blackwell.

LAVE, J., MURTAUGH, M., and DE LA ROCHA, O. (1984) 'The dialectic of arithmetic in grocery shopping', in Rogoff, B. and Lave, J. (Eds.) *Everyday Cognition: its development in social context* pp. 66–94, Cambridge, Harvard University Press.

MENNINGER, K. (1969) *Number Words and Number Symbols: A Cultural History of Numbers*, Cambridge, Mass: Harvard University Press. {Original German edition dated 1957–58.}

NEWMAN, J.R. (1988) 'The Rhind Papyrus', in Newman, J.R. *The World of Mathematics: a Small Library of the Literature of Mathematics from Ahmose the Scribe to Albert Einstein*, New York, Tempus, pp. 163–171.

PIAGET, J. (1952) *The child's conception of number*, New York: Norton.

PIAGET, J. (1973). 'Comments on mathematical education', in Howson. A. G. (Ed.) *Developments in Mathematical Education*, Proceedings of the Second International Congress on Mathematical Education, Cambridge; Cambridge University Press. pp. 79–87.

RESNICK, L. (1986) 'The development of mathematical intuition', in Perlmutter M. (Ed.) *Minnesota Symposium on Child Psycholgoy*, Vol 19. Hillsdale, NJ: Erlbaum.

RESNICK, L. (1987) 'Learning in school and out', *Educational Researcher*, 16 (9) pp. 13–20.

RESNICK, L. and OMANSON. S. F. (1987) 'Learning to understand arithmetic', in Glaser R. (Ed.) *Advances in instructional psychology*, Vol 3. pp. 41–95. Hillsdale, NJ: Erlbaum.

SAXE. G. (1988). 'The interplay between children's learning in formal and informal social contexts', in Gardner M., Greeno, S., Reis. F. and Schoenfeld, A. (eds.) *Toward a scientific practice of science education*, pp. 151–194. Hillsdale: Lawrence Erlbaum Associates.

SAXE, G. (1988) *Mathematics in and out of school*, Unpublished manuscript. School of Education. University of California, Los Angeles.

SAXE, G. (1988a) 'Candy Selling and Math Learning', *Educational Researcher*, (August-September) pp. 14–21

SCHLIEMANN, A.D. and ACIOLY, N.M. (1988) 'Mathematical knowledge developed at work: The contribution of practice versus the contribution of schooling', *Cognition and Instruction*. 6 (3) pp. 185–221.

SCHLIEMANN, A.D. and ACIOLY, N.M. (1989) 'Numbers and operations in everyday problem solving', in Keitel, C., Damerow, D., Bishop, A. and Gerdes, P. (Eds.) *Mathematics, Education, and Society*, Paris, UNESCO.

SCRIBNER, S. (1984) 'Cognitive Studies of Work', *The Quarterly Newsletter of the Laboratory of Comparative Human Cognition*, 6, pp. 1–49.

VERGNAUD, G. (1983) 'Multiplicative Structures', in Lesh, R. and Landau, M. (Eds.) *Acquisition of mathematics: Concepts and process*, 127–174. New York, Academic Press.

VERGNAUD, G. (1988) 'Multiplicative Structures', in Hiebert, M. and Behr, J. (Eds.) *Number Concepts and Operations in the Middle Grades*, National Council for Teachers of Mathematics.

17
Theories of Practice

Jeff Evans and Mary Harris

Introduction

Traditional theorists, when contemplating the idea of learning transfer are surprised by two related phenomena: the 'dart-scorer syndrome' and the 'no maths here, we're practical people' syndrome. The first appears as different arithmetic behaviour on the part of the same people in different settings, for example, young people at school fail in mathematics or people in work demonstrate a lack of 'numeracy' yet these same people are competent at scoring at darts or in the quantitative or geometrical aspects of their hobbies. The second syndrome appears in the fact that many people fail to 'recognize' the 'mathematics' in their work.

In this paper we examine a group of theorists who, rather than being tempted by the easy assumption of the ubiquity of learning transfer from school to practical contexts that characterizes behavioural and cognitive psychology, are rather sceptical of the possibility of transfer happening very much at all. In response to what the transfer theorists assure us will be an easy application of properly learned, sufficiently general mathematical ideas, Lave for example proposes discontinuities in the actor's perception of the activities (dart-scoring versus mathematics), in the strategies used (see also, Carraher, Chapter 16), and in 'performance' levels in activities that might (or might not) be seen to be 'the same' (See Carraher on 'tasks correct' in street maths versus 'questions correct' in school maths, in Chapter 16).

For any discussion of the practice of mathematics we need to confront the idea of context. This implies a study of everyday life *in situ* (in context) and an emphasis on the routine 'lived experience' of people's lives, rather than focusing exclusively on people's behaviour in relatively 'unusual' or artificial settings such as experimental laboratories or test situations in schools.

In so doing we need to emphasize the importance of social structures. Social practice theorists attempt to integrate social action with social structures. In attempting this integration different researchers adopt different approaches; for example Jean Lave identifies schemes of interacting structures and analyzes the dialectical relations between them. Valerie Walkerdine chooses to use ideas from linguistics (below). The characteristic of social practice theorists is that they focus on the organization of social activity in social practices.

In this section we discuss recent work of Sylvia Scribner (especially Scribner, 1984), Jean Lave (1988 and Lave, *et. al.*, 1984) and Valerie Walkerdine (1988 and Walkerdine, *et al.*, 1989)

Mathematics and Social Practice

Previous work in trying to understand the formative role of culture in cognition (for example Cole and Scribner, 1974) had illustrated the dependency of particular 'skills' on the social organization of the activities in which they were practised, and this led Scribner and her colleagues to use the term *practice* to emphasize the culturally organized nature of the activities. Since cognitive skills are so closely tied to the intellectual demands of the practices in which they occur, a study of their characteristics must entail a study of the practices themselves. 'What intellectual tasks do these practices pose? What knowledge do the various tasks require, and what intellectual operations are involved in their accomplishment?' (Scribner, p. 4) To a certain extent these questions are those that are asked in laboratory studies of cognition where the conditions and performance of the tasks, however irrelevant to practices outside the laboratory, can be rigorously controlled. In moving to a factory for the research, Scribner was largely motivated by attempts to validate research method. 'Can we derive models of cognitive tasks empirically from a study of ongoing activities that will help us understand the characteristics of practical thinking?'

But Scribner's study (like Carraher's), was also motivated by egalitarian social concerns. It is well known that children from lower socio-economic groupings often achieve lower attainments in school than the children of middle- and professional-class parents, yet they often go on to perform competently in jobs that require technical knowledge and skill. If the researchers could specify in detail how cognitive tasks such as arithmetic are carried out in the workplace then they would have a basis for the comparison of cognitive demands and outputs in school and work, and educational implications could follow.

Factories are very fruitful places for such research. Although they do not exist for the purpose, workplaces 'offer many occasions for the development of expertise in tasks involving complex intellectual skills' (Scribner, p. 15). From a research point of view, factories impose tight constraints on activities which are shaped both technically and socially by the requirements of production. A factory can therefore be viewed as a culture that takes particular care to make required performance and outcome explicit; researchers do not have to begin by defining discrete units of behaviour since this has already been done. 'Initially the investigator can accept the social system's definition of a work task as a unit of behaviour and allow the evolving research to test its adequacy' (Scribner, p. 16).

The factory chosen for Scribner's research was a milk processing plant employing about three hundred people. After an ethnographic study of the whole plant, four common blue-collar tasks were chosen for cognitive analysis; all of these were essential for the performance of the job and all of them involved operations with written symbols including numerals. The tasks were product assembly, counting product arrays, pricing delivery tickets and using a computer form to represent quantities of the various products. The objective was to describe the skilled performance and identify the systematic characteristics of each task. Only one of these tasks, product assembly, will be discussed here, but very briefly.

Product assembly is carried out by 'pre-loaders' whose task is to make up orders for delivery drivers. It is classed as an unskilled job, is low-paid and is carried out in the low temperatures required for the storage of dairy products. The pre-loaders' work is to collect the various dairy products required by the route drivers who deliver wholesale orders consisting of a range of products specified in gallons, quarts, pints or half-pints. In the plant however, all products are packed in standard-sized cases which contain different quantities of the different products depending on how the individual product items are packed. Although the drivers specify their orders in units, the firm's computer that handles all the paper work, specifies them in case lots. If the number required for an order is not a round case load, then the computer expresses the number as 'case + number of units' for numbers up to and including half a case or as 'case − number of units' for numbers of more than half case. On the print-out, such numbers would appear as '1 + 3' or '2 − 4' for 'one case and three units' or 'two cases less four units' respectively. Detailed observation and carefully controlled intervention with the actual numbers revealed that, in the unpleasant working conditions, the pre-loaders mostly did not work literally from the computer instructions by adding or subtracting items exactly as instructed but had a large repertoire of strategies involving the saving of physical effort for dealing with what appears to be on the surface such unskilled and repetitive work.

> For example, if an order is 1 − 6 (10) quarts and a preloader has the option of using a full case and removing 6 quarts (the literal strategy) or using a case with 2 quarts already in it and adding 8, the literal strategy is optimal from the point of view of physical effort: it saves 2 moves. If the partial case, however, has 8 quarts and only 2 quarts must be added, filling the order 8 + 2 is the least-physical-effort solution (the saving is 4 quarts). (Scribner, *op cit*, p. 22)

The solutions were not specific to products but involved cognitive transformations of the numerical features triggered by visual arrays in the incomplete cases (not all of whose contents would have been visible). As a result of the observations, the researchers proposed an explanatory 'law of mental effort' by which the workers expended mental work to save physical work. The amount of physical work involved could be checked immediately from observational records. It was clear that in the majority of cases where the pre-loaders did use literal strategies, these strategies were also the ones that required least physical effort. On every occasion where non-literal strategies were used, they were also least-physical-effort strategies.

The researchers then went on to determine by means of laboratory studies of simulations in which 'pre-loader' performance was compared with 'novice' performance of clerks and students, whether such solution strategies would transfer to different situations. The inclusion of office clerks provided the most marked novice-expert contrast (Scribner, p. 18) for although they shared a common 'cultural knowledge' of dairy products with the pre-loaders they did not actually handle the products themselves. Laboratory-based simulations were devised, the outcome of which provided evidence that transfer of solution strategies from the natural to laboratory context did occur and that even when pre-loaders did fail to use optimal strategies, they were never as inefficient as those of the comparison groups.

> Clerks showed little tendency to adapt strategies to the properties of the problem at hand, using nonliteral solutions in less than half of the instances in which there were strategies of choice. [School] Students by and large were single algorithm problem-solvers; they were overwhelmingly literal . . . their use of optimal strategies continued to be dependent on the nature of the problem and the type of conversion required (Scribner, *op cit.*, p. 24).

In short the performace of the 'unskilled' workers was revealed to be more efficient than that of the 'skilled' workers and it was the latter groups who provided all the examples of the most inefficient solutions. As Scribner reports (*op. cit* p. 26):

> In product assembly, we have a suggestion as to one possible defining characteristic of practical thinking which might warrant the use of the qualifier 'intelligent', that is, the extent to which thinking serves to organize and make more economical the operational components of tasks . . . The issue is not accuracy or error but rather modes of solution. Strategy analysis demonstrated that experience makes for different ways of solving problems, or to put it another way, that the problem-solving process is restructured by the knowledge and strategy repertoire available to the expert in comparison to the novice. Other studies have amply

demonstrated these effects of experience in pursuits such as chess and music . . .
(Scribner p. 38)

'What is called the "same operation" is done now in one way and now in another, but each way is, as we say "fitted to the occasion"'. (Scribner, p. 39). Practical thinking is goal-directed and varies adaptively with the changing properties of problems and changing conditions in the task environment. 'In this respect, practical thinking contrasts with the kind of academic thinking exemplified in the use of a single algorithm to solve all problems of a given type.' (Scribner, p. 19) It is of course the use of single algorithms in this way that is held up by formal mathematics to be an example of the highly valued property of generalizability.

The suggestion that academic thought may not be a proper yardstick with which to measure, diagnose and prescribe remedies for 'everyday thought' is one of the points taken up by Lave (1988) in her work with the *Adult Mathematics Project* (AMP) in which arithmetic practices in various settings were investigated with the aim of developing a different perspective on problem solving from that of school or laboratory. Lave's investigations arose from concern about the ecological validity of experimental cognitive research in the light of 'speculation that the circumstances that govern problem solving in situations which are not prefabricated and minimally negotiable differ from those that can be examined in experimental situations', (Lave, Murtaugh and de la Rocha, 1984). One such investigation was of arithmetic decision-making processes during grocery shopping. A supermarket, appears to be like a factory, in that it is a highly structured environment geared to support a specific task[1].

The study began with simple questions about the place of arithmetic in such settings, how it is used, whether it is a major or minor part of the activity and what are the procedural differences between arithmetic in such situations and in school. It chose deliberately to conduct its investigations in places other than the 'academic hinterland' and with 'ordinary people' in 'everyday activity'. The research involved 'extensive interviewing, observation, and experimental work with twenty-five adult, expert grocery shoppers' from a wide range of ages and incomes and from both sexes. Field data was obtained by participant observation. Shoppers wore a tape recorder and talked with an accompanying research worker, a procedure that was found less uncomfortable than talking in monologue; the researchers took care not to interpret the situation, but simply to clarify the shoppers' behaviour for the record. The shoppers' arithmetic in the supermarket was then compared with their performances on pencil and paper tests of arithmetic.

Arithmetic problem solving in the test and shopping situations was quite different. The shoppers' scores averaged 59 per cent on the arithmetic test 'compared with a startling 98 per cent — virtually error free — arithmetic in the supermarket'. (Lave, *et. al.* p. 82). Not only was the 'practical workplace arithmetic' almost error-free but it came at the end of decision-making processes, implying to Lave and her colleagues that the shoppers had already assigned 'rich content and shape to a problem solution by the time arithmetic becomes an obvious next step.' Far from being applied to the problem from the outside, the form and content of the arithmetic used so accurately, grew out of it. The arithmetic practice was quite specific to the situation and appeared as what Lave calls a 'gap-closing process' that draws the problem and the already anticipated form of problem solution closer together. It was the iterative use and the monitoring of these processes by the shoppers that accounts for the extraordinarily high level of successful problem solving observed.

For Lave 'the specificity of arithmetic practice within a situation, and discontinuities between situations, constitute a provisional basis for pursuing explanations of cognition as a nexus of relations between the mind at work and the world in which it works' (Lave, 1988 p. 1). To characterize this 'nexus of relations' Lave uses the term 'structuring resources', the different resources which combine to structure the relations. These have as their sources not only activities but also social relationships, subjective experience of problems as dilemmas that produce motivation, and standard crystallized forms of quantity, such as money and mathematics (Lave, Chapter 6).[2] This is the basis of Lave's ideas about context.

Workplace Research and Schools

The concern of both Scribner and Lave's research is with the development of an understanding of cognition in practice. A major strategy in this and much other research is the choice of arithmetic behaviour as a focus of study. As Lave points out (Lave, 1988, Chapter 1) arithmetic was originally used for purposes of methodology since it provides an excellent tool for comparative studies for school and workplace learning. It does however have further advantages: it is an accepted research topic in cognitive psychology; its 'incorrigible lexicon', tight stucture and easy recognition makes it a sympathetic medium for the study of open-ended situations. From the point of view of mathematics education however, the special position of arithmetic in such studies gives it a significance it does not deserve. Arithmetic may have widespread respectability as a research tool but as a measure of mathematical thinking it is grossly distorting to mathematics' own view of itself. Certainly in the public mind which includes employers, the terms 'arithmetic' and 'mathematics' are virtually synonymous and inability to perform out-of-context algorithms in test situations is still widely regarded as a measure of cognitive deficiency which cannot be disturbed by evidence of the same person's abilities in for example practical geometry, strategic planning or skill in the betting shop.

Although the social practice theorists stress the significance of context, there is a number of ways in which they use the idea of context itself. (Evans discusses 'context' in more detail in part 2.3 of Strässer *et. al.*, in Chapter 15). In her discussion, Lave proposes a number of dialectical relations that are meant to transcend the polarities of cognition versus culture. She distinguishes the *arena*, a durable public context such as the supermarket from the *setting* which is malleable and is experienced by the person-acting. At a macro level, there is a relation beween the 'experienced world', as described so far and the 'constituent order' relating 'culture' (semiotic systems) and 'material and social organization' (including political economy and social structure). Lave calls this 'constitutive order', the 'context of the context', and indeed the whole structure seems to be a valuable way to characterize activity in context and its relationship to larger structures.

This relation however, would seem to require a systematic way to relate 'culture' and 'material and social organization' to the 'person-acting-in-context' which Lave does not really provide. It is provided however, by considering activities or structuring resources as 'discursive' practices, and by using the linguistic terminology of signified and signifier (see Dowling above) as in the work of Walkerdine (1988 and Walkerdine, *et. al.*, 1989).

An example of the potential size of the problem, and of the possibility of solving it in the framework of Walkerdine's approach, can be given by considering the difficulties young children may have with the way the term 'more' is used in primary school discourses, as contrasted with the way the term has been used as a signifier at home.

> Mathematical meanings — indeed, the development of language word meanings in general — cannot be separated from the practices in which the girls grow up. The mother is positioned as regulative in these practices, in which desires, fears and fantasies are deeply involved. So 'mathematical meanings' are not simply intellectual, nor are they comprehensible outside the practices of their production. Yet in school . . . children have to learn that there are special meanings to these terms which are not necessarily those used at home. These meanings lead to the generation of mathematical statements of enormous power . . . We analyzed the transcripts of recordings of thirty mother-daughter pairs . . . While there were many examples of 'more', 'less' did not occur once . . . all instances of 'more' come from mother-daughter exchanges where the daughter's consumption of scarce or expensive resources and food is regulated by the mother . . . The opposite of 'more' in food regulative practices is something like 'no more', 'not as much', and so on. Here these terms, for the girls . . . carry strong emotional . . . content and act as signifiers in very different [ways] from the word pair 'more'/'less' as used in school mathematics . . . Shifts from these practices and emotions to understanding 'mathematical terms' — in this case the 'more'/'less' pair in . . . the comparison of quantities — have to be

accomplished. (Walkerdine, and Girls and Mathematics Unit 1989, pp. 52–53)

Thus different contexts, such as home and school, are characterized by — indeed are constituted by — different practices and related sets of terms and meanings: we therefore call them 'discursive practices'. Subjects are put into 'positions' by the practices in which they are engaged, as well as structurally (for example by gender, ethnic or social class origins).

Thus Walkerdine *et al.* (1989) are also not optimistic about the possibility of easy transfer. However illustrations of how transfer or translation between discourses can be accomplished through careful attention to the relating of signifiers and signifieds in particular chains of meaning, can be given. Walkerdine (1988, Chapter 6) shows how a primary teacher accomplished this in a lesson on number for infants and Evans (1988) shows how a sensitive adult student can 'speak' the discourses of both 'college maths' and 'money and market maths'. Research with mother-daughter interactions provides ample evidence that mathematical meanings cannot be separated from the practices in which they are produced. Thus special meanings of school mathematics are not necessarily those of the home. Further, as Dowling also suggests (Chapter 11), the discourses and special meanings of workplace mathematics will not be, indeed cannot be the same as either those of the home or of the school.

Conclusion

In comparison with the workplace research of Scribner and of Lave and their attempts to build a workable theory of practice; in comparison with the detailed discussion of the conditioning of mathematical learning by the context discussed by Walkerdine and in comparison with the systematic analyses of differences in performance in school and work mathematics reported by Carraher (Chapter 16), the research that has been done in England and Wales in the field of mathematics and employment is underdeveloped and uninformative. The conceptual similarity of its approaches is revealed in both its categories, the work done by school mathematics educators with a view to influencing teaching in schools (the Cockcroft research) and in studies done on behalf of industrial and further education interests with a view to influencing teaching and learning in further education (the skills projects noted above). The fundamental similarity in the approach conceals questions of the differing aims and purposes of schools as a place for a broad education and of further education as a place for specific work-focused training. That a different perspective can be revealed from within the very body of the data such research generates is discussed by Harris (Chapter 13).

Research based on counting the sums people do now is, of course, concerned with maintaining an essential *status quo*. No amount of counting of current skills of itself, can explain what a person might be capable of doing in other contexts or indeed what they might do in changed circumstances. On the other hand, the research described here does attempt to understand the generation and development of 'mathematical' thinking and performance in different contexts and promises to keep open possibilities not only of explanation but also of improvement.

Notes

1 Of course, the extent to which a factory and a supermarket are 'similar' contexts for arithmetic in activity, requires further analysis. See Lave (1988, especially pp. 68–71).
2 Thus, when Lave speaks of 'proportional articulation of structuring resources' she means the relative predominance of different practices in shaping actual activity in the particular situation. This allows her to conceive of a different mix of 'shopping' and 'maths' say, in the supermarket and in her best-buy simulation experiments. Put another way, Lave's concept of 'proportional articulation of structuring resources' implies that a person is 'positioned', almost always, in several practices, rather than one.

References

COLE, M and SCRIBNER, S. (1974) *Culture and Thought*, New York, Wiley.
EVANS, J. (1988) 'Contexts and Performance in Numerical Activity among Adults', in Borbas, A. (Ed) *Psychology of Mathematics Education Conference XII*, Hungary 1988, 20–25 July, Proceedings, Veszprem. OOK Printing House pp. 296–303.
LAVE, J. (1988) *Cognition in Practice*, Cambridge, Cambridge University Press.
LAVE, J. MURTAUGH, M. and DE LA ROCHA, M. (1984) 'The Dialectic of Arithmetic in Grocery Shopping', in Rogoff, B. and Lave, J. (1984) *Everyday Cognition: its Development in Social Context.*, Harvard. University Press.
SCRIBNER, S. (1984) 'Studying Working Intelligence', in Rogoff, B. and Lave, J. 1984, *Everyday Cognition: its Development in Social Context*, Harvard University Press.
SCRIBNER, S. and COLE, M. (1973) 'Cognitive Consequences of Formal and Informal Education', *Science* Vol 182. pp. 553–559.
WALKERDINE, V. (1988) *The Mastery of Reason.*, London. Routledge.
WALKERDINE, V. and The Girls and Mathematics Unit, Institute of Education. University of London (1989) *Counting Girls Out*. London, Virago in Association with the University of London Institute of Education.

Part 3: Introduction
Mathematics in the Workplace:
User Views

Part 3 looks at the status of mathematics and work from the point of view of an employing organization, a training organization and a substantial part of the workforce.

In an article entitled 'Women who are a Mixed Blessing' in the *Times Educational Supplement* of November 3rd 1989, Bob Finch, returning to education from industry, remarks on how the odds are still stacked against women in most walks of life. 'Paid less for doing the same work, offered fewer training opportunities and promoted more slowly than their male colleagues' the situation of women at work is 'morally indefensible and commercially inept.' 'In industry' he reminds us and in contrast to education, 'with few exceptions, women tend to work for men rather than with them'. 'It will be next to useless to make it easier for young women to join [industry ...] if what they are joining ... is still a male-dominated organisation which does not offer them training, experience and promotion on an equal footing with men.'

Industry is indeed still much more highly gendered than the maintained educational system and blatant sexual (and racial) discrimination is commonplace in many workplaces. The soft-porn pin-up, out of sight of young children on the top shelf of the corner shop, is in full view on the factory wall when they go on their industrial visit.

The relative positions of women and some of the mathematics they do in their work features in the paper by Ingham, written as a contribution to the volume from Courtaulds Textiles. The hierarchical valuing of different sorts of mathematics shows clearly in the perception of mathematics 'merely as the skill that helps the machinist to calculate her bonus-related pay packet'. On closer examination this mere skill, in an industry that employs mainly women at the labour intensive and low status end of the textiles chain, turns out to be rather complex and not by any means the only skill of the 'girls' who are in fact 'bombarded with mathematics all day'. The compliment that their ability to deal with it is assumed, may be some compensation for its lack of reward. The mathematically rich activity of the 'girls' that is ignored in both recruitment and training presents a paradox explicable only in terms of power and control. For many industries, the perception of mathematics is related first and foremost to its use in recruiting, itself only sometimes related to its use in work. In many too, the fact that young entrants appear disabled mathematically in the training room or test situation yet 'get by' in work is noticed, but not noted. (See Mathews, Chapter 14).

Pye in his article on mathematics in the training activities of the Clothing and Allied Products Industry Training Board (CAPITB) blames the lack of recognition of machinists' skills on the 'image which has dogged the clothing industry' rather than on the industry's own lack of recognition or reward for the machinists. Recognition of the effects on recruitment of the demographic time bomb has caused some industries to tinker with their image in place of analyzing its origin, so that the focus of schools-indsutry work has shifted from attempts at practical help and real co-operation (now dismissed as altruistic and luxury that can no longer be afforded) to glossy career literature at least one of whose effects in school is to raise questions of cost and extravagance.

When state education and the use of arithmetic for curricular control began, women's work was limited to textile manufacture and domestic work and the rate for it was fixed with little regard for the skills involved. The 'natural role of women' was hedged about with the same moralizing that defined the role of the working classes and some of the old arguments remain. When women are needed for textiles labour their nimble fingers are praised. When their jobs are taken over by male-controlled technology, the same work becomes boring and repetitive. (Cockburn, 1985).

In her review paper specially written for this volume, Holland examines gendering at work from a historical perspective and offers a review of explanation for the gendering process. In general the organization of social relations across widely differing social, cultural, geographical and historical settings can be seen as based on gender hierarchy, an unequal distribution of power between women and men, with women subordinated to male power, authority and control. The term used to characterize this distribution of power in society is patriarchy. It is at points of change in the content of gender divisions that the mechanisms through which gender hierarchy is maintained can be examined. Holland's paper reviews a range of studies which have examined and described the ways in which work is and has been gendered, and assesses a number of explanations for gender divisions and inequality in the workplace.

As both Pye and Ingham show, however, mathematics is there in work at all levels of employment and there is no doubt that more enlightened training policies which Pye describes could exploit it for the development of the workforce, if that was really wanted. It is however the level of employment that determines the perception of mathematics.

If mathematics in schools is what the academics say it is, then mathematics in work is what the employers say it is. However different the definitions, both come from organizations in which hierarchical power is often of more moment than the cognitive growth of individual members of the silent majority.

Reference

COCKBURN C. (1985) 'A Wave of Women: New Technology and Sexual Divisions in Clothing Manufacture, in *Machinery of Dominance*, London, Pluto Press.

18
An Industry and Mathematics: A View from Courtaulds

Sarah Ingham

Introduction

From the shopfloor, through all levels of recruitment and management, mathematics has its place in one of the United Kingdom's most people intensive industries — textiles. Sometimes this serves merely as the skill that helps the machinist to calculate her bonus-related pay packet. More often it helps manage effectively an industry that is increasingly embracing new computer-controlled technology.

Courtaulds Textiles, one of the largest textile businesses in Europe, has a near one billion pound turnover and employs 30,000 people at 165 manufacturing locations, 140 of them in the United Kingdom. It is a major supplier of clothing to top high street stores, has internationally based fabrics operations, especially strong in lace and elastomerics, and is a major manufacturer of spun yarn.

A few examples of the part that mathematics plays within Courtaulds Textiles' fast-moving manufacturing operations can be gleaned from the qualifications needed and the training received by graduates and shopfloor employees.

Graduate Entry

The degree of numeracy possessed by graduates joining Courtaulds Textiles (CT) is obviously largely dictated by degree discipline and will be high in the case of mathematics, engineering, computing and most sciences. The numerical abilities of those with little or no mathematical content to their degree courses will be a function of the extent to which 'O' level or GCSE mathematics was an entry requirement to any degree course that was taken. The numerical abilities required of graduates joining CT is a function of their career path within CT and there is a wide variety of routes for graduates to progress through to middle and senior management levels. However, there are no examples of either a general route or of specific roles in which a lack of basic numeracy and ability to work with and interpret numbers would not be a distinct disadvantage.

For example, production management in garment making, a role in which a large proportion of new graduates is employed at first, requires an ability to think constantly in terms of ratios, proportions and percentages and the ability to estimate and calculate them quickly and easily in order both to plan the flow of work and maintain its smooth passage along production lines. The many variables affecting their functioning (absent operatives, material shortages, machines breaking down, customers changing their minds and so on), mean that managers are faced daily with numerous estimations and calculations.

Marketing and sales are areas in which graduates of any degree discipline may spend some of their training. Naturally, these too, require mathematical ability every day in dealing with orders, sales volumes, costings, sales margins, market areas and so on. The same applies to those involved in purchasing fibres, yarn, fabric and trims, who will be concerned not only with prices and volumes but also technical specifications and quality requirements of purchased material, all expressed in a manner requiring an understanding of and ability to manipulate mathematical concepts.

Some graduate selection boards use a test of basic numeracy and mathematics as selection criteria, although these probably require less than the old 'O' level ability. It is therefore clear from both the roles and training undertaken by graduates within CT, that without a better ability in mathematics than such a test requires, a graduate would be at a great disadvantage.

Eighteen-plus Entry

Joining CT at the age of eighteen or nineteen with 'A' levels is not a usual entry route into the group and would result only from individual applicants for specific positions. One example is the recruitment of trainee Work Study Officers or Industrial Engineers. This role requires a high degree of numeracy and the initial training course lasting between six and ten weeks has a high mathematical content. Thus in this particular example, at least the old 'O' level mathematics is likely to be a requirement.

Sixteen-plus Entry

Today, most 16-year-old school leavers go into pre-vocational and vocational training schemes rather than straight into employment. Courtaulds Spinning has two entry levels for 16-year-olds. The first is into their Youth Training Scheme to train as machine operators and requires no specific entry qualifications. The second is entry into a three year Technician Training Scheme to train to set and maintain plant machinery and which is potentially for management. The entry requirements for the scheme are the equivalent of four 'O' levels of a minimum grade C, including mathematics and engineering and/or science subjects.

The YTS typically requires a period spent off-the-job on basic numeracy and literacy. Operatives must be able to carry out some basic calculations and be able to record and interpret numerical information correctly. Many YTS trainees are unable to do this even with a calculator, and certainly not without one and are unable to express more than simple ideas in writing. However, once these requirements are placed in context within a specific job, the majority 'get by'.

The level of numeracy required for most of the jobs for which school leavers or older teenagers could be employed in fabric companies is quite low. However, although some ability is required, only a small proportion of those companies questioned during the preparation of this article actually tested specifically for any basic skills at selection. Tests that are carried out tend to be very basic and to vary according to the specific job in question. Trainee knitting mechanics in one knitted fabric manufacturer have to pass a simple mathematics test at selection while trainee knitters take a test of comprehension, manual dexterity and a test of their ability to recognize and record numerical information correctly, since no actual calculation is required in the job. Dying and finishing operatives were in some cases tested for ability to count, weigh and measure correctly. Thus in general, selection tests, where carried out at all, are very specific to the job requirements and test for a bare minimum of necessary ability.

Spinning and fabric manufacture employ a largely male workforce, whereas garment manufacture, by far the most labour intensive end of the textile chain, employs mainly women below management levels. In contrast to spinning and fabrics, the clothing businesses also employ a larger proportion of young people. Garment manufacture is an area where the demand for flexibility from customers is forcing a change in attitudes to training and multi-skilling production workers. The traditional piecework payment system where workers are paid in proportion to the volume of work they produce promotes inflexibility, as operatives want only to do what they are fastest at in order to earn a minimum wage. The need for flexibility in skills to meet rapidly changing customer requirements and an effort to enhance the quality of production line jobs in order to retain labour, are both factors applying pressure to move away from piecework as a method of payment.

Selection testing for the main production line jobs in clothing manufacture is in most cases limited to tests for comprehension of verbal instructions and to manual aptitude and dexterity testing. In fact, clothing workers face a variety of mathematical tasks every day for which many are ill equipped. The most frequently recurring calculations that they need to carry out relate to the piecework payment system itself. In order to know in advance exactly how much money they would receive in next week's pay packet they would have to perform a series of reasonably complex calculations. For each job they do, for example closing the side seam on a vest, attaching the rib to the neck of a sweater or examining garments for faults, there is a rate expressed in 'standard minutes' provided by the industrial engineers, expressed in up to two decimal places. According to their performance, expressed in percentage form, they earn standard minutes which are converted to money by multiplying by a rate expressed in pence in up to four decimal places. If their machine breaks down or if they are moved to another job by their supervisor, they may be paid at a percentage of the average or some other rate for that period. If they are working on a new style they will earn a 'top up' expressed in pence or percentage of standard minutes. Thus, calculating their wages can be a complex mathematical task. In reality most do not do this but develop an ability to estimate their wages based on experience and a feel for how well they are performing. Many in fact do not fully understand how their wages are calculated while others who claim no ability at 'sums' pick it up remarkably quickly. As well as the long multiplications of decimals and percentages of their wage packet calculations, they also need to be able to count in base twelve as all quantities are still expressed in dozens. They need an ability to estimate seam allowances and tolerances in centimetres or millimetres but stitch at densities to the inch. Increasing emphasis on quality has introduced displays of bar charts, pie charts and line graphs on fault rates and repair rates around the factory.

The 'girls' are in fact bombarded with mathematics all day. Apart from the basic literacy and numeracy training required on the YTS scheme, the ability to cope with all this is assumed. They have to 'pick it up' on the job or from one another. Many never really do know how to perform the calculations or indeed which calculations to carry out, but as long as they can cope with the required recording systems, this would rarely be a consideration when deciding whether to employ or retain them.

Production lines are managed on a day-to-day basis by supervisors, who are almost exclusively female and in the majority of cases, ex-production workers themselves. The supervisor's role involves a great deal more mathematics in theory than that of the production workers themselves. A supervisor needs to be able to manage the piecework payment system as it applies to each of her operatives, allocating allowances, or different rates of payment in different situations. She also has to be able to maintain a 'balanced line' to ensure the smooth flow of work and ensure all her 'girls' are fully occupied. She therefore has to be able to integrate individual operative performances which can vary daily, with the differing standard-minute work content of each operation involved in making-up the finished garment. Thus long multiplication of decimals and percentages should in theory be a daily requirement. In practice, many of the decisions made are based on experience and the constraints of the workers available to her. Many of today's supervisors were promoted to this position not because of their superior numeracy or even their people management abilities but because they were good all round experienced machinists. Thus it is true that although many are not constantly armed with a calculator because they are sufficiently experienced and adept at mental arithmetic, others would not be able to do the calculations any way and have got by without. This is however another area where the increasing rate of change and complexity required in garment manufacture is forcing a change in management attitudes to what is required of a supervisor. As the role becomes broader and more demanding, the selection criteria for the role will be forced into present day reality.

Thus the reality of some shop floor workers is that, although numeracy is not considered at selection and, until the advent of YTS they were never trained, their daily activities do in fact naturally involve a lot of 'mathematics'. The manner in which this is carried out has not really changed much for a production line worker for many years. However, the introduction of automated handling systems which can potentially be linked to a computer providing up-to-the-minute information both on the position and flow of work along the line and on individual worker's current performance, could potentially revolutionize a supervisor's role. This is only one of the many ways in which information technology is affecting and in some cases fundamentally changing many different jobs within textiles.

New Technology

As high street retailers demand a faster response to changes in consumer demand and as the traditional two seasons become four or five, so the complexity of the textile and clothing industry has increased. More yarns, fabrics and clothing styles used for shorter runs mean that the textile supply chain has to be managed more closely. Unsold garment 'markdowns' and unused materials result in direct financial losses and so the balance between supply and demand has to be monitored. Computers and their associated technologies excel in the collection, processing and dissemination of the large volumes of data and can thus provide any information needed very quickly. Computer aided manufacturing, planning and control has increased the emphasis on the production control functions in the industry and offers considerable scope for those who are numerate, logical and disciplined.

The application of IT to the manufacturing processes themselves varies considerably through the industry. In clothing, sewing remains highly labour intensive with such technologies as robotics and vision systems still very much in the development stages. In such areas as spinning and weaving however, programmable logic controllers (plc's) have been controlling large machinery complexes for some time. This technology has had two employment effects; firstly, it has reduced the number of people employed and secondly it has changed the nature of the remaining jobs. People skilled in electronics and numerical part programming are now in demand in the textile industry.

Electronic data interchange (EDI) now means that orders processed on the retailer's computers can be transmitted directly on to their suppliers' machines in a matter of minutes replacing the slow, error prone and expensive, paper (even computer paper) procedures. EDI is becoming the way for routine business communications to be transacted between companies and for automatic storage and retrieval systems within them. The textile and clothing industry is leading the way in these developments and once again the number and nature of the jobs involved also changes.

The Future

From the examples detailed above it is possible to see that IT is now at last permuting all areas of the textile and clothing industry. There is no doubt that many jobs have been displaced by the use of technology but these have been mainly in repetitive, boring areas. New jobs have been created both for the users of IT and those who define, design and implement the systems. The process is inexorable and technology will continue to change the industry at all levels. Some jobs will be lost but for those who are numerate and disciplined and who can use a VDU as part of their normal working lives there will be many opportunities.

19
An Industry and Mathematics: A View from a Training Board

Keith Pye

The Clothing Industry Today

Introduction

The clothing industry is the fourth largest manufacturing industry in Britain. Quite apart from the large numbers of people involved in the textile industry on the upstream side and the retailing sector on the downstream, the clothing industry employs around a quarter of a million people in almost 6,000 enterprises throughout the British Isles. The range of products is huge – from the highest of high fashion to the toughest of nuclear protection clothing and from underwear to yachting gear. The range and sophistication of the fabrics used is similarly wide.

This said, the image of the industry remains one of stark contrasts between high fashion houses in London producing garments for royalty, to the back-street sweatshop of popular fiction and television. As with most industries, both images represent part of the reality but neither presents anything like a true picture. The majority of clothing companies are well run, industrial manufacturing enterprises whose success depends upon their skill at marketing a product which the public want at a quality and price which meets their needs.

More than any other commodity perhaps, clothes are taken for granted by consumers who expect high quality and fashionable products but rarely consider the British firms which have manufactured them.

Industrial Training Boards

The Labour Government of the late 1960s under Harold Wilson, decided that the only way to ensure adequate training was to establish Industrial Training Boards for every sector of industry and commerce, to be controlled by tripartite committees of educationalists, manufacturers and trade unionists. Twenty-four Training Boards were established between 1967 and 1969. They were permitted to raise a levy from their industries and to supply training, manpower planning and skills investigations in return. During the late 1970s, the Boards came under considerable pressure from industrialists who believed that they were not receiving value for money. Most Boards in those days operated on a system of levy-exemption by which companies which could prove that they were doing adequate training would be exempt from all or part of the levy payable to their particular Industrial Training Board. This system generated considerable resentment amongst companies which believed that they were training correctly but who, in the eyes of their Training Board, remained liable to pay all or part of the levy. Their resentment peaked in 1979 with the election of a Conservative Government pledged to principles of the free market. By 1981 the new government had abolished all the Training Boards but had offered each industrial sector the option of maintaining theirs or of establishing a voluntary training organization in its place. In the clothing industry the decision was one which fell in favour of the Industrial Training Board as a statutory entity.

Since 1982 the Clothing and Allied Products Industry Training Board has maintained the levy upon its industry at an extremely low level (0.1 per cent of emoluments for firms having salary bills of greater than £68,000 per year) but has massively increased its services to the industry through the sales of its products, training and services. The 1988 Employment White Paper laid down the stipulation that the remaining seven Industrial Training Boards will lose their statutory basis — and their right to raise a levy — within the next three to five years. CAPITB, however, had already been moving towards a less dependent position and by 1987, had reduced its dependence on levy to just 7 per cent of its turnover.

The Work of CAPITB

The reconstitution of the new-style Training Boards in 1982 meant that the seven remaining Boards including CAPITB had to identify the means by which they could continue to assist their industries while depending less and less upon them for their resources. CAPITB's activities now extend into a vast number of areas all of which are concerned with supporting and assisting the British clothing industry to meet the challenges which face it. Since 1982 the Board has developed a wide variety of services for its industry as well as promoting and becoming involved in the Government's various training initiatives. What follows is a brief explanation of some of the major areas of its work. Although it should give the reader an idea of the scope and scale of modern Training Boards's activities, it is by no means a comprehensive list.

Youth Training Scheme

A large proportion of CAPITB staff are engaged in delivering what is the largest Youth Training Scheme in manufacturing industry and the second largest scheme in the country. In 1988 CAPITB had almost 8,000 YTS trainees in around 700 companies nationwide. These young people are engaged on the full two-year YTS programme and a very high proportion of them are taken on as full-time employees at the end of their training period. Each trainee is supervized by a CAPITB Youth Training Officer whose role is to ensure that they receive all the necessary on-the-job and off-the-job training.

Most of the young people enter the industry as sewing machinists, pressers, cutters or other employees at the operative level. The extent and depth of YTS training together with the flexibility which it engenders, is such that a large number of the YTS trainees are found eminently suitable for promotion within the industry to the supervisor level. Many of such trainees received their initial training in CAPITB workshops in Glasgow and Birmingham. The workshops provide trained operatives for the smaller companies within the industry and particularly for those which have no means of effective in-company training.

Qualifications; the new Clothing Skills Award

A new and expanding role of the Training Boards is the provision of National Vocational Qualifications within the clothing industry. CAPITB for example has developed, tested and launched three qualifications for sewing machinists which establish standards for the workbased performance of machinists for the first time. These require the machinist to complete a total of nine modules of which five are introductory and four of which are sewing modules selected from a variety of those available. The modular structure of the qualifications means that most companies can adopt the system, can have their own internal staff trained as assessors and can offer young people the chance to achieve qualifications for their extremely skilled performance in their place of work. In England and Wales the CAPITB qualifications are jointly certificated by City and Guilds and in Scotland by SCOTVEC. Both of these latter institutions assist CAPITB to maintain the quality of the scheme by providing a second level of moderation within the system. CAPITB's own team of national moderators provide the first level.

In common with many other manufacturing industires, the modern clothing industry is one in which quality, speed of production and flexibility are paramount. The current qualifications for sewing machinists attempt to emphasize all of these aspects of modern production requirements by setting standards for the performance of various sewing tasks which emphasize both accuracy and speed. The result has been qualifications which are directly relevant to the performance required from operators by the employers in their everyday work. Embedded within the performance of their everyday work however, are a whole series of other skills which, while not directly recognized are nevertheless essential for the performance of the job task. Accurate measurement, understanding of decimal systems, comprehension of forms and graphical displays, familiarity and competence with computer assisted machinery, together with high levels of personal and communications skills are all an integral part of the sewing machinist's job.

Engineers

The British clothing industry is no different from any other manufacturing industry in the requirements that it has for skilled engineers to maintain and repair its machinery. Modern clothing factories contain advanced machinery which can include computerized cutting machines, computer assisted overhead distribution systems, automatic, semi-automatic and operator managed sewing machines and complex steam pressing machinery, as well as the normal machinery which any modern factory requires for the handling, packing, storage and distribution of its raw materials and products. The industry is also no different from other British manufacturing industries in suffering from the shortage of trained engineers and the attention which it is currently giving to remedying that shortage. CAPITB, together with the Business and Technician Education Council has been at the forefront of these attempts at developing and promoting a range of courses for clothing machine technicians and engineers which range from the BTEC First Level up to BTEC Higher Level. CAPITB has also identified four centres of excellence for the provision of such training and has been instrumental in promoting and guiding these courses. The young engineers of today require a wide range of skills in electronics, mechanics and systems evaluation which place increasingly high demands upon the individuals and their education and training. For its part CAPITB is developing and writing a textbook for clothing machine engineers and technicians designed to be the first in the world to support formal training courses with a comprehensive and detailed background to the profession.

Production management

Machine operations are controlled by production management staff, that is production managers, work study engineers, production controllers and supervisors. It is their role to ensure that a smooth and controlled system of production occurs throughout the company.

The production department of any clothing company is the heart of the organization and for this reason requires the staff involved to be developed effectively to deal with the pressures of their job. Production staff are concerned with the quantity, quality and cost of production and should manage all three areas efficiently and effectively. CAPITB provides companies with the expertise to help identify scope within the company for improving production output and quality. There is a number of specialist training avenues through national courses, in-company programmes and on-the-job coaching.

Supervisors

Supervisors within clothing companies constitute the first line of management and are very responsible individuals. Their tasks can range from the day to day management and control of a small group of operators through to the complete handling, management and guidance of a line producing a specific product for a major clothing company. In recent years the pool of supervisors has been used for promotion or development into other areas of the company's management structure, especially those of instructing, of quality management and of qualifications assessment. CAPITB is currently working on a system of National Vocational Qualifications for supervisors and instructors within the industry, based on nationally set standards for supervisory management and including clothing-specific competences.

Management

In addition to supervisory training, the Board now undertakes a wide range of management training across the entire industry. These courses are usually of short duration so that executives are away from their companies for as short a time as possible, but they are intensive and provide the managers with the skills necessary to enhance their performance in a variety of areas.

Currently CAPITB offers a wide range of management courses in such subjects as personal effectiveness, financial awareness, production management, quality management and communication skills. These, as well as a large number of other courses, are now in the process of being expanded with the addition of specific clothing related marketing and sales training designed to improve the industry's capacity to meet the challenges of the introduction of the single European market in 1992.

The Board, in conjunction with the Training Agency and the Council for Management Education and Development, is also directly concerned with the developments in management training which are designed to provide a wide range of standard modules of management competence in skills directly related to the requirements of management today. Supervizing this project is Management 2000, a Committee of Enquiry chaired by the Baroness Cox. It is hoped that the Management 2000 Committee will lead the way in defining management competences for the development of National Vocational Qualifications for clothing managers as well as helping the industry to develop management development strategies.

Consultancy and assessment of training needs

CAPITB not only concerns itself with the training needs of the clothing industry (though this is often the first stage in any work in a company) but also with developing business improvements within a company through a range of consultancy services. The need for an independent consultant to be brought into a company need not be because of problems but rather because of business opportunities that the company would like to implement. Consultancy covers a wide spectrum of operations from operator training systems, factory layouts, production methods, quality control and work measurement to whole company re-organization.

The assessment of training needs often plays a large part in consultancy work within a company. Many firms accept the need for a professional approach to marketing, work study and so on but fail to implement training in a systematic way, so that benefits are not felt to their best advantage. Systematic assessment of training needs by CAPITB training staff looks for potential improvements in the cost of labour turnover, operator performance and manager effectiveness. Part of today's consultancy is the provision for the industry of computerized management information and production control systems which CAPITB staff are now offering to both large and small companies.

Design

For many years, the Design Council has been at the forefront of increasing the awareness in British industry of the importance of design. Since 1987 the Board has had a national committee investigating the problems of design and design training within the industry. The initial investigative committee under the chairmanship of the international menswear designer Paul Smith reported in 1988 on an interim basis and this has resulted in the establishment of a further committee, the Design 2000 Development Committee, which has the role of investigating the exact needs of the clothing industry for designers, their education and training requirements and the competences which will need to be assessed if clothing industry designers are to achieve National Vocational Qualifications.

One of the most important results of the Design 2000 Committee has been the establishment by the Committee of five definitions of designer which categorize the design function within clothing industries in terms of the very different roles which a designer fills. Among these categories are design technician, technical designer, product developer, garment technician and creative designer.

Both the Committee and CAPITB regard it as vital that designers who have real competences in a wide variety of areas apart from the creative design function, are produced by the educational and training system. The committee has identified such areas as pattern cutting and grading, management and financial skills. The demand in the industry for skilled pattern cutters and graders, especially those with knowledge and experience of computer aided design systems is such that CAPITB is currently lauching a national initiative to create training schemes, to upgrade training and to supply retraining for pattern cutters and graders.

Schools

The clothing industry is well aware of the demographic problems which face it and of the problem of the image of the industry amongst the general public. On the whole there is a lack of public awareness of the existence of a British clothing industry on any large scale and even where such awareness exists it tends to be of an industry which is small, unsophisticated, poorly paid and non-technical. It was in order to counter this lack of awareness that CAPITB has developed the first clothing industry *Schools' Pack* launched in early 1989. It was designed to increase the awareness of the clothing industry amongst students of the 14—16-year-old age group by providing posters of illustrations of the production process and careers within the industry as well as providing GCSE relevant material for use within a variety of subjects. The first series of eight worksheets in such areas as history, geography, English and mathematics was distributed with the original Schools' Pack but further work cards are currently being developed. CAPITB's Training Officers are also involved in helping to support careers initiatives and local industry-schools links throughout the country. To further assist careers officers within schools, CAPITB has produced a video illustrating some of the opportunities to be found by school leavers in the clothing industry. In particular CAPITB has been active in promoting the *Compact* system as well as the Education 2000 initiative and a variety of other industry-schools links.

Mathematics and the Clothing Industry

The clothing industry relies on the use of a wide range of mathematical skills every day that its factories are operating. The use of mathematics concerns all its employees whether in counting how many garments have been made or in complex budget planning for the next year. Departments employ mathematics in everyday life in a number of different ways.

Managers

The wide variety of management roles to be found in the clothing industry includes financial, marketing and sales among others. Managers coming into the industry today tend to have formal qualifications, for example a degree or Higher National Diploma possibly relating to the clothing industry. So far as mathematical competences are concerned, a manager would need to have a range of abilities upon which he could rely depending upon his role in the company; for instance the marketing/sales manager would need to interpret statistical data regarding sales, demand trends, income and expenditure and also be able to translate this data into a form he could use to anticipate the demand for any particular product. A financial manager would need excellent accountancy skills to assess the performance of the company and project budgets for the future. The production manger is required to calculate the rate of work through the factory, the raw materials used and required and project output for the future. All managers need to directly concern themselves with costs at each stage of production, both direct costs such as raw materials and indirect costs such as quality maintenance. The level of mathematical and statistical ability required is high and involves the basic arithmetical and algebraic skills together with geometric knowledge in some cases. Specialist managers, as in many other industries, require specific attributes in financial and engineering terms.

Designers

A designer will usually come from a further education background with a degree or Higher National Diploma in clothing design. This training usually does not concentrate wholly on the artistic development of the student but also on their awareness of production constraints as well. Production constraints are very often mathematically based; costing of garments for example, depends on how much fabric is used, the time required to produce the garment and the complexity of the production process. The designer must also be able to manipulate sizes and scales as a garment will very often not be produced in one size only. Analysis of three-dimensional shapes into two-dimensional forms underlies a major part of the work.

Pattern cutters and graders

Pattern cutting and grading is a highly skilled operation usually taught at further education institutions or through on-the-job training. It requires a great deal of accuracy in translating the designer's work into a series of patterns for cutting the different sizes of garments out of the cloth. A pattern cutter and grader must have just as good skills in manipulating sizes and scales as those of a designer.

Engineer

The technology to be found in the clothing industry has increased greatly over the last few years. This has led to an increase in the need for highly qualified engineers to service this machinery. Traditionally training was on-the-job with perhaps day release to a local college, however the increase in technology requires more time at college and training is now available in company with longer periods of block release to colleges. Mathematics and engineering go hand in hand as many of the principles of the working of a particular piece of machinery are mathematical. A good engineer needs a wide spectrum of mathematical competence in order to complete his or her training, including for example, algebra, calculus and statistics as well as the more basic mathematical procedures.

Machinists

The image which has dogged the clothing industry has tended to obscure the levels of skill displayed by industrial sewing machinists. However, anyone who has attempted to join two pieces of fabric at home will understand the elements of their problems. Sewing machinists in a modern clothing factory are required to work at high speed and to incredible levels of accuracy (in some sectors the tolerances are measured in millimetres) with flexible fabrics which have so far defied all the efforts of electronics and engineering companies to automate. The simple act of inserting a sleeve on to the body of a shirt involves the forming of an ellipse in which gather has to be inserted by eye in an exact and evenly spaced manner. The tolerance for the lines of stiches is extremely small and the speed with which the process has to be completed, is extremely demanding. Increasingly machinists are required to use computer-aided sewing systems and to carry out some programming of their machines.

Supervisors

The education and training of supervisors must provide them with the skills to cope with 'man' management as well as the more technical aspects of their job which can include the ability to 'balance' lines of operators, each of whom will be working at a specific and different speed to others, as well as a comprehension of the ability to use elementary statistics in the company's record keeping system (usually these statistics will include those on bonus, performance, time-keeping and so on). In addition, of course, supervisors must understand this system of British Standard Institute rates for performance within the company.

The use of mathematics is widespread in all levels of the clothing industry and is part and parcel of the whole context of the training initiatives of CAPITB.

20
The Gendering of Work

Janet Holland

Introduction

'A woman's work is never done' — well-worn clichés can grate in their familiarity, but of course gain their status as a cliché by offering a succinct rendering of a known reality. The statement seems to refer to the ubiquitous and all embracing nature of women's work in the household or family; to their domestic labour. But women constitute more than 40 per cent of the labour force, and paid work in the labour market is a crucial part of women's lives, their work intersects and connects the workplace and the home (Daune-Richard, 1988; Holland, 1988).

In our current social and cultural context, most women do work which is done mainly by women. The ideologically saturated, common sense category 'women's work' refers to two types of sexual division; a division in the home, in which most of the work is seen as women's work, and a division in the labour market in which most of the work is seen as men's work. This is not to deny the paid work which is done by women, but to point out that the range of jobs in which they can be found is radically limited. Jobs are gendered and there is a strict division between women's work and men's work which individuals cross at their peril. The specific content of these categories varies both historically within cultures and across cultures at any given time but a crucial constant is gender hierarchy.

Gendered Work in Historical Perspective

'Gender has been a basic principle of occupational division and inequality from as far back as the thirteenth century' Middleton (1988). Middleton's and other studies also attest to the ubiquity and tenacity of income differentials between men and women. He uses a number of examples from agriculture to demonstrate how the sexual division of labour was transformed over the course of several centuries by the processes of (a) *segregation*, intensified by technological change and (b) the *exclusion* of women from the harvest fields. These two processes tirelessly based on gender hierarchy, can be seen as operating to create and maintain gendered divisions in the labour market both through time and across different industries, trades and professions. Historians have mapped the processes for particular time periods and types of work and have uncovered patterns of which the following examples are a necessarily cursory review.

Many such analyses, looking at the period of the development and consolidation of the capitalist mode of production, have addressed the issue of the interests which were being served by the specific sexual division of work which emerged at particular moments and in particular periods. It is generally agreed that capitalism took over or built upon pre-existing sexual divisions, and patriarchy as a descriptive or explanatory concept is frequently employed in discussing the nature of those prior divisions. Control and authority over women's labour (in field, home or labour market) was exercised by men, with variations related to the level of power in the social system held by men in different social classes or groups. The inter-relation between class and patriarchal power and interests, forms the focus for many analyses of sexual division and segregation in work.

Women have always worked in home or household, the type and extent of their work varying in relation to its historical and class location. Compare the work in the home done, for example, by a yeoman farmer's wife (Middleton, 1988), a master craftsman's wife (Hall, 1982), an early nineteenth century collier's wife (Mark-Lawson and Witz, 1988), a domestic servant (and a wife) in a middle-class home in the late nineteenth century and a twentieth century working-class working mother (Glendinning and Millar, 1987). But prior to the growth of factory based manufacturing industry, the household and workplace were more closely integrated — separation generated changes in the type of work women (and men) could do in the home and an examination of this separation demonstrates the processes of segregation and exclusion referred to above.

Pinchbeck (1981) has argued that the productive work of women was of greater importance in textiles than in any other trade. In the case of textiles, the pre-industrial family economy (household based) although often organized by mercantile capitalists rather than as independent production units could experience considerable autonomy, for example over hours and pacing of work. In the family economy of textile production, it was the father/male household head who retained the supervisory role, organizing and disciplining the labour of other household members. Weaving was a male task and spinning was a female task. But technological change from the late eighteenth century onwards brought radical changes to this sexual division of labour. The invention of Crompton's mule and the increased importance first of water, then of steam power moved spinning from the home to the factory and in this move, in the space of one generation, spinning became a male occupation (Pinchbeck, 1981; Hall, 1982).

Lazonick (1976) suggests that men established their dominance as spinners at the moment when spinners became supervisors of work. Men were able to carry over the authority they had in the family economy into the factory, taking on the more skilled occupation of mule-spinning and frequently employing their own wives and other family members, including children, as their assistants or 'piecers'. This control of men over the spinning occupation was then consolidated by the development of unions, and the exclusion of women from membership, training and jobs in the early nineteenth century (Hutchins, 1915; Walby, 1986).

Weaving, a traditional male occupation in the family economy, stayed in the home for longer than spinning, but in the late 1830s and 40s power loom weaving began to be introduced on a large scale and it was women and children who were employed on these looms in the factories as men were displaced. But as Hall (1982) points out this challenge to men as head of household and breadwinner did not last very long. A combination of successive factory acts restricting children and women's labour, and a pressure on married women not to work (outside the home) originally emanating from the middle class but adopted by the working class and unions in the notion of the 'family wage'[1], led to a gradual reorganization of the labour force and the increasing employment of men. The ideal, or indeed ideology of the family wage, which was not financially realistic for most working class families, legitimated low pay for women ('pin money') as it does today (Barrett and McIntosh, 1982).[2]

Walby's (1986) trenchant analysis of the factory acts (which limited the hours and types of work which women [and children] could perform) argues that the legislation, far from being benign and progressive, was an important attempt to maintain and reinforce the patriarchal structuring of society. This structuring mobilized the interests of and pressure from various groups in the society, including 'male workers, female workers, different sets of employers, landowners, bourgeois philanthropists, and the women's rights lobby'. Walby outlines the way in which patriarchal and capitalist structures and interests coincided at some points and were in contradiction at others in the process of the gendered restructuring of the labour force. It is in the interests of individual capitalists for example, to employ cheaper, female labour; it is not in the interest of men nor patriarchy that women should (a) attain relative independence from men by working long hours away from home in the factory for a wage; (b) displace men in factory work which increasingly came to be seen as offering a power base for organization; (c) be rendered less able to perform their reproductive functions in the home.[3]

While Walby's analysis offers a broad sweep on forces operating to construct gendered work in the nineteenth century, supported by a wealth of empirical information on the textiles, engineering and clerical trades, Lown (1983) gives a detailed analysis of patriarchal power and paternalism at work in the setting up of a particular silk mill by Courtaulds in 1825. She sees paternalism as a legitimizing form of social interaction which occurs in a hierarchical, patriarchal system and argues that women cushioned the effect of industrialization for men, enabling them to secure better conditions for themselves out of the economic and social changes taking place. In the Courtaulds mill most workers were female, but two distinct hierarchies of labour existed, one male and one female. Most of the men were in jobs involving maintenance of machinery and supervision of the workforce. Women operated the power looms at half the wage rate of the males who attended the machines. The employers were uncomfortable about women's employment in factories to the detriment of their training in domestic skills and performance of domestic services (letter written by George Courtauld II, 1846). Courtaulds developed a number of strategies which enabled them to employ cheap female labour, whilst allowing women to combine this work with domestic duties. They set up nurseries, an evening school and a boarding house, for example, to keep young women from lodging in 'immoral' surroundings. They also allowed women to move in and out of the workforce with ease in response to the demands of household life cycles. Lown suggests that 'running through all their paternalistic activities was a persistent attempt to construct femininity and masculinity in their own image' (p. 40). The men who worked in the factory in their separate, segregated, more highly paid hierarchy, were under no threat from the women who worked there; the husbands and fathers of these women, predominantly agricultural labourers, benefitted from both the economic contribution of their wives (married women had no rights to their earnings till the 1870s) and the ability of the women to stay at home, or adopt employment for short periods as circumstances required. What Lown sees as central to the Courtauld experience was the realignment of patriarchal interests, in which a gender and occupational hierarchy was negotiated which 'structurally marginalised women's economic status and ideologically restricted them to the realm of the "domestic"'. (p. 44) Patriarchal interests were maintained despite women's work outside the home in the factory.

Lewis (1984) describes the characteristics of women's work through the period 1870–1950. She suggests that the range of opportunities for women to work in the nineteenth century was limited largely to textile manufacture and domestic services, and that in general, employers, trade unions and women themselves shared the dual concept of 'a woman's job' and 'a woman's rate'. But although there were strong patterns of sex segregation, these were not fixed throughout the country. Brickmaking for example was a woman's job in the Black Country, but a man's in Lancashire. The 'woman's rate' resulted in a tendency to fix women's pay with little regard for the skill[4] required, but in relation to this notional rate, or as a fixed percentage of the male rate (Lewis 1984, Routh 1965).

Alexander (1976) describes women's work in London in the mid nineteenth century, attempting to retrieve the lives of working-class women from the 'relentless scrutiny of the middle class'. Revelations about the conditions under which women worked in the mid nineteenth century 'shattered middle class complacency and aroused the reformatory zeal of the Evangelical and Utilitarian philanthropists' (p. 60). But she regards the middle-class enquiries into the working class which ensued as coloured by moralism and mystification and the ideology of the 'natural' role of women.

Alexander argues that women's work, especially when it took place in the home (e.g. washerwomen, domestic servants, working with their husband in his trade) passed almost unnoticed, was overlooked and unmentioned, for example in the 1851 census. Women were excluded from the heavy or skilled industries in London (shipbuilding and engineering), public service (gas, building), transport, factories, professions and clerical work during this period. Their work fell largely into four categories: (1) all aspects of domestic and household labour; (2) childcare and training (including teaching); (3) distribution and retail of food and (4) specific skills in manufacturing based on the sexual division of labour in the family economy prior to industrialization. Women did undertake other work — George (1930) wrote that there is no work 'too heavy or disagreeable to be done by women provided it is also low paid' (p. 170) and Mayhew (1861) and Munby (Hudson, 1972) give examples of other types of work that women were found to be engaged in (flower girl, mudlark, entertainer, sackmaker). Alexander describes the processes of change in the labour market arguing that each step in the development of capitalist production 'the breakdown of the handicrafts system, division of labour, exclusion from the skilled craft guilds, separation of workshop and home, formation of trade societies' undermined women's labour market position. With others we have noted above, she argues that it was the transfer of the sexual division of labour from the family into social production which ensured that women remained hierarchically subordinate in the labour market. She describes in detail some of the types of work which women did undertake and concludes that 'Throughout the nineteenth and twentieth centuries, amid all the technical changes within trades as well as the industrial transformation of London as a whole, a survey of women's work reveals the tenacity of the sexual division of labour — a division sustained by ideology not biology, an ideology whose material manifestation is embodied and reproduced within the family and then transferred from the family into social production' (p. 110/111).

Glucksman (1986) examines the role of women in the development of the labour process in the new industries which emerged in the inter-war period. Although it is difficult to gather the statistics, Glucksman reviews the material that does exist[5] to demonstrate both the expansion of women's employment in the new industries and a link between the introduction of mechanized production processes and the use of women as semi- and unskilled labour. (See too Liff (1986) for a similar analysis of new industries in the post-World War II period). Glucksman uses the electrical engineering industry to illustrate the growth of the female labour force relative to men, the high proportion of young women[6] employed and a differing age structure for the male and female workforce, sex segregation and women's employment as operatives and in semi- and unskilled, repetitive and assembly work. Similar calculations were made for the food processing, rayon and ready made clothing industries. Women's labour was adaptable, interchangeable and temporary in the eyes of employers and contemporary writers reflected, justified and never questioned this stance. As automatic machinery and mass production was introduced into electrical engineering, a clear sexual division of labour was established. Men acquired or retained control over the technical process and 'know-how'; women operated the machinery with no knowledge of how it worked. This process is one whereby gender subordination is created within the labour process, a point we shall return to later.

The impetus for the incorporation of this subordinate gender relation can come from male workers or from employers, or indeed from both.[7] Loveridge (1987) has examined technological change in a number of industries over a longer time span and illustrates the way in which employers can use gender relationships in the transformation of manufacturing and service delivery processes, sometimes with the collusion of women themselves. He gives the example of male laboratory technicians (Harvey, 1987) who attempted to use strategies of closure and appropriation to retain the judgmental elements of their work, when their routine tasks were being automated. In contrast the female administrative workers in Loveridge's case studies were often willing collaborators in programming their tacit knowledge and skills. The impact of micro-electronics on service-sector performance revealed in the international study from which his examples were taken, Child *et. at.* (1985) was seen to produce and reproduce the gender hierarchy with which we have become familiar. Men were retaining or moving into both supervisory and 'technical' positions, women remained in junior, subordinate positions.

Gendered Work Today

We have reviewed some evidence outlining the processes which have produced gendered work in the past; how can we characterize gendered work today? Starting from the woman's perspective, we can see that work is a crucial part of women's lives and their labour, as ever, is a major component in the formal, the informal and the household economies. Women's economic activity rates[8] range from 21 per cent at ages 60–64 to 70 per cent at ages 20–24. A small proportion of the average female lifespan is occupied by child bearing and full-time rearing, which interrupts their employment for about 7 years (Joshi and Owen, 1981). The proportion of married women entering the labour market has risen from under 10 per cent in 1901 to over 50 per cent in the 1980s and many studies have shown that without the women's wage a large number of families would fall below the poverty line and that this dependence is increasing (Land, 1976; Royal Commission, 1978; Walker, 1988).

The labour market in the UK is highly sex segregated, both horizontally and vertically. Horizontal occupational segregation means that men and women work in different types of occupation. The vast proportion of both manual and non-manual women workers are each concentrated in only three occupational categories.[9] Vertical segregation means that men are more usually found working in higher grade occupations, and women in lower grades (Hakim, 1979). Gender hierarchy operates within occupations in which women are present or in the majority. In school teaching, for example, at both primary level, where women predominate and secondary level, where there are almost equal numbers of women and men, the familiar pyramid emerges. Higher scale teaching posts and managerial and authority positions are filled by men, women cluster in lower grades, rarely ascending the heights of the profession.

Most professions are dominated by men and present women with a 'discriminatory environment' (Bourne and Wikler, 1978). In a recent collection of essays describing the position of women in a range of professions including the law, medicine, engineering, science, the civil service and academia (Spencer and Podmore, 1987) processes of exclusion and marginalization are identified which channel the minority of women who attempt to enter these areas into certain types of work. In the civil service, for example, Walters (1987) identifies an ambivalence in the handling of women's employment (despite an explicit commitment to sex equality) which leads to the situation where the service 'opens itself to women but squeezes them out, ... integrates them and yet marginalises them, ... both facilitates and impedes its women employees' efforts to realise their full potential' (p. 14). An inspection of any high status, high paid work reveals a singular absence of women.[10] The power to make decisions which radically affect people's lives and work is very rarely in the hands of women, while more than half of the people about whom men make such decisions are women.

The past three decades have seen a dramatic growth in part-time work in the UK, mirroring the increase in the participation of married women in paid work. Most part-timers are women (83 per cent, see Robinson and Wallace, 1984; Robinson, 1988) and conversely a large proportion of women are in part-time work (44 per cent of female employees (Robinson, 1988)). Women often take part-time work as it becomes available whatever it may be, in order to accommodate their domestic responsibilities. In most cases this work is poorly paid, characterized by poor working conditions and extreme insecurity and is not covered by protective employment legislation[11]. It also represents under-utilization of qualification and skill level for many women (Martin and Roberts, 1984) and there is considerable evidence of downward occupational mobility in women returning to part-time work after a period of absence from the labour market (Dex, 1987). It has often been argued that the growth of part-time work and the fact that it is largely women who are employed in it is a labour supply phenomenon in which employers create part-time work in response to the availability of a 'marginal' supply of married female labour wanting to work short hours to meet their domestic responsibilities. Robinson (1988) argues that it is a demand-side phenomenon and that employers have expanded and used part-time work and shift systems to increase their flexibility in response to market demand, both in those manufacturing industries in which women predominate and particularly in the service industries in which 90 per cent of part-time employment is located and which has been a growth area for female employment. Robinson suggests that 'At firm, organisation or establishment levels, the points at which employment decisions are taken, policies which result in the growth of the part-time labour force reflect the responses of employers (whether in the private or public sector) to pressures to reduce and contain operating costs in an increasingly competitive or constrained environment' (p. 131).[12]

Homeworking has also increased since the early 70s (Hakim, 1984) and most of those working *at* home rather than *from* home are women (Cragg and Dawson, 1981). Most homeworkers are found in manufacturing, office and clerical work and needlework, with the largest concentrations in occupations that are themselves predominantly female. The work itself is primarily characterized by 'extremely poor rates of pay and the hidden costs and hazards for the women who do it and their families' (Lonsdale, 1987). One of its major attractions for employers is that it enables them to displace the risks of business more easily than they can with a conventional workforce (Rubery and Wilkinson, 1981). Homeworkers, uncompensated for their greater insecurity and costless in terms of overheads, provide a genuine low-cost labour alternative for employers, quite apart from any actual income differentials between homeworkers and other categories of workers which exist (Low Pay Unit, 1984). In a study of the clothing industry Mitter (1986) argues that there has been a decline in officially registered employers in the industry and an increase in 'invisible' workers in undeclared workplaces, including the home, which partly explains reported increases in productivity, in the face of falling profits and competition from the Third World (Mitter, 1986; Humphries and Rubery, 1988; Greater London Council, 1985). Paradoxically it is often women from black and other minority ethnic groups who are most likely to be employed as homeworkers. But the labour market situation and the experience of women in minority groups, suffering as they do from the discriminatory effects of institutionalized racism in British society, is even more difficult than that of women as a group (Carby, 1982; Giddings, 1984; Amos and Parmar, 1984; Bryan *et. al.*, 1985; Bhavani and Coulson, 1986; Cook and Watt, 1987; Thorogood, 1987; Phizaklea, 1988). Clearly there are different groups with differing experiences and chances within black and ethnic minority groups, as there are of course amongst women as a whole (Ramazanoglu, 1989). And Parmar (1982) has pointed out the error of applying an additive model of disadvantage to the situation of black, working-class women. The combination of these attributes may not necessarily relegate such a woman to the bottom of the social pile, but coalesce to generate a collective identity and solidarity which enables her to overcome some of the more obvious forms of discrimination. An example here is the instrumental use by young black women of the educational system to provide them with the necessary qualifications which will enable them to improve their labour market chances, in the face of low and negative expectations of their performance and capacity by teachers (Fuller, 1980; Mac an Ghaill, 1988).

Humphries and Rubery (1988) discuss the growth of part-time work and homeworking in the context of the 1979–85 recession, and its impact along with 'disabling measures'[13] on the situation of women. They point out that 'The additional domestic labour imposed upon women by the alterations in government policies in conjunction with the greater importance of women's earnings, must have left many women increasingly desperate for some earnings while simultaneously increasingly unable to work full-time' (p. 97). They conclude that disabling measures have had their effect on the terms and conditions of women's employment rather than their participation rate and that many British women have suffered an increase in exploitation during the period of recession covered by their study.

Women have also suffered straightforward job loss in the recent recession. A dramatic example of job loss during a rationalization programme in the textile industry (which as we have seen is a large employer of women) in the face of the 1980s recession is an example. In a process in which the weakest amongst their operations were eliminated, Courtauld's workforce of 100,000 was halved by 'industrial surgery' in three years.

All of the characteristics of women's work outlined above contribute to the tenacity of income differentials between women and men. There are difficulties in calculating this figure precisely but the Equal Opportunities Commission report that after an initial rise in the mid 70s when the Equal Pay Act (1975) became law, women's gross hourly earnings as a proportion of men's have settled at 73–75 per cent, comparing gross weekly earnings lowers the figure to 67 per cent. Snell (1979) has demonstrated that employers, often in collusion with trade unions and no doubt fearful that the structural constraints of women's labour market positioning were insufficient to maintain the existing imbalance of female and male earnings, used the implementation period from 1970 until the Act became law to reclassify work in such a way that women remained in clearly separated, lower paid grades of work.

The Construction of Gender

Factors contributing to the gendering of work and women's labour market position have been emerging throughout this discussion and one major contributory fact is girls' and women's educational experience. We briefly mentioned teachers' expectations in the context of young black women's attitude towards school; teachers' expectations of girls and young women in terms of their educational performance and future life and work experience in general support traditional gender patterns. Clarricoates (1980) and Stanworth (1981) have demonstrated for primary and secondary school respectively, that teachers value boys more highly than girls and regard the most important aspect of girls' future to be related to their domestic situation rather than work. The school itself, through its structure, organization and practices has been seen as reinforcing and reproducing traditional gender roles to an even greater extent than other institutions in the society (Delamont, 1980). The evidence for this proposition has been provided by a proliferation of feminist research in education over the past fifteen years (see for example Wolpe, 1977; Deem, 1978, 1980; MacDonald, 1980, 1981; Arnot, 1982, 1986; Kelly, 1981, 1985, 1986; Chisholm and Holland, 1986; Blackman and Holland, 1989).

The groundwork for gendered work is laid in the school which structures pupils' expectations and provides girls and boys and different class and ethnic groups within those gender categories with widely different types of educational experiences and qualifications, leading seemingly inexorably to widely different ranges and types of work. Boys go into mathematics, science and technical subjects; men into science, engineering and technology. Girls go into arts, languages and domestic subjects; women are unprepared and untrained for lucrative and progressive careers particularly in science and technical areas — they are trained indeed largely for their domestic role (Pratt *et. al.*, 1984; Whyte, 1986; Ormerod, 1975). This process of reproduction of class and gender relations in the school has been variously theorized, for class by Althusser (1971) Bowles and Gintis (1976), Bernstein (1977), Bourdieu and Passeron (1977) and for gender by MacDonald (1980, 1981), Arnot, (1982), Kessler *et. al.*, (1985).

Gender difference is constructed through the processes of schooling, including the interactional practices of all the participants in the educational experience and the structure and organization of the education system at all levels.[14] In this process, occupational destiny becomes mapped onto sexual identity and that linkage is notoriously hard to break. Cockburn (1986), describing YTS schemes, has elaborated on the very real problems at the personal and practical level which confront a young woman who wants to step outside the expected mould, to take up for example a traditionally male working-class occupation such as plumber or electrician.

But the workplace itself does not merely present ready-made gendered positions for women and men to slot into; gender difference is constructed in and through the work process itself. We have seen in some of the historical examples given earlier the importance of both the enabling part played by changing technology in the process of gendering work, used both by employers and male workers and of men as individual and collective agents. There are many examples of the part played by men in trade unions in constructing and maintaining gender difference and segregation in the workplace (Hartmann, 1979, Walby, 1988). Cockburn (1988) suggests that there are two mechanisms which men as individuals employ in this process. First, they exploit horizontal and vertical differentiation in the workplace: if women move into an area of work, men move out of the sphere 'contaminated' by women, either horizontally into work at a similar level but still 'male', or vertically into higher graded work. They can also create new vertical subdivisions. Cockburn (1985) gives examples of women moving into medical physics (responsible for high-tech equipment in hospitals[15]) and changes in the clothing industry, where men are even prepared to lose their jobs and take redundancy pay rather than work in a 'female' area.

The second mechanism which Cockburn identifies is the active gendering of both people and jobs — this applies to the cultural processes involved in the construction of gender differences and the hegemony of the male gender ideology. As Cockburn puts it 'masculine equals superior, active and powerful; feminine equals subordinate, submissive and directed' (p. 39) and 'It is very difficult to stop ourselves seeing things in this way, because gender is part of our cultural tools for thinking, for ordering and understanding the world' (p. 37, 38). Behind occupational segregation and the gendering process lies male power and it is this which has to be confronted if there is to be change.

Game and Pringle (1984), who provide detailed case studies of a number of industries and services in Australia, placing them in the context of trends in advanced capitalist societies, concur that gender is about power. It is about the domination of men and subordination of women which is produced in the labour process, as well as in other institutional sites. They examine manufacturing industry, banks, retailing, 'computer-land', the health industry and housework in the Australian setting. Like Cockburn they look at both the construction of the division of labour in particular worksites, in their case in relation to reproduction in the private sphere and at the role of the symbolic in constructing female and male experiences of work. Points of change in technology and the labour process and concomitantly in the sexual division of labour represent points of challenge for Game and Pringle, 'since gender relations, like power relations are characterized by struggle and resistances' (p. 141).

A. M. Scott (1986) discusses the gendering of work at a more general level, arguing that the social relations of gender are intertwined with the social relations of production and that this is demonstrated in the sex-typing of jobs, which occurs when gender segregation has crystallized in an occupation. The fact that the job is gendered exerts a strong pressure on the 'rational' economic market and substitution of the other sex is resisted even when economic conditions of supply and price change. Sex-typing affects recruitment by stressing the appropriateness of men or women for a job rather than their actual abilities to do it. And the stronger the influence of sex-labelling on particular occupations, the more occupational identities merge with sexual identities. Scott examines four very different societies — Britain, Peru, Egypt and Ghana to demonstrate to two basic points in her argument. The first is that even in advanced industrialized societies the market is subject to political and ideological influences; the second that political and ideological influences on the market derive from the high degree of linkage between economic and non-economic structures and institutions. 'It shows how, as a result of these linkages, gender is incorporated into the division of labour and manifests itself in gender segregation and the structure of gendered occupations . . . Patterns of gender segregation are therefore constructed socially and historically and cannot be deduced from a universal economic and biological logic. (p. 155)'

We have some examples of the ways in which employers have used technological change to restructure work processes and their employment of women in this connection (e.g. Hall, 1982; Glucksman, 1986). We have seen, too, collusion between male workers, trade unions and employers in the restructuring and gendering of work to male advantage.[16] Male power realized through the workplace can also be seen operating in the gendered construction of technological and other skills. A number of social analysts have discussed the ways in which technology and technical know-how are constructed, maintained and reproduced as male in all of the areas in which they impinge on our lives (Faulkner and Arnold, 1985; McNeil, 1987) but especially in the workplace (Game and Pringle, 1984; Cockburn, 1985; Faulkner and Arnold, 1985).

Two of the definitive male attributes which underpin the gendered definition of work are skill and strength (Holland, 1988), both deeply ideologically saturated, socially constructed concepts. Skill has become increasingly identified as a social rather than a technical phenomenon, with women's jobs classified as unskilled and men's as skilled or semi-skilled, regardless of the amount of training or ability required to perform them. It is the sex and power of the workers who do the work which dictates the skill level accorded to it (Phillips and Taylor, 1980; Beechey, 1988).

Phillips and Taylor give a fascinating example in the clothing industry from an unpublished paper by Birnbaum. Birnbaum argues that machining has been done by both men and women throughout this century, but when it was done by men it has been classified as skilled, when by women as semi-skilled and this distinction has survived numerous re-definitions of the basis for skill classification. In Birnbaum's view the origin of the distinction lay in the struggle of male immigrants to retain their status within the family when excluded as immigrants from 'skilled' work which they would otherwise have done. When forced to undertake the 'semi-skilled' work of machining, more usually done by women, these men preserved their masculinity and their social status within the family by struggling to re-define and maintain *their* machining as skilled labour.

Historical, current and cross-cultural analysis have demonstrated that gender divisions and hierarchy separate women and men in the home and in the workplace.[17] In the material which we have reviewed so far we have seen certain explanations for these phenomena put forward. We now turn to a consideration of some of these explanations.

Explanations for Gender Divisions

A range of explanations has been put forward for women's labour market position and more generally for gender inequality. Economic theories, such as deskilling or dual labour market theories, tend to focus on production and the workplace. Feminist explanations, placing women at the forefront of the analysis, are concerned with change, and have taken the following forms:

1. They concentrate on either patriarchy (gender relations) *or* capitalism (class relations) as the fundamental area in which these inequalities are created, maintained and reproduced.

2. They suggest two systems of social relations, which are sometimes seen as different modes of production: patriarchy and capitalism. These two systems are regarded either as (a) totally integrated; *or* (b) separate, autonomous but inter-related. (Holland 1989).

Dual market theory (see for example Barron and Norris, 1976) postulates two distinct sectors in the labour market which have the following characteristics:

1. a primary sector, with high wages, fringe benefits, skilled work, opportunities for training and promotion, employment stability and high unionization;

2. a secondary sector with low wages, few fringe benefits, unskilled work with no opportunity for promotion, employment instability and lack of unionization.

Women and various ethnic groups are found in the secondary sector. More recent versions examine the new forms of flexibility used by employers to reduce costs, using the notions of core and peripheral workers rather than primary or secondary workers.

Dual labour market theory has been roundly criticized — it is descriptive, provides no explanation for the growth of segmented labour markets and the actual organization of the labour process. Versions which focus on employer strategies ignore the impact of organized labour on rational market strategies which has contributed to the positioning of particular groups in the secondary sector. The explanations for women's position in the secondary sector are inadequate, indeed the definition of secondary sector workers is inadequate and the allocation of women to a heterogeneous category of secondary workers, submerges important differences between predominantly female occupations and the nature of horizontal and vertical segregation. But the major problem in relation to gender is that the dynamics of the labour market are assumed to be the main factor determining the position of female labour and no link is made between the structure of the labour market and the sexual division of labour, between production and reproduction. In general, deskilling and dual labour market theories largely conceptualize production from the perspective of skilled male workers and ignore gender.[18] It will have become clear from the earlier discussion that these theories also have problems with their conceptualization of skill, regarded as a technical, measurable attribute. This ignores the ideological and power components in the construction of skill categories and classifications which many feminist analysts have identified. These theories attempt to operate with a notion of production and the labour market as sexually neutral, a postulate which the weight of evidence reviewed in this paper indicates to be demonstrably untenable (see Scott, 1986; Beechey and Perkins, 1987).

Some feminist explanations which concentrate on patriarchy pay little attention to production and the labour market. Radical feminist arguments regard patriarchy as an autonomous system with the emphasis on male power and control over women. Particular aspects of women's oppression in the area of reproduction are emphasized — for example childbirth, abortion, motherhood. This emphasis on biological reproduction is taken one stage further by revolutionary feminists who, following Firestone (1970), develop a theory of patriarchy and sex-class based on male ownership of and control over women's reproductive capacities. They argue that there are two systems of social class, one the economic class system based on relations of production and the other the sex-class system, based on relations of reproduction. Patriarchy refers to this second system where women are subordinated to male control over their reproductive power (Jeffreys, 1977) and it is the constancy of this power which provides the unchanging basis of patriarchy.

More recently, radical feminists have argued that male power and control over female sexuality is a crucial mechanism of women's oppression, that male violence against women is an essential instrument in maintaining that control and that both of these form an essential part of the construction of masculinity. The extent and ubiquity of male violence against women and the power that the threat of violence — from wife battering to rape — has to control women's lives has been detailed, and the role of the state and the law in maintaining male control over women through male threat and use of violence described, particularly in the study of rape, and wife battering (Brownmiller, 1976; Dobash and Dobash, 1980; Stanko, 1985; Hanmer and Saunders, 1984). Many radical feminist formulations have been criticized for basing the explanation of women's subordination in biology and for claims of universality, which limit the potential for change. But more recent versions incorporate notions of the social construction of sexuality (Mahoney, 1985).

Socialist feminists have been concerned to make the link between the labour market and the sexual division of labour and draw on a number of Marxist concepts to elucidate the way in which the sexual division of labour is functional for capital in terms both of production and of their domestic labour. The domestic labour debate for example, which was an attempt to conceptualize the material position of women as housewives, drew attention to the links between work inside the home and outside, indicating the contradictory demands on women as domestic labourers and as workers selling their labour power in the market place.[19]

Feminists who see patriarchy and capitalism as autonomous systems and analytically independent, specify the ways in which the two systems inter-relate. One form argues that different spheres of society are determined by either patriarchal or capitalist relations. Capitalist relations are usually assigned to production or the economy; patriarchy may be confined to the sphere of ideology, culture and sexuality (Mitchell, 1975) or to reproduction and the family (Delphy, 1977; O'Brien, 1981). Another form argues that patriarchal and capitalist relations articulate at all levels and in all spheres of society (Hartmann, 1979, 1981; Walby, 1986, 1987).

Hartmann for example (1979, 1981) provides a powerful analysis of gender inequality in terms of the inter-relationship of patriarchy and capitalism. Patriarchy and capitalism should be seen as separate structures which have had important effects on each other historically but which can be analytically separated, although in contemporary western society they operate in partnership. Hartmann provides a historical analysis of the interaction of patriarchy and the processes of capitalism focusing on the role of male workers in restricting women's participation in the labour market, using examples from both the UK and USA. For Hartmann the crucial factors in this process are job segregation by sex and the family wage, which have been used by patriarchy and capitalism in interaction, combining to exclude women from much paid work. Occupational sex segregation has lowered the wages of women in the jobs which do remain open to them and forced them to remain dependent on men within the family.

Walby (1986) pushes the specification of the relationship between patriarchy and capitalism further, arguing that it should be seen as historically and spatially variable and riddled with conflict. Her thesis is demonstrated in a detailed analysis of gender relations in paid work in the UK from 1800 to the present day to which we have referred earlier and which focuses on three industries: cotton textiles, engineering and clerical work. Overlapping rounds of restructuring in employment are examined, indicating the two main strategies by which men and patriarchy retain control: the exclusion of women from paid employment; and confining women to jobs which are graded lower than those of men.[20] Walby reverses the usual linkage made between the labour market and family situation of women, which suggests that their subordinate position in the labour market derives from their role in the family, by arguing that it is because women are excluded and downgraded in the labour market that they are forced into dependence in the family.

Whichever way the link is made, this is the crucial connection which has to *be* made. The position of women in production and reproduction, the material and ideological components which contribute to the construction and maintenance of gender division and hierarchy in production and reproduction in a patriarchal capitalist society, are what have to be described and explained. All of the material from feminist scholarship which has been reviewed here contributes to this description and to the refinement of the analysis of gender divisions. It looks ultimately to change and remove gender inequality; but what are the chances?

Where do we go from here?

The current situation could be interpreted as one of flux; flux at the level of understanding women's position and gender divisions, with feminist theory standing on the threshold of a break through the barrier of the relationship between patriarchy and class into an embracing theory of gender; and perhaps more importantly and not unrelated, flux in the actual situation of women. We could see the analyses and descriptions which have been discussed in this chapter as endlessly depressing, a constant reminder of the durability of women's subordinate, oppressed and exploited condition. On the other hand we could see the descriptions of the points of change in the content of gender division as not only revealing of the ways in which gender hierarchy can be maintained, but also of how it can be changed.

Hagen and Jensen (1988) argue on the basis of a collection of articles describing women's position in seven countries[21] that 'women's situations are neither eternal facts of nature nor social constructs given once and for all. Rather, there is a state of flux' (p. 13) and that 'The final outcome will be the product of choices made by all the actors involved, including women themselves and the institutions within which they act and which help to structure their everyday lives' (p. 11), although they do admit that the current situation is currently bleak. But economic history since the Second World War[22] has demonstrated an increasing feminization of the labour force, coming out of a period of sustained economic growth and stability followed by one of turmoil, economic crisis and restructuring. Three basic themes emerge, (a) that there is considerable diversity in women's actual work experience, but that increasing labour market participation is occurring in these countries regardless of whether the context encourages or discourages women's participation, (b) that there is a state of confusion because increasing female participation does not occur alone, affecting other relatively stable factors, but in a situation where all factors are changing and interacting including structural economic changes in the countries involved, (c) that there is a need to rethink the basic categories with which women's work has been analyzed in the past which have focused on men and relegated women to the margin or the family.

Beechey (1988) building on the analyses of gender made by Harding (1986) and J. W. Scott (1986) has thoughtfully outlined what a theory of gender would need to address. These themes have resonated through the discussion in this chapter and lightly paraphrased they are:

1. The separation of public and private spheres. What is its basis and consequences for women and men?
2. What are the reasons for job segregation and for its persistence in different cultural and social settings?
3. How do workplace cultures operate and how do they contribute to the maintenance of inequality between women and men?

4. How are people's identities bound up in the public-private split and gender relations, and how are subjectivities created?
5. What are the links between changes in the organization of work and sexual definition of jobs, and changing representations of work?
6. What is the relationship between gender division and other forms of division, for example race, ethnicity and class?

In constructing this theory feminists too must follow J. W. Scott's (1986) advice to historians and scrutinize our methods of analysis, clarify our assumptions, and look for the source of change in interconnected processes which may not be amenable to disentanglement. We must recognize that power is not necessarily unified, coherent and centralized, but 'dispersed constellations of unequal relationships' which leaves a space for human agency, in contrast to the overdetermined conceptualization of women that some of the material we have considered here has suggested.

Notes

1 A wage sufficient to keep a wife at home, not engaged in paid work outside the home.
2 Humphries (1977) defends working-class demands for a family wage as a defensive strategy against capitalist exploitation.
3 This work involves reproducing the labour force physically, emotionally and ideologically both inter-generationally (childbearing and rearing), and intro-generationally (servicing husband and children).
4 See below for a discussion of skill in relation to women's work.
5 Government statistics are available in the 1931 Census of Population, the 1935 Census of Production and in Ministry of Labour employment figures.
6 Adult pay rates started at 18, female youth labour was particularly cheap and employers preferred 14–18 year olds.
7 See Coyle (1982) for an analysis of deskilling, technical change and the rationalization of the labour process and the use of women's cheap labour in the clothing industry. Coyle argues that management strategies to exert downward pressure on wages combined with union strategies to resist that, reinforce sexual divisions in the labour process.
8 The economic activity rate is the number of economically active (employed, or unemployed but seeking employment) as a percentage of the total population in that age group.
9 60 per cent of women manual workers are in 'catering, cleaning, hairdressing and other personal services'; 15 per cent in 'painting, repetitive assembling, product packaging and related; and 11 per cent in 'making and repairing (excluding metal and electrical)'. Of non-manual women workers, 53 per cent are in 'clerical and related'; 27 per cent in 'professional and related work in education, welfare and health'; and 12 per cent in 'selling' (Holland et. al., 1985; Holland, 1981). These proportions have remained relatively stable over time (see successive EOC Annual Reports for statistics on women's employment, including teaching).
10 Examples at random are doctors, 23 per cent women, chartered accountants, 6 per cent and architects, 4 per cent, professors in higher education 2 per cent (figures from professional associations, (Blackman et. al., 1987)). Of 160 top jobs in the BBC, 6 are held by women (radio discussion, 1988); of 635 Members of Parliament, 41 are women.
11 The Women and Employment Survey (see Martin and Roberts, 1984) demonstrated that there

was a large peak in women's part-time hours at 16 hours per week, above which level employers become liable for more overhead costs associated with the work (Dex, 1987) and employment rights accrue to the worker. Robinson (1988) demonstrates that a growing proportion of women is below the threshold for entitlement for such employment rights as maternity benefits, redundancy payment, and protection against unfair dismissal.

12 Beechey and Perkins (1987) found that employers only created part-time jobs when they hired women. Where men were employed, they used other strategies, for example the use of overtime, short-time working and temporary employment contracts. See too Buswell (1987, 1988) who argues that employers use the pattern of women's labour force participation (with a break for childrearing) and the availability of cheap youth labour through YTS in a process of restructuring labour markets, drawing young unmarried women and older married women into the labour market as needed.

13 Disabling measures are 'Changes in firms' or the State's behaviour which make it more difficult for women to combine paid work and domestic family responsibilities' (Humphries and Rubery, 1988 p. 85, 86). These authors outline legislative changes and government expenditure cuts in areas such as education and health which have had a disabling effect on women's capacity to work.

14 Gendered differences in education and training continue through YTS, Further and Higher Education (Marsh, 1986; Wickham, 1985; Cockburn, 1987; Stoney and Reid, 1981; Stoney, 1984; Tomlinson, 1983). Grass roots and other attempts to construct and promote non-sexist or even anti-sexist education have grown in the last decade. See for example Cornbleet and Libovitch, 1983; Whyte et. at., 1985; Weiner, 1985; Blackman et. al., 1987; Burchell and Millman, 1989).

15 As women move into medical physics they are allocated to jobs in 'nuclear medicine' concerned with contact with patients; men fill the newer slots which are involved with the technology of the machines.

16 See Ellis (1988) for recent trade union attempts to counter gender segregation in work.

17 For examples of cross cultural data see Holland (1983), Davidson and Cooper (1984).

18 Dex (1987) has developed a more elaborated model of the segmentated labour market for women, in which women appear in a primary non-manual sector (as teachers) and in sexually segregated women's primary and secondary sectors. She describes women's mobility between these sectors over the working life on the basis of the Women and Employment Survey.

19 For a discussion of the domestic labour debate and a collection of articles from it see Molyneux (1979) and Malos (1980).

20 See Cockburn (1985) for a discussion of the interaction between the processes of gendering and subdivision of work during technological change, which reproduces male dominance. As McNeil (1987) puts it, when women appear to be replacing men, you can almost always be sure that the job has been redesigned and devalued (in pay, status, conditions) in some way.

21 Britain, Canada, France, the Federal Republic of Germany, Italy, Sweden and the US.

22 In the western industrialized countries covered in their volume: Jensen, Hagen and Reddy, (1988).

References

ALEXANDER, S. (1976) 'Women's work in nineteenth-century London: a study of the years 1825–50', in Oakley, A. and Mitchell, J. (Eds) *The Rights and Wrongs of Women*, Harmondsworth, Penguin.

ALTHUSSER, L. (1971) *Lenin and Philosophy and other essays*, London, New Left Books.

AMOS, V. and PARMAR, P. (1984) 'Challenging imperial feminism', *Feminist Review* 17, pp. 3–19.

ARNOT, M. (1982) 'Male hegemony, social class and women's education', *Boston University Journal of Education*, 164, pp. 64–89.

ARNOT, M. (1986) 'State education policy and girls' educational experiences', in Beechey, V. and Whitelegg, E. (Eds.) *Women in Britain today*, Milton Keynes, Open University Press, 1986.

BARRON, R. D. and NORRIS, G. M. (1976) 'Sexual divisions and the dual labour market', in Barker, L. D. and Allen, S. (Eds.) *Dependence and exploitation in work and marriage*, London, Longman.

BARRETT, M. and McINTOSH, M. (1982) *The anti-social family*, London, Verso.

BEECHEY, V. (1988) 'Rethinking the definition of work: Gender and work', in Jenson, J., Hagen, E. and Reddy, C. (Eds.) *Feminization of the labour force: Paradoxes and promises*, Cambridge, Polity Press.

BEECHEY, V. and PERKINS, T. (1987) *A matter of hours: women, part-time work and the labour market*, Cambridge, Polity Press.

BERNSTEIN, B., (1977) *Class, codes and control: Vol. 3. Towards a theory of educational transmissions*, 2nd ed. London, Routledge and Kegan Paul.

BHAVNANI, K. K. and COULSON, M. (1986) 'Transforming socialist-feminism: The challenge of racism', *Feminist Review*, 23, pp. 81–92.

BIRNBAUM, B. *Women, skill and automation: A study of women's employment in the clothing industry 1946–1972*, unpublished paper.

BLACKMAN, S. J., CHISHOLM, L. A., GORDON, T., HOLLAND, J. (1987) *Hidden messages: An equal opportunities teachers' pack*, Oxford, Blackwell, 1987.

BLACKMAN, S. J. and HOLLAND, J. 'Equal opportunities in education and training', *Comprehensive Education*, No. 51, Spring, 1989.

BOURDIEU, P. and PASSERON, J. C. (1977) *Reproduction in education, society and culture*, London, Sage.

BOURNE, P. G. and WIKLER, N. J. (1978) 'Commitment and the cultural mandate: Women in medicine', *Social Problems*, 25, pp. 430–440.

BOWLES, S. and GINTIS, H. (1976) *Schooling in capitalist America: Educational reform and the contradictions of economic life*, New York, Basic Books.

BROWNMILLER, S. (1976) *Against our will. Men, women and rape*, Harmondsworth, Penguin.

BRYAN, B., DADZIE, S. and SCAFE, S. (1985) *The heart of the race: Black women's lives in Britain*, London, Virago.

BURCHELL, H. and MILLMAN, V. (Eds.) (1989) *Changing perspectives on gender: New initiatives in secondary education*, Milton Keynes, Open University Press.

BUSWELL, C. (1987) *Training for Low Pay* in Millar and Glendinning qv.

BUSWELL, C. (1988) 'Flexible workers for flexible firms?' in Pollard, A., Purvis, J. and Walford, G. *Education training and the New Vocationalism: Experience and policy*, Milton Keynes, Open University Press.

CARBY, H. V. (1982) 'Schooling in Babylon', in CCCS *The Empire strikes back*, Hutchinson University Library & CCCS.

CHILD, J., LOVERIDGE, R., HARVEY, J. and SPENCER, A. (1985) 'Microelectronics and the quality of employment in services', in Marstrand, P. (Ed.) *New technology and the future of work and skills*, London, Frances Pinter.

CHISHOLM, L. A. and HOLLAND, J. (1986) 'Girls and Occupational Choice: Anti-sexism in action in a curriculum development project', *British Journal of Sociology of Education*, 7, 4.

CLARRICOATES, K. (1980) 'The importance of being Ernest...Emma...Tom...Jane...The perception and categorisation of gender conformity and deviation in primary schools', in Deem, R. (Ed.) *Schooling for women's work*, qv.

COCKBURN, C. (1985) *Machinery of dominance: Women, men and technical know-how*, London, Pluto Press.

COCKBURN, C. (1986) 'Sixteen: Sweet or sorry?' *Marxism Today*, December, pp. 30–33.

COCKBURN, C. (1987) *Two-track training: Sex inequalities and the YTS* London: Macmillan.

COCKBURN, C. (1988) 'The gendering of jobs: Workplace relations and the reproduction of sex segregation', in Walby, S. (Ed.) *Gender segregation at work*, Milton Keynes, Open University Press.

COOK, J. and WATT, S. (1987) 'Racism, women and poverty', in Millar, J. and Glendinning, C. (Eds.) *Women and Poverty in Britain*, Sussex, Wheatsheaf Books.

CORNBLEET, A., and LIBOVITCH, S. (1983) 'Atni-sexist initiatives in a mixed comprehensive school: a case study', in Wolpe and Donald (Eds.) *Is there anyone here from education?* London, Pluto Press.

COYLE, A. (1982) 'Sex and skill in the organisation of the clothing industry', in West, J. (Ed.) *Women, work and the labour market*, London, Routledge and Kegan Paul.

CRAGG, A. and DAWSON, T. (1981) *Qualitative research among homeworkers*, Research Paper No. 21, London, Department of Employment.

DAUNE-RICHARD, A-M, (1988) 'Gender relations and female labour: A consideration of sociological categories', in Jensen, J., Hagen, E., and Reddy, C. (Eds.) *Feminization of the labour force: Paradoxes and promises*, London, Polity Press.

DAVIDSON, M. J. and COOPER, C. L. (Eds.) 1984 *Working women: An international survey*, London, Wiley.

DEEM, R., (1978) *Women and Schooling*, London, Routledge and Kegan Paul.

DEEM, R. (1980) *Schooling for women's work*, London, Routledge and Kegan Paul.

DELAMONT, S. (1980) *Sex roles and the school*, London, Methuen.

DELPHY, C. (1977) *The main enemy: a materialist analysis of women's oppression*, London, WRRC Explorations in Feminism No. 3.

DEX, S. (1987) *Women's occupational mobility: A lifetime perspective*, London, Macmillan.

DOBASH, R. and DOBASH, R. (1980) *Violence against wives*, London, Open Books.

ELLIS, V. (1988) 'Current trade union attempts to remove occupational segregation in the employment of women', in Walby, S. (Ed.) *Gender segregation at work*, Milton Keynes, Open University Press.

FAULKNER, W. and ARNOLD, E. (Eds.) (1985) *Smothered by invention: Technology in women's lives*, London, Pluto Press.

FIRESTONE, S. (1970) *The dialectic of sex: The case for feminist revolution*, New York, Morrow.

FULLER, M. (1980) 'Black girls in a London comprehensive school', in Deem, R., (Ed.) *Schooling for women's work*, London, Routledge and Kegan Paul.

GAME, A. and PRINGLE, R. (1984) *Gender at work*, London, Pluto Press.

GEORGE, D. (1930) *London life in the eighteenth century*, London, Kegan Paul.

GIDDINGS, P. (1984) *When and where I enter: The impact of Black women on race and sex in America*, New York, Bantam Books.

GLENDINNING, C. and MILLAR, J. (Eds.) (1987) *Women and Poverty in Britain*, Sussex, Wheatsheaf Books.

GLUCKSMAN, M. (1986) 'In a class of their own? Women workers in the new industries in inter-war Britain', *Feminist Review*, 24, pp. 7–37.

GREATER LONDON COUNCIL, (1985) *The London industrial strategy*, London:

HAGEN, E. and JENSON, J. (1988) 'Paradoxes and promises: Work and politics in the postwar years', in Jenson, J., Hagen, E. and Reddy, C. (Eds.) *Feminization of the labour force: Paradoxes and promises*, Cambridge, Polity Press.

HAKIM, C. (1979) *Occupational segregation: A comparative study of the degree and pattern of the differentiation between men and women's work in Britain, the United States and other countries*, Research Paper No. 9, Department of Employment.

HAKIM, C. (1984) 'Homework and outwork', *Employment Gazette*, 92, 1, pp. 7–12.

HALL, C. (1982) 'The home turned upside down? The working-class family in cotton textiles 1780–1850', in Whitelegg, E. *et. al.* (Eds.) *The changing experience of women*, Oxford, Martin Robertson.

HANMER, J. and SAUNDERS, S. (1984) *Well-founded fear: A community study of violence to women*, London, Hutchinson.

HARDING, S. (1986) *The science question in feminism*, Milton Keynes, Open University Press.

HARTMANN, H. (1979) 'Capitalism, patriarchy and job segregation by sex', in Eisenstein, Z. R. (Ed.) *Capitalist patriarchy and the case for socialist feminism*, New York, Monthly Review Press, 1979.

HARTMANN, H. (1981) 'The unhappy marriage of Marxism and feminism: towards a more progressive union', in Dale, R., Esland, G., Fergusson, R. and MacDonald, M. (Eds.) *Education and the state Vol. 2: Politics, patriarchy and practice* Lewes, Falmer Press and Open University Press.

HARVEY, J. (1987) 'New technology and the gender divisions of labour', in Lee, G. and Loveridge, R. *The manufacture of disadvantage: Stigma and social closure*, Milton Keynes, Open University Press.

HOLLAND, J. (1981) *Work and women*, Bedford Way Paper 6, London: Institute of Education.

HOLLAND, J. (Ed.) (1983) *Bibliographic guide to studies on the status of women: Development and population trends*, Paris: Bowker/UNIPUB/Unesco.

HOLLAND, J. (1988) 'Girls and occupational choice: In search of meanings' in Pollard, A., Purvis, J. and Walford, G. *Education, training and the New Vocationalism: Experience and policy*, Milton Keynes: Open University Press.

HOLLAND, J. (1989) 'Gender in Britain today', in Cole, M. (Ed.) *The social contexts of schooling*, Falmer, 1989.

HOLLAND, J., BLACKMAN, S. J., GORDON, T. and TEACHER TEAM, (1985) *A woman's place: Strategies for change in the educational context*, GAOC Working Paper No. 4, Institute of Education, London.

HUDSON, D. (Ed.) (1972) *Munby, man of two worlds*, London, J. Murray.

HUMPHRIES, J. (1977) 'Class struggle and the persistence of the working class family', *Cambridge Journal of Economics*, 1, 1, pp. 241–258.

HUMPHRIES, J. and RUBERY, J. (1988) 'Recession and exploitation: British women in a changing workplace, 1979–1985', in Jenson, J., Hagen, E. and Reddy, C. (Eds.) *Feminization of the labour force: Paradoxes and promises*, Cambridge, Polity Press.

HUTCHINS, B. L. (1915) *Women in modern industry*, London, G. Bell and Sons.

JEFFREYS, S. (1977) 'Sex-class — Why is it important to call women a class?' (with J. Hanmer, C. Lunn and S. McNeill); 'The need for revolutionary feminism', 'Male sexuality as social control', all in *Scarlet Women Five*, North Shields, Tyne and Wear.

JENSON, J., HAGEN, E. and REDDY, C. (Eds.) (1988) *Feminization of the labour force: Paradoxes and promises*, Cambridge, Polity Press.

JOSHI, H. and OWEN, S. (1981) *Demographic indicators of women's work participation in post-war Britain*, London, Centre for Population Studies.

KELLY, A., (Ed.) (1981) *The missing half*, Manchester, Manchester University Press.

KELLY, A. (1985) 'Changing schools and changing society: some reflections on the Girls Into Science and Technology project', in Arnot, (Ed.) *Race and gender: Equal opportunities policies in education*.

KELLY, A. et. al. (1986) 'Gender roles at home and school', in Burton, L. (Ed.) *Girls into maths can go*, London: Holt, Rinehart and Winston.

KESSLER, S., ASHENDEN, D. J., CONNELL, R. W. and DOWSETT, G. W. (1985) 'Gender relations in secondary schooling', *Sociology of Education*, 58, pp. 34–48.

LAND, H. (1976) 'Women: supporters or supported', in Leonard Barker, D. and Allen, S. (Eds.) *Sexual divisions and society: Process and change*, London, Tavistock.

LAZONICK, W. H. (1976) *Historical origins of the sex based division of labour under capitalism: A study of the British textile industry during the Industrial Revolution*, Harvard Institute of Economic Research Discussion Paper No. 497.

LEWIS, J. (1984) *Women in England 1870–1950*, Sussex, Wheatsheaf Books.

LIFF, S. (1986) 'Technical change and occupational sex-typing', in Knights, D. and Willmott, H. (Eds.) *Gender and the labour process*, Aldershot, Gower.

LONSDALE, S. (1987) 'Patterns of paid work', in Glendinning, C. and Millar, J. (Eds.) *Women and Poverty in Britain*, Sussex, Wheatsheaf Books.

LOVERIDGE, R. (1987) 'Social accommodations and technological transformations: The case of gender', in Lee, G. and Loveridge, R. *The manufacture of disadvantage: Stigma and social closure*, Milton Keynes, Open University Press.

LOW PAY UNIT, (1984) *Sweated Labour*, London.

LOWN, J. (1983) 'Not so much a factory, more a form of patriarchy: gender and class during industrialisation', in Gamarnikow, E. *et. al.* (Eds.) *Gender, class and work*, London, Heinemann Educational Books.

MAC AN GHAILL, M. (1988) *Young, gifted and black*, Milton Keynes, Open University Press.

MACDONALD, M. (1980) 'Socio-cultural reproduction and women's education', in Deem, R. (Ed.) *Schooling for women's work*, London: Routledge and Kegan Paul.

MACDONALD, M. (1981) *Class, gender and education*, Milton Keynes, Open University Press.

McNEIL, M. (Ed.) (1987) *Gender and expertise*, London, Free Association Books.

MAHONEY, P. (1985) *Schools for the boys? Co-education reassessed*, London, Hutchinson and The Explorations in Feminism Collective.

MALOS, E. (1980) *The politics of housework*, London, Allison and Busby.

MAYHEW, H. (1981) *London labour and the London poor*, 4 volumes.

MARK-LAWSON, J. and WITZ, A. (1988) 'From family labour to family wage? The case of women's labour in 19th century coalmining', *Social History*, 13, 2.

MARSH, S. (1986) 'Women and the MSC', in Benn, C. and Fairley, J. (Eds.) *Challenging the MSC on jobs education and training*, London, Pluto Press.

MARTIN, J. and ROBERTS, C. (1984) *Women and employment: A lifetime perspective*, London: HMSO.

MIDDLETON, C. (1988) 'Gender divisions and wage labour in English history', in Walby, S. (Ed.) *Gender segregation at work*, Milton Keynes, Open University Press.

MILLAR, J. and GLENDINNING, C. (1987) 'Invisible women, invisible poverty', in Glendinning, C. and Millar, J. (Eds.) *Women and Poverty in Britain*, Sussex, Wheatsheaf Books.

MITCHELL, J. (1975) *Psychoanalysis and Feminism*, Harmondsworth, Penguin.

MITTER, S. (1986) *Industrial restructuring and manufacturing homework: Immigrant women in the UK clothing industry*, Capital and Class, Winter.

MOLYNEUX, M. (1979) 'Beyond the domestic labour debate', *New Left Review*, 116 (July-August), pp. 3–27.

O'BRIEN, M. (1981) *The politics of reproduction*, London, Routledge and Kegan Paul.

ORMEROD, M. B. (1975) 'Subject preference and choice in co-educational and single sex secondary schools', *British Journal of Educational Psychology*, 45, p. 257–267.

PARMAR, P. (1982) 'Gender, race and class: Asian women in resistance', in Centre for Contemporary Cultural Studies, *The Empire strikes back*.

PINCHBECK, I. (1981) *Women Workers and the Industrial Revolution*, 1750–1850, London, Virago.

PHILLIPS, A. and TAYLOR, B. (1980) 'Sex and skill: Notes towards a feminist economics', *Feminist Review*, 6, pp. 79–88.

PHIZACKLEA, A. (1988) 'Gender, racism and occupational segregation', in Walby, S. (Ed.) *Gender segregation at work*, Milton Keynes, Open University Press.

PRATT, J., BLOOMFIELD, J. and SEALE, C. (1984) *Option choice: A question of equal opportunity*, Oxford, NFER/Nelson.

RAMAZANOGLU, C. (1989) *Feminism and the contradictions of oppression*, London, Routledge.

ROBINSON, O. (1988) 'The changing labour market: Growth of part-time employment and labour market segmentation in Britain', in Walby, S. (Ed.) *Gender segregation at work*, Milton Keynes: Open University Press.

ROBINSON, O. and WALLACE, J. (1984) *Part-time Employment and Sex Discrimination Legislation in Britain*, Research Paper No. 43, London, Department of Employment.

ROYAL COMMISSION ON THE DISTRIBUTION OF INCOME AND WEALTH, (1978) Research Report 6, *Lower Incomes*, London, HMSO.

ROUTH, G. (1965) *Occupation and pay 1906–1960*, Cambridge, Cambridge University Press.

RUBERY, J. and WILKINSON, F. (1981) 'Outwork and segmented labour markets', in Wilkinson, F. (Ed.) *The dynamics of labour market segmentation*, London, Academic Press.

SCOTT, A. M. (1986) 'Industrialisation, Gender Segregation and Stratification Theory', in Crompton, R. and Mann, M. (Eds.) *Gender and stratification*, Cambridge, Polity Press.

SCOTT, J. W. (1986) 'Gender: a useful category of historical analysis', *The American Historical Review*, 91, 5, December.

SNELL, M. (1979) 'The Equal Pay and Sex Discrimination Acts: Their impact in the workplace', *Feminist Review*, 1, pp. 37–57.

SPENCER, A. and PODMORE, D. (1987) *In a man's world: Essays on women in male-dominated professions*, London, Tavistock.

STANKO, E. A. (1985) *Intimate intrusions: Women's experience of male violence*, London, Routledge and Kegan Paul.

STANWORTH, M. (1981) 'Gender and Schooling: A study of sexual divisions in the classroom', *Explorations in Feminism* No. 7, WRRC.

STONEY, S. M. (1984) *Girls entering science and technology: The problems and possibilities for action as viewed from the FE perspective*, paper given at the EOC/Manchester Polytechnic Conference, Girl-Friendly Schooling, September.

STONEY, S. M. and REID, M. I. (1981) *Balancing the equation: A study of women and technology within FE*, FE Unit, London.

THOROGOOD, N. (1987) 'Race, class and gender: The politics of housework', in Brannen, J. and Wilson, G. (Eds.) *Give and Take in Families: Studies in resource distribution*, London, Allen and Unwin.

TOMLINSON, S. (1983) 'Black women in higher education — case studies of university women in Britain' in Barton, L. and Walker, S. (Eds.) *Race, class and education*, London, Croom Helm.

WALBY, S. (1986) *Patriarchy at work*, Cambridge, Polity Press.

WALBY, S. (1987) *Gender and inequality*, Oxford, Basil Blackwell.

WALBY, S. (Ed.) (1988) *Gender segregation at work*, Milton Keynes, Open University Press.

WALKER, J. (1988) 'Women, the state, and the family: The case of the UK', in Rubery, J. (Ed.) *Women and recession*, London, Routledge.

WALTERS, P. A. (1987) 'Servants of the crown', in Spencer, A. and Podmore, D. *In a man's world: Essays on women in male-dominated professions*, London, Tavistock.

WEINER, G. (Ed.) (1985) *Just a bunch of girls: Feminist approaches to schooling*, Milton Keynes, Open University Press.

WHYTE, J. (1986) *Girls into Science and Technology*, London, Routledge and Kegan Paul.

WHYTE, J. et. al. (Eds.) (1985) *Girl-Friendly Schooling*, London, Methuen.

WIDDOWSON, F. (1980) *Going up into the next class: Women and elementary teacher training 1840–1914*, London, Hutchinson.

WICKHAM, A. (1985) 'Gender divisions, training and the state', in Dale, R. (Ed.) *Education, training and employment: Towards a new vocationalism*, Oxford, Pergamon Press.

WOLPE, A-M. (1977) 'Some processes in sexist education', *Explorations in Feminism* No. 1, London, WRRC.

Part Four: Introduction
School Mathematics in Context

There is no doubt that a major factor in learning mathematics is the context of activity of which mathematics is perceived to be a part.

The increasing trend towards vocationalism in school mathematics curricula has been traced through various papers in this volume. Such a trend, allied to vocationally-oriented mathematics courses carries with it assumptions and prejudices already present in mathematics, in mathematics education and in employment. Some initiatives in vocational education take on the differentiated nature of educational mathematics by the wholesale adoption of a concept of numeracy; others are concerned to build good mathematics from within an existing vocational context or from within a context that recognizes workplaces as mathematics resources. The authors of the papers in Chapters 21 to 24 discuss some of the issues raised when a particular focus on vocationalism comes into contact with the broad aims of mathematics education.

Sullivan describes the origin and evolution of the Certificate of Pre-Vocational Education (CPVE) in reactive initiatives deriving from the unemployment of the 1970s and the assumption that the ills of the British economy can be blamed directly on inappropriate school curricula. She illustrates the controversy that surrounds the formal introduction of vocationalism into school curricula and the competitive activities of validating bodies and other vested interests in what is a highly politicized field. CPVE has its enemies in the academic world, for example its attempt at bridging the academic/vocational divide by profiling both academic and non-academic attainment has been charged with being no more than a structuring of the bottom rung of a new tertiary education tri-partism. CPVE however does take on the idea of curriculum negotiation, relevant experience and in the curriculum, planning has taken account of educational concerns in both content and process of teaching and learning. In some courses at least, some of the problems raised by Drake in Chapter 22 have been recognized.

Drake points to danger in the workplace practice of vocationalism when the origin, content and pedagogy of the numeracy aspect of courses are planned, taught and assessed without reference to those engaged in mathematics education. The situation is made worse by the low status of numeracy and work-related mathematics within the profession of mathematics education itself. Drake's project responded to the need to supply good learning resources both for the use of Youth Training students and for developing the diagnostic and training skills of their trainers. Her research again emphasized the effects of context on mathematics performance. In the workplace context, mathematical performance was not problematic but out of context the same mathematics became extremely hard. Drake also notes that there is a clear relationship between the mathematics tests given to trainees and the status of jobs into which they are fed as a result, in contrast with the lack of relationship between the mathematics tested (and the method of testing it) and the mathematics that really goes on in the job; some higher status jobs actually involved less mathematics than some lower status jobs.

The Technical and Vocational Education Initiative (TVEI) was another vocational programme launched at about the same time as CPVE but by the Manpower Services Commission, a government funded body linked to the Department of Employment not the Department of Education. It pushed the age level of school vocational courses from the sixth form down to the age of 14 and the lavish funding it offered was a source of friction in schools more used to educational budgets. The *Maths + TVEI* project described by Grey however has its base firmly in mathematics education and its aim is to exploit the cross-curricular situation offered by TVEI to enrich problem-solving in mathematics. The project uses cross-curricular possibilities already in school to develop mathematics at its most challenging, when students learn to construct mathematical models in problem-solving situations. Grey again illustrates the tension between mathematics teaching and some technology teaching whose perception of mathematics is the application of skills learned out of context. The mathematics education emphasis of the *Maths + TVEI* Project challenges the skills applications model and its accepted implication that mathematics in vocationally-oriented work is low level.

This challenge is taken up by Harris and Paechter who reject the assumptions that 'non-academic' experiences are relevant only to low class and low ability students and that the intellectual content of such experience is also low. They discuss classroom activities in which mathematics is revealed in the work of low status groups; it is shown both through stereotyped non-academic work in the mathematics class and by revealing mathematics in low-status cross curricular activities in Design and Technology.

Part 4 ends on a positive note. In spite of the differential esteem attached to theoretical and practical work in both mathematics and employment, there *are* possibilities for emancipatory mathematics learning within the national curriculum.

The final paper in the volume is a postscript by Harris on some of the work of the Maths in Work Project a part of which is the origin of this book and its related in-service course. In it she offers the experience of the Project to anyone whose concern with *Schools, Mathematics and Work* is strong enough for them to want it to have an effect in practice.

21
The Certificate in Pre-vocational Education

Sheila Sullivan

CPVE in Action

The origins of the Certificate in Pre-Vocational Education (CPVE) lie in the Schools Council[1] Working Papers of 1972 and 1973 which identified the *New Sixth Former*. This was the 'non-academic' young person staying on at school and challenging the education system to provide a relevant curriculum. The Certificate of Extended Education (CEE) was developed for people so defined but rendered obsolete by later policy changes. Developments in further education in the 1970s gave rise to the City and Guilds of London Institutes[2] (CGLI) General Vocational Preparation (365) courses. Within the education system there was growing dissatisfaction that a single client group was being provided for by two different curricula and two forms of certification. A debate over what constituted the most appropriate solution was expressed in the recommendations of the *Keohane Report* (1979) which favoured the single subject examination approach of CEE, and *A Basis for Choice* (1989) which proposed a framework for the integrated provision of general and pre-vocational education. The Government, preferring the more vocational course, adopted the proposals of *A Basis for Choice* in its White Paper *17 + : a New Qualification* (1982) which described an embryonic version of CPVE. There followed a period of intense lobbying by interest groups, in particular the certificating bodies, CGLI, BTEC[3], and RSA[4], at the end of which CGLI and BTEC formed the Joint Board for Pre-Vocational Education. In 1984, the Joint Board produced its consultative documents and piloted CPVE. The following year CPVE was launched nationally and by 1988 over 100,000 students had received certification. Changes in CPVE took place from 1989 as a result of a national evaluation carried out in 1987/88.

CPVE in Action

Students who embark on a CPVE programme are usually sixteen-years-old and the majority of them are in a school sixth form. They want to continue in education because their age, qualifications, experience or state of certainty about their future careers prevents them from starting work, training or the further education course most suitable for them. While they may be interested in qualifications such as GCSE that can be taken alongside CPVE, they will be attracted by the vocational element in CPVE which includes a period of work experience. Sixteen-years-olds often find this vocational bias to be a source of motivation. These young people are a relatively new client group in the post compulsory phase of education in England and Wales. In the past, they would have gone into work maintaining their education, if at all, through day release or the adult education system.

Every school or college offering CPVE can determine its own organization. The time-table shown below (Figure 21.1) demonstrates the variety of experience in one student's work. In determining an individual time-table, the tutor and the student have to agree on targets for the student in the light of his or her particular needs and aspirations. Clearly the resources available constrain student choice.

	am	pm
Mon	Mini-company	GCSE English
Tue	Computing	Core Studies/Tutorial
Wed	Work experience	
Thur	Clerical Services	Core Studies
Fri	Numeracy Workshop	Book-keeping

Figure 21.1.

Within a CPVE scheme, activities such as mini-company, residentials, visits and projects abound. Such activities are designed to be stimulating and realistic while providing the opportunity for students to encounter and practise new skills and enter new fields of knowledge or experience. An activitiy such as running a fast-food service in the school or college has all the pressures of a real service industry but also delivers in an integrated way a wide range of vocational studies: catering, customer service, nutrition and health and safety at work. Core studies, which are always difficult to justify to young people, occur naturally in this sort of activity. The students appreciate the importance of working in a group, communicating and calculating. They can see science and technology in action in hygiene checks and they have to make plans and solve problems. They can also reflect upon the aptitudes they have shown during the project.

The basic structure of CPVE is quite simple. Every programme must contain vocational studies, core studies and additional studies.

Vocational studies are selected from a comprehensive list organized into five categories:

Business and Administrative Services
Technical Services
Production
Distribution
Services to People

Those most commonly offered are Services to People and Business and Administrative Services. Within each category there are three levels: introductory, exploratory and preparatory. At each level there is a more specific focus on vocational skills.

With Core Studies there are no options. All students must have the opportunity to experience the whole of the core. The CPVE was revised in 1989 and now comprises five areas:

Communication and Social Skills
Applied Numeracy
Problem Solving
Science, Technology and Information Technology
Social, Industrial and Economic Awareness

Within the general area of applied numeracy there is a further sub-division into three skills:

Collecting and Presenting Data
Analyzing and Interpreting Data
Calculating

The level of difficulty at which these skills are practised is provided by the context in which they are found. 'Calculating' for instance may refer to the simple addition of the prices of goods or to finding out the cost of decorating a room. Details of the core and its assessment are available from the Joint Unit for CPVE and Foundation Programmes. The CPVE core is now less comprehensive than before it was revised.

Additional studies are optional and may occupy up to 25 per cent of course time. Examples of options are; GCSEs[5], workshops in communication or numeracy, keyboarding, computing, sport and artistic or craft activity. The purpose of additional studies is to enable the student to pursue individual needs and interests outside the CPVE experience. They are separately certificated.

While vocational and core studies describe much of the content and skills covered in CPVE, the Joint Board for Pre-Vocational Education (now replaced by the

Joint Unit for CPVE and Foundation Programmes) prescribed further characteristics for CPVE programmes which deal largely with the processes of teaching and learning considered to be essential to CPVE. At the time this was novel although it was building upon good practice in both pre- and post-16 education. The CGLI 365 General Vocational Preparation course was an example of a curriculum that pre-figured many of the CPVE developments.

There are five prescribed characteristics of CPVE programmes.

1. Activity-based learning is the most important. It develops understanding through experience and enables the practical application of skills and knowledge. It is also an approach that encourages student autonomy.
2. All CPVE students must have real or simulated work experience. This will differ from the work experience of many fourth and fifth year pupils because it will be relevant to the students' vocational preferences.
3. Negotiation of programmes and assessment must be built into each student's programme through a system of guidance and support.
4. Assessment has to be continuous and formative, setting targets and reviewing progress throughout the course. The final reporting of student achievement is centred around a portfolio which demonstrates the students' best work.
5. In the organization of programmes, centres must ensure that core and vocational studies are integrated and are not perceived as different 'subjects'.

CPVE: Origins

The events which marked the evolution of CPVE have been noted above but what were the underlying social and educational forces which produced them? Throughout the 1970s in response to growing unemployment more young people were staying on in schools and colleges. For many, the most popular course was to 'repeat O level'. GCEs had an established currency with parents and students and could easily be provided by schools and colleges. The pass rate on such courses however was dismal, at around 39 per cent and amongst teachers there was growing concern for their value. At the same time research by NFER[6] (1982) showed that a vocationally focused curriculum was more motivating for students and led to more success for them. Students found subjects easier to cope with partly because of the increased relevance gained from the vocational focus. A 'pre-vocational' methodology was emerging in practice which included the concepts of *negotiation* of the curriculum and of assessment, *guidance* education within each student's programme, *relevance* of content to the student's interests and learning arising from *experience*. This methodology was seen as progressive by the teachers involved but there can be no doubt that colleges in particular found it difficult to adjust to the notion of student centred curriculum design.

As youth unemployment grew, there was a rapid increase in courses to meet the needs of this client group, from the GCSE boards, CGLI, BTEC and RSA. Given the impending decline in the cohort and the confusion regarding the boundaries between these courses among students, parents and employers (not to mention teachers) the proliferation was unsatisfactory and some rationalization was necessary.

These educational pressures existed in an economic and political landscape in which the transformation of education was to be an important feature. A major strategy in this transformation was the vocationalization of the curriculum by a variety of methods such as mini-enterprises, education-industry links and the more complex TVEI[7]. The assumption underlying this strategy is that the performance of the British economy had suffered because education was inappropriate for the young in terms of the skills and attitudes which they took from school to employment. This is an essentially ideological stance which ignored economic and political issues such as technological change, regional policy, the oil crisis and low investment by both the private and public sector in continuing training and education. From this, the argument moved on to place the blame for industrial decline and unemployment upon the educational system. The government was therefore in favour of a curriculum which had a vocational focus.

CPVE: The Future

Controversy attended the birth of CPVE and during its short life, the changing environment has created further pressure upon it. Since its launch, the market for CPVE has shrunk because of increasing levels of employment and declining numbers of 16-year-olds. Given these conditions, it is surprising, to say the least, that new awards, most notably the BTEC First Diplomas have been devised. These are available to full- and part-time students and last for one year. Each First Award covers a single vocational area such as Art and Design, Business Studies or Engineering. Entry is usually open and successful completion of the course offers progression onto the prestigious BTEC National courses. First Awards recruit students who in the past would have taken CPVE. Considering that BTEC has responsibility for both First Awards and CPVE it is difficult to understand why it is promoting such competition. Indeed, the existence of First Awards actively undermines one of the aims of CPVE which was to rationalize the range of post-16 qualifications. It does appear that CPVE is declining in colleges where First Diplomas are expanding.

Other developments within the educational system also threaten CPVE. As TVEI begins to influence the post-16 curriculum, CPVE-type experiences will be extended to all students. Records of Achievement could replace the formative and summative systems of assessment in CPVE. It is possible that NCVQ[8] could accredit vocational competence gained in CPVE courses. The original flexible framework of CPVE may become established practice without the need for certification when this costs around £33 per student.

Important changes to CPVE were announced in 1989 arising from the national evaluation of 1987/1988. CPVE may now be a two year course as well as one year thus giving it a role in the certification of vocational elements in GCE A level courses and of special needs programmes. The CPVE core has been altered by compressing the present ten core areas into five. Students and other lay people will find that the language has been considerably revised and simplified. Vocational achievement receives greater emphasis by two means. It is recorded more prominently but more importantly, negotiations are under way to obtain National Vocational Qualification status for preparatory modules. The purpose of this is to tie CPVE vocational achievement into some recognized national system to ensure progression.

In spite of these changes there is still doubt over the future of CPVE arising from its past difficulties, its identification as a low-level course, the difficulty of organizing it especially in further education colleges across departmental boundaries, the proliferation of other awards and the unresolved problems of progression.

CPVE: An Assessment

So what is the worth of CPVE? The most recent HMI[9] report on CPVE (1988) notes a wide range of different styles of programmes in terms of their emphasis on GCSE or vocational studies, whether they are exploratory or assume that students have already chosen their career paths, the attitude and aptitude of the students and so on. The report comments favourably on the experiences of work, the local environment, working in groups and with adults and practical activity. Increased confidence and maturity seem to be important personal outcomes. Assessment and guidance are central features of CPVE and the HMI note that arrangements are good and the profiling interviews provide the opportunity for pulling together the various elements of a student's programme. With the growth of Records of Achievement, teachers are not so isolated in their use of profiling. Problems of moderation between teachers and centres remain.

The HMI conclude:

> The CPVE provides a valuable opportunity for students to discover or confirm their vocational orientation and to develop their knowledge, understanding, skills and personality . . . The learning experiences afforded by the CPVE have been generally enjoyed and appreciated by the students

and through them many have gained, sometimes for the first time, a sense of achievement. (HMI, 1988).

Some would say that this HMI report gave the seal of approval to CPVE therefore helping to assure its future. More support came in a speech made by the Secretary of State for Education and Science to the Association of Colleges of Further and Higher Education. (Times Educational Supplement 17 February 1989, No. 3790) He gave emphasis to the Joint Boards' intention to improve progression by the accumulation of credits towards other qualifications. This probably means that CPVE preparatory modules will have NCVQ status and be accepted for progression purposes by notably BTEC and CGLI. So far progression routes have been locally negotiated and this process has been severely hampered by the lack of national acceptance of CPVE modules. Elsewhere in his speech the Secretary of State lamented the 'academic/vocational divide' which is so characteristic of education in England and Wales. He proposed the idea of a core curriculum for FE and gave illustrative rather than prescriptive examples which are so clearly related to CPVE that some have suggested that he was proposing a 'super CPVE'. He also supported the practice of integrating core skills into the students' mainstream studies.

CPVE may have friends in the HMI and the Secretary of State for Education and Science but it has many enemies within the world of academic educationalists. Although CPVE was a Department of Education and Science initiative and despite support for it by many progressive teachers, it is frequently grouped with measures such as YTS[10] and TVEI as State interventions to vocationalize the curriculum. Writers such as Holt (1987) and Ranson (1984) seem to reject the legitimacy of a vocational element in the curriculum. Ranson argues that CPVE was part of a conscious strategy at DES when he quotes an official who said 'If we can achieve things with this new 17 + , that will give us an important lever to vocationalise or re-vocationalise the last years of public schooling'. (Ranson, 1984, p. 24)

Seemingly progressive aspects of CPVE such as profiling are characterized by Ranson using Bernstein's work on classification and framing as reactionary: 'In this way, more of the student is available for social control.' (Bernstein, 1975 p. 109)

Precisely by assessing more than academic ability alone and for recognizing a wide range of capability, CPVE is condemned by Ranson. His mistake lies in part in assuming that the typical features of élite education are 'good'. Academic approaches, traditional subject boundaries and the obscure conventions of the examinations system may now in fact be giving way, even for the élite, to a more relevant curriculum which through coursework is being assessed in ways which the clients feel are fairer. It is interesting to note that the experiments in student-centredness took place in lower status courses such as CPVE before being extended to higher status courses.

Ranson also coined the term *tertiary tri-partism* to describe the way in which different courses post-16 have different value and status. CPVE and YTS schemes are the *tertiary modern* provision, BTEC courses are *tertiary technical* and GCE A levels form the *tertiary grammars*. He is right to observe that the post-16 system is indeed stratified. Any observation of an FE College would reveal that point but eliminating courses such as CPVE would not change the situation. It would merely serve to exclude a range of young people from the system altogether. The question of the differentiation and stratification of the curriculum is an important one which does require further exploration and analysis.

Hostility to a vocational element in the curriculum is not reflected in the work of prominent curriculum thinkers such as Peters (1981), White (1982) and Hargreaves (1982). We need to evaluate vocational elements according to their role in the development of students. As educators our focus should be the student, not the welfare of the economic system and it seems reasonable that we should respond to young people's natural interest in the adult world of work.

It is not clear at this time what the future of post-16 education and training will be. In many of the initiatives being taken to improve the experience of young people in schools and colleges there are echoes of CPVE. Curriculum change is occurring all around the system where teachers can improve the service by their own efforts. Some, notably the team at the University of London Institute of Education Post-Sixteen Education Centre, argue that there is a need for more radical structural reform within the curriculum and look towards modularization as a solution. Its advantage would be that the students could select a personalized package of modules formed from the present range of CPVE, BTEC First Awards, GCSE and other courses. There are however, powerful vested interests who have not helped CPVE carry out its task of rationalizing provision, and these would have to change if we are to see reform. Perhaps what is most needed is a clearer, unified view of what the post-16 curriculum should be.

Notes

1 The Schools' Council was set up in 1964 as an educational body to deal with curriculum development and examinations. It was given a life of twenty years and came to an end in March 1984. The Secondary Examinations Council (SEC) was set up by Government in 1983 to deal with the new General Certificate of Education. SEC was replaced by SEAC, the School Examination and Assessment Council in August 1988 as part of the Government's Education Reform Act that introduced the national curriculum.

2 The independent and self-financing City and Guilds of London Institute was founded in 1878 by the Corporation of the City of London and the City Livery Companies. It is concerned with education and training for occupations in industry, commerce and public services, developing criteria and tests and providing internationally recognized qualifications in technical and vocational skills.

3 The Business and Technician Education Council (BTEC) was formed from the merging in 1983 of the former Technician Education Council (founded in 1973) and the Business Education Council (founded in 1974). The latter were established by the Secretary of State for Education and Science

respectively as independent bodies to develop schemes of education for people in technician and business type occupations to study at non-degree levels. BTEC is a financially independent company whose main source of income is fees paid by and for students. It is ultimately accountable to government through the Secretary of State for Education and Science.

4 The RSA (the Royal Society for the Encouragement of Arts, Manufactures and Commerce) is a voluntary body that was founded in 1754. Its original aim was to award prizes for commercially useful ideas or inventions and it was involved in initiating the Great Exhibition of 1851. It pioneered public examinations in 1856 and has been an examining board for general and vocational qualifications ever since. It initiated the Education for Capability movement in 1979 to counteract perceived academic bias in British education. The RSA is particularly well-known for its clerical and secretarial courses.

5 General Certificate of Secondary Education. This certificates single subjects and most students enter for it at the end of their secondary schooling. Subjects cover the whole of the secondary curriculum including traditional ones such as English and mathematics as well as some newer subjects such as community studies or business studies. Post-16 students can also take GCSEs in order to enhance their programmes or improve their qualifications. GCSE is moderated by a number of Examination Boards which are private companies.

6 The National Foundation for Educational Research is an independent body undertaking research and development projects on issues of current interest in all sectors of the public education system. It was set up in 1946. Its membership includes all the local education authorities in England and Wales, the main teachers' associations and a large number of other organizations with educational interests.

7 The Technical and Vocational Educational Initiative is funded and administered through the Training Agency, a Government funded body. Its aim is to stimulate curriculum development in schools and colleges with a view to increasing the relevance of the curriculum to life outside school.

8 The National Council for Vocational Qualifications (NCVQ) was set up by government in 1986 for the reform of the system of vocational qualifications in England, Wales and Northern Ireland. It covers all sectors of employment and types of occupation from basic to senior professional and aims to improve performance and personal progess at work through qualifications based on competence and valued by employer and employee. Standards are set by industry.

9 Her Majesty's Inspectors of Education are a body of educationalists responsible to reporting on the quality of education in schools and colleges. They are independent of the local education authorities and the Department of Education and Science.

10 The Youth Training Scheme was developed by the Manpower Services Commission in response to the increase in youth unemployment in the 1980s. Its aim is to encourage training for young people in work by subsidising their employment and training. It is a scheme of the Training Agency, a government funded but independent body.

References

BERNSTEIN, B. (1982) *Class, Codes and Control Vol 3*, London Routledge and Kegan Paul.

DEAN, J. and STEEDS, A. (1982) *17 + : The New Sixth Form in Schools and Colleges*, Slough, NFER Nelson.

DES (1980) *Examinations 16-18*, London, HMSO

DES (1982) *17 + A New Qualification*, London, HMSO.

FURTHER EDUCATION UNIT (1979) *A Basis for Choice*.

FURTHER EDUCATION UNIT (1981) *Vocational Preparation*.

HARGREAVES, D. (1982) *The Challenge for the Comprehensive School*, London, Routledge and Kegan Paul.

HMI (1988) *A Survey of Courses Leading to the Certificate of Pre-Vocational Education*, London, HMSO.

HOLT, M. (Ed.) (1987) *Skills and Vocationalism: The Easy Answer*, Oxford, Oxford University Press.

JOINT BOARD FOR PRE-VOCATIONAL EDUCATION (1984) *Consultative Document for CPVE*, London, Joint Board for Pre-Vocational Education.

JOINT BOARD FOR PRE-VOCATIONAL EDUCATION (1985) *The CPVE Framework*.

KEOHANE (1979) *Proposals for a Certificate of Extended Education*, The Keohane Report, London, HMSO.

PETERS, R. (1981) 'Democratic Values and Educational Aims' in *Essays on Education*, London, Unwin.

RANSON, S. (1984) 'Towards a Tertiary Tri-partism: new codes of control and the 17 + ', in Broad, P. (Ed.) *Selection, Certification and Control*, Basingstoke, Falmer Press.

SCHOOLS COUNCIL (1972) *Working Paper 45*, London, Schools Council.

SCHOOLS COUNCIL (1973) *Working Paper 46*.

WHITE, J. P. (1982) *The Aims of Education Re-stated*, London, Routledge and Kegan Paul.

22
'Maths in the Workplace': Some Issues Arising out of the Development of a Resource Pack

Pat Drake

Introduction

In Britain it is now most unlikely that 16-year-old school leavers will move straight into employment. The majority go into pre-vocational and vocational training schemes, based in school, or college, or industry. The Youth Training Scheme, (from May 1990 called just Youth Training) has expanded dramatically since its inception in 1983. It is now true to say that YT improves job opportunities, if only because most job opportunities are linked to YT schemes.

Youth Training

The Youth Training scheme combines 'off-the-job' with 'on-the-job' training over a period of one or two years. Delivered by Training and Enterprise Councils, schemes must now show that their training satisfies given criteria before funds are forthcoming. These criteria are outcomes-based[1] and include demonstrating skill in problem solving and numeracy. They also emphasize 'skill transfer', that is developing the ability to apply what has been learned in one work context in a different context.[2] The rationale for skill transfer is that in Britain what is said to be needed is a flexible, skilled workforce prepared to change occupation and geographical location according to national needs.

This increasingly utilitarian climate within which we operate makes demands for numeracy teaching which are responded to by a system which almost completely by-passes those actually engaged in mathematics education. Even within Colleges of Further Education, numeracy is rarely in the domain of mathematics but is delivered by General Studies departments. Rarely is this work seen as anything other than extremely low status and almost nowhere is there any in-service training for teachers engaged in it. Even though mathematics, or numeracy, or whatever you want to call it, is now a facet of the continued educative/training experience of most young people in Britain, they and their teachers are getting little or no support. They are marginalized from the work of mainstream mathematics education. The courses which must be implemented are determined directly by employment concerns. Training is to serve the needs of that employment, and is being implemented at all levels with little regard to the considerable experience of mathematics educators. Thus we have a situation where young people, having spent eleven years of compulsory schooling doing compulsory mathematics, find themselves in training doing yet more compulsory mathematics.

Maths in the Workplace

The project 'Maths in the Workplace' was funded in 1985 by the then Manpower Services Commission (now renamed the Training Agency)[3] to address the difficulties faced by young people on YTS schemes in doing mathematical activity, and the difficulties faced by their trainers in teaching them. The project was based in England at Sheffield City Polytechnic from 1985–87 and was carried out in conjunction with the BBC, Milton Keynes, and the Institute of Education University of London. The aims of the project were stipulated in advance by the Training Agency:

> to provide materials which will enhance the abilities of young people and adults to use mathematical skills in a variety of different work settings. The sets of materials will each contain:
>
>   a video cassette
>   worksheets for supervisors
>   worksheets for trainers
>
> and will be designed for use on YTS schemes. The material will be designed for use by trainees, to reinforce and extend their work-based learning, and to provide clear diagnositc information on their skills. It will also provide assistance to supervisors in understanding good training procedures, especially in identifying and developing transferable skills.

Clearly in order to even begin to fulfil these aims, considerable thought had to be given to the relationship between the video and the written materials, as well as to the actual content of both.

At the start of the project we visited many YTS schemes in order to identify mathematics activities done at work by sixteen-year-olds across à range of occupations. We also needed to find places in which to film. Working with the BBC gave us a cachet which we suspect gained us admittance to and offers of co-operation from a number of YTS work schemes, and perhaps allowed us to ask for more detailed information than we might otherwise have obtained. We were alerted to many different issues, as well as to the more general response that there were too few resources to support the teaching and learning of mathematics to trainees, and certainly none which helped trainers pick up on the issue of transferability. Resources of 'appropriate' mathematical content are often obviously directed either at younger children, or at the adult education market.

It is relevant to mention particular themes which recurred:

a) Trainees are poorly motivated. One reason for this is previous lack of success in mathematics at school. This leads many to fear mathematics and avoid doing it if they can. The mathematics required for school examinations is beyond that required for everyday life by most people, so school pupils see a lot of mathematics as being irrelevant. It takes considerable effort on the part of supervisors to overcome these attitudes in young people when they arrive on YTS.

b) Trainees lack confidence, but so do supervisors. Supervisors are expected to teach mathematics, very often with little other than their own school experience to go on.

c) Supervisors are expected to develop skill transfer with trainees. In mathematics this is often difficult for supervisors to do, skilled as they are in specific work areas. How can they know when the same mathematics occurs in a different occupational context?

d) The mathematics courses followed by trainees at school do not always contain specific skills required in real life. For example, supervisors are dismayed to meet trainees who can't use feet and inches, and fractions of an inch. However, school syllabuses no longer include these imperial units, even though they remain a facet of British industrial life.[4]

e) Trainees arrive at YTS with widely different experiences and at different levels of mathematical achievement. The difficulties experienced within a particular group of trainees are correspondingly diverse.

f) There are few text books or other mateirals to help supervisors and trainees relate mathematics to specific occupations and yet to encourage skill transfer.

g) There is little, if any, in-service training for supervisors, to develop and support their mathematics teaching.

We found that the actual mathematics done at work by young people in training fell into a number of broad categories.

dealing with money and time,
understanding numbers used as codes, references, or in putting things in order,

measurement of length and area,
reading scales to measure length, area, volume, electricity,
mixing things in proportion,
stock control,
following plans, diagrams or other instructions to make things,
making decisions within constraints of money, time, and space.

These activities occur in a wide range of occupations including agriculture, building, catering, clerical, community care, hairdressing, leisure and recreation, motor vehicle maintenance, painting and decorating, retail, textiles, travel, warehousing and wood trades.

Generally speaking, anything more mathematically sophisticated, was done in relation to a particular aspect of a particular occupation area. We observed first that more sophisticated mathematical activity was very rare in early stages of training, second, was performed in a specific way according to the occupational constraints and therefore not all transferable and third, did not pose problems for young people working in that context. This last point was true of the more general 'elementary' mathematics too; people did not find mathematical activity difficult when it was embedded in specific practical contexts. When it was removed to the training room and called mathematics, it became immensely hard.

For example, in the weaving trade, an analysis of the way pattern repeats across a piece of fabric are incorporated into the width of material, reveals the necessity for sophisticated understanding and application of principles of multiples and factors, in order to calculate the numbers of threads of different colours required on each row of woven pattern. Weavers have developed a notation and vocabulary to describe the process they adopt, which is specific to this activity, and descriptive of it. For instance they refer to a 'handful' of threads. This 'handful' is precisely the lowest common multiple of the numbers of each thread in each of the colours used in one pattern, without repeats. A decision as to how large a 'handful' should be is made each time the pattern changes. A trainee weaver would gradually acquire the expertise required to make this decision from an experienced weaver at a loom — not in a classroom.

Meanwhile trainers bewailed the inability of trainees to grasp the simple (to them) principles of Imperial units, percentages, fractions, decimals, operating with time, timetabling, money . . . the list goes on, and they were at a loss to know what to do about it. Sometimes they managed to help trainees towards an understanding, provided they avoided the word 'mathematics' at all costs — a sorry comment on the confidence of young people meeting these ideas at work for the first time. Sometimes they resorted to pages of sums, finding that these were of little help because the next time the trainee met the real problem they did not know which sum to do. We found trainers who spent the little time and energy they have for generating resources in the production of these drill and practice exercises, conscious that they do not really work, but helpless in thinking of alternative approaches.

In many places, tests are applied to discriminate between those who have 'basic skills' and those who do not. This discrimination then determines the occupation into which young people are directed. Those whose test results indicate possession of 'basic skills' are then directed into high status occupations. The trouble is that the tests always include sums, yet so far as we could tell, these kinds of sum are not often required in any workplace by any young person on YT; multiplication of mixed numbers is an example. Secondly, as Wolf (1984) points out, the mathematics required in specific occupations is not linked to the status of the occupation. For instance, carpenters who have high occupational status have to measure, yet bricklayers whose occupational status is low have to both measure *and* estimate big numbers.

These observations presented us with some difficulties. We were not sure of the value of producing resource materials for 'off-the-job' training. We proceeded with some caution, conscious that we were balancing what we met as a declared need of trainers for resource material, with questions about the effectiveness of such material out of context.

We decided to base the materials absolutely on what mathematics was required by young people on YTS, no more and no less. We also decided that we were not going to let the inexperienced supervisor 'off the hook' by providing a teacher-proof pack of what in effect would be distance learning materials. We attempted to build into the materials an element of negotiation between trainers and trainees, and we tried to acknowledge the varied backgrounds of trainees by deliberately omitting a hierarchy of difficulty in the activities we offered. So the materials are not presented with a presumed order of difficulty, and there are far more activities than it is possible to do in the available time. This implies an element of choice — trainees and trainers select the activities that they consider to be the most suitable for their needs. Neither are the activities all paper and pencil ones. They are presented in not only differing occupational contexts but also as requiring different styles of working, for instance they contain small and large group activities, role plays and scripted conversations. Throughout, the trainers are encouraged to act as consultants to the trainees, rather than didactic teachers, so that trainees are encouraged to take control of their own learning and progress. One concern of the project team was the blatant sex-stereotyping of occupation which we saw around the country. We made an attempt to counter this by illustrating young people in different occupational roles, independent of their gender.

After consultations with YTS practitioners and trainees, it was decided to use the ninety minute videotape to make nine ten-minute programmes, each relating to a particular aspect of mathematical activity done at work, and to accompany each with a pack of written materials. The video was not intended as a replacement maths teacher. The strength of video is that it can enlarge the walls of the training room: through the medium of television, the viewer can travel to workplaces other than his or her own. The video is therefore used primarily to illustrate a range of actitivies, in different work contexts, which involve mathematics. In this way the problem of skill transfer can at least be addressed.

A pilot video module plus written materials was produced after six months, and was handed over to the Training Agency for comment, as well as being tried out in YTS schemes around the country. Subsequent packs of written materials were also tested and tried as they were produced, but it was not possible to trial the remainder of the video, as it was not produced sequentially.

Each of the nine programmes then showed young people, most of them YTS trainees, engaged in different working environments, engaged in mathematical activity. Each programme is intended as a starting point for the written materials which follow. It is a motivator, a stimulus for discussion, an opportunity to recognize and acknowledge the skills that they already possess, and so to empower by giving trainees status.

Conclusion

The whole set of materials including the completed video, was handed over to the Training Agency at the end of the two year period, in August 1987. By this time, negotiations with them had resulted in a licence to publish 'Maths in the Workplace' being issued to Sheffield City Polytechnic publishing house, PAVIC. There was also further funding for a short period of dissemination, four months spread over one year, starting September 1987.

During the dissemination phase of the project, since September 1987, publication of the materials was suspended pending negotiation of a compromise to accommodate Training Agency sensitivity to Health and Safety at Work issues which have emerged out of the illustrative nature of the video. A debate on the relative merits of this case is inappropriate here, as at the time of writing the matter has been satisfactorily resolved. However management of this suspension draws our attention to the fundamental point made at the beginning of this article relating to mathematics education of young adults in Britain. We are experiencing tangible evidence of forms of control which visibly mould the mathematics experience of post-sixteen-year-olds in Britain.

This is not only the fault of those responsible for training programmes. As we have seen, control by the bureaucratic paymasters fails to address the very real difficulties faced by those whose job it is to implement policy and teach. I would suggest that because of the traditional split between mathematics and numeracy, between academic schooling and the world of work, because of the low status attributed to genuinely applied mathematics within the school curriculum, and because of current political rhetoric and increasing hegemony of 'training', there has hitherto been little incentive for experienced mathematics educators to pick up the baton. Surely it is now the responsibility of those educators in Britain to acknowledge the trend of an extended mathematics curriculum beyond sixteen and to widen horizons beyond the school gates to include the world of sixteen plus, so that individuals may become empowered by their lengthened mathematics education, not just functionally trained.

Notes

1 See the article by Mathews in Chapter 14 of this volume. (Editor)
2 See the articles by Harris and Evans, Chapter 12, and Strässer *et. al.* in Chapter 15 of this volume. (Editor)
3 The Manpower Services Commission, now the Training Agency, is a government agency which operates independently from the Department of Education and Science.
4 Imperial units now feature in Attainment Target 8 Level 6 of the National Curriculum. (Editor)

References

DRAKE, P., MARDELL, J. and WOLF, A. (1988) *Adding Reality to Numeracy*, YTS News, No. 43. Sheffield, Manpower Services Commission.

SHEFFIELD CITY POLYTECHNIC AND THE BBC (1989) *Maths in the Workplace* Sheffield, PAVIC Publications.

WOLF, A. (1984) *Practical Maths at Work*, MSC Research and Development Series, No. 21. Sheffield. Manpower Services Commission.

23
Mathematics and TVEI

Annette Grey

Introduction

With the extension of the Technical and Vocational Education Initiative (TVEI), problem solving activities become central to all areas of the curriculum. TVEI courses are expected to meet the following criteria: they must be relevant to the world of work, be accessible to pupils of all abilities, make good use of new technology such as computers, develop links with industry and involve active learning and problem solving.

The project *Maths + TVEI*, based at the Department of Mathematics, Statistics and Computing of the University of London Institute of Education, aims to integrate the teaching of mathematics with the curriculum areas which are to become *Technology* under the National Curriculum: these are the subjects of Craft, Design and Technology; Food Technology and Business Studies/Information Technology. By using authentic technology problems in mathematics lessons, and increasing the use of mathematical approaches in technology classrooms, it is hoped that pupils will start to apply their skills and knowledge more effectively.

Practice

In the Craft, Design and Technology room, a student is making a piece of Flat Pack Furniture. Working from an isometric drawing of her chosen design, she is trying to draw up a cutting list, giving the dimensions in millimetres of each piece of chipboard she requires. Her teacher eventually has to help her. In the Maths Room, the class are doing the latest piece of GCSE practical coursework, set by the exam board. Pupils are to design and sketch a bookcase, which must be made from one piece of board measuring 150cm × 100cm × 1.5m. In all probability, the teachers of these two classes are unaware that they are tackling similar tasks. In recent years, mathematics teachers have been laboriously constructing problems and contexts through which to deliver the mathematics syllabus, seemingly unaware of the teaching resources represented by their colleagues down the corridor. Is it any wonder that pupils fail to transfer their mathematical skills?

Problem solving in mathematics lessons can be a very strange affair. All the background information needed is supplied on an A4 sheet of paper. The problem itself is selcted for the amount of 'maths' (calculation, measurement, ratio, algebra) involved, rather than for any reality or relevance to the pupils. Very often, artificial constraints are introduced in order to involve more 'maths'; after all, pupils will learn more that way, won't they? The numbers however, are neatened up so as not to appear off-putting. One wonders if pupils are fooled by this window dressing. They may find such problems more interesting than a page of sums, but do they become more effective problem solvers as a result?

The difficulty in solving real problems lies not in the complexity of the mathematics (which is often fairly straightfoward) but in the selection of the appropriate operation that is required. From the tangle of information, constraints, and objectives involved in the task, pupils must construct the underlying mathematical model which is capable of solution. This is mathematics at its most challenging and powerful, even when the model is a simple one.

Technology teachers are not unaware of the mathematical demands of their subject, although their perception of mathematics seems to be limited to surface elements such as percentages, rearranging formulae and so on. Yet a great deal of progress can be made within a school simply by comparing the demands of different subjects in terms of such 'skills', and particularly their timing.

Learners experience great difficulties in applying skills in an unfamiliar context at the best of times. Here the problem is compounded by the way mathematics appears in other lessons.

> Frequently a technique is required before it has been developed sufficiently in the maths classroom. Pupils will be confused by different methods, language and notation.

> Non-maths specialists may be unaware of the factors which influence the difficulty of a particular type of problem, and may unwittingly over-complicate the task by the way they present the question.

> The mathematics occurs suddenly, with no preliminary revision, and it demands instant recall on the part of the pupil.

> Frequently the mathematics is not an end in itself, so if the pupil does not succeed first time, difficulties are not explored but rather the correct answer is supplied and the lesson moves on.

> Out of the context of the maths lesson, pupils often find it difficult to

select the appropriate technique, or even to acknowledge the need for mathematics.

The following areas of mathematics are easily identified by looking at syllabus demands of other subjects.

Ratio and Proportion

CDT: working with scale models and diagrams; gear ratios

Food Studies: scaling up recipes; composition of foods such as pastry, syrups; nutritional balance

Percentages

CDT: tolerance levels; factors of safety; efficiency

Food Studies: nutritional values; survey results; business costs

Business: interest, taxation, profit, growth; invoices with discounts and VAT; case study data

Units

CDT: length, area, volume, mass (SI units); time; money; also common units in electronics, structures and mechanisms

Food Studies: metric and imperial units of mass and volume (including fluids); temperature; time

Graphs and Charts

CDT: graphic communication; force-extension and stress-strain graphs; distance-time charts

Food Studies: display and interpretation of nutritional information and survey data;

Business: sales figures; expenditure; demographics; market shares profit/loss figures

Use of Formulae

CDT: energy (KE = $\frac{1}{2}mv^2$); power (W = IV); velocity ratio; gear ratio; mechanical advantage; etc.

Reading Scales

CDT: use of multimeter to measure current, voltage and resistance; use of measuring instruments in technical drawings and construction

Food Studies: use of a variety of weighing machines and thermometers

The practical situations encountered in TVEI lessons provide an ideal context in which to reinforce the mathematical skills, and most importantly *estimation and approximation*, getting a *feel* for number. There are times in every problem solving activity when mathematics can provide the most efficient way forward, but all too often it is avoided by using trial and error, or by choosing numbers which are easy but not the best solution to the problem.

Potential

The *Maths + TVEI* project involves mathematics and technology teachers from seven West London boroughs. Eight cross curricular resource packs are being produced, which are based largely on current GCSE syllabus requirements. The aim is not to turn mathematics teachers into technology teachers and vice versa, but to make everyone more aware of the opportunities for collaboration, for the benefit of pupils. Topics include gears and bicycles; choosing the site for a new leisure centre; design, production and packaging of a new biscuit; rewiring a room; food maths; conducting a survey; constructing a flat pack disaster shelter, and collecting a portfolio on design themes across the curriculum. Preliminary trials have been very successful, but we have been made increasingly aware for the need for full training and support. Simply giving out written teaching materials does not have much effect on classroom practice. When the packages are published they will include a variety of strategies for overcoming the barriers to cross curricular work, and they will be accompanied by in-service training recommendations.

The traditional organization within schools has made it difficult for teachers in different departments to communicate effectively. The sheer effort involved in finding time to be together prevents most departments from attempting to match their syllabuses. A faculty system (for examples Maths, Science and Technology) ought to improve matters but it is evident that this sort of activity is low on the list of priorities. One of the aims of the *Maths + TVEI* project is to show people that it is a good investment: that it can save time by cutting down on overlap between subjects, it can enhance pupil motivation and learning, and it can broaden the skills and confidence of teachers. Another powerful incentive for co-operation between departments is the possibility of submitting one piece of coursework for more than one GCSE subject.

The national curriculum has many references to cross-curricular work, indeed some see this as the only possible method of survival. Mathematics teachers must all be aware of the emergence of Technology as a powerful new curriculum area, with many resources and newly designed teaching schemes. Active learning and problem solving activities must have rigour and precision in order to stretch older pupils; in this respect mathematics has a unique and vital contribution to make.

24
Work Reclaimed: Status Mathematics in Non-élitist Contexts

Mary Harris and Carrie Paechter

In the early 1990s mathematics teachers in England and Wales are proceeding with the task of reconciling the demands of vocationalism with the aspirations of academism through the national curriculum. Historically, vocational or work-related education has been seen as aimed at working-class students or those of below average ability. This has led to its low status in the curriculum and a concomitant denial of its academic content. In order for teachers of vocationally-related subjects to obtain the teaching facilities that accompany academic status, this status has to be demonstrated, usually through the sacrifice of relevance:

> ... it is ... possible to discern three stages in the development of school subjects. In the first, the callow intruder stakes a place in the time-table, justifying its presence on grounds such as pertinence and utility. During this stage learners are attracted to the subject because of its bearing on matters of concern to them. The teachers are rarely trained specialists, but bring the missionary enthusiasm of pioneers to their task. The dominating criterion for the selection of subject matter is relevance to the needs and interests of the learners.
>
> In the second stage, a tradition of scholarly work in the subject is emerging along with a corps of trained specialists from which teachers may be recruited. Students are still attracted to the study, but as much by its reputation and growing academic status as by its relevance to their own problems and concerns. The internal logic and discipline of the subject is becoming increasingly influential in the selection and organization of subject matter.
>
> The third stage is characterised by a further advance along the road of specialisation and expertise. The teachers now constitute a professional

body with established rules and values. The selection of subject matter is determined in large measure by the judgments and practices of the specialist scholars who lead inquiries in the field. Students are initiated into a tradition, their attitudes approaching passivity and resignation, a prelude to disenchantment. (Layton, 1972. See also Goodson, 1988).

Giroux has pointed out that it is usually black and working-class students that are encouraged to take vocationally based courses with relatively little academic content, by teachers operating within the logic of cultural deprivation theory, 'which defines education in terms of cultural enrichment, remediation, and basics' (Giroux, 1985).

The assumptions behind this view of work-related education are twofold. The first is that it is only necessary to provide learning experiences that are relevant to the non-academic world for working-class students and those of lower ability. This ignores the problems that arise as a result of more able students' taking on board frames of reference that operate as if educational texts were context-free (Keddie, 1971). The second assumption concerns the processes and content involved in work-related tasks, which are seen as having low academic status. We wish to challenge both these assumptions.

The exploitation of contexts of work for mathematical development, making explicit the mathematics involved, provides a promising way out of an impasse. The Maths in Work Project (Harris, Postscript, this volume) has attempted to make a reconciliation between school ideal and work practice by bringing mathematically rich activities of daily working life into school as resources to be developed mathematically by teachers. Such an approach is related both to ethnomathematics (D'Ambrosio Chapter 2 this volume) in that it aims to start from the context of the activity of the person at work rather than the context of school mathematics, and to the work of Grey for example (Chapter 23) who demonstrates cross-curricular mathematical development through the Maths and TVEI Project. At the same time it has become necessary to show that the mathematics learned through cross-curricular, context-based work can be assessed within the academic framework of the National Curriculum Statements of Attainment. Work is currently being done in this general area at GCSE level by Paechter and others working on the Cross-curricular Assessment Project at King's College, London.

In cross-curricular frameworks such a TVEIE[1] there is no doubt that motivation amongst students increases when they see the relevance of their school work to work outside school (Millman, 1985). Like other workplace influences however some of the first effects of vocationally oriented courses were to demonstrate the extreme of vocational curricula in confirming gender prejudice, with girls being channelled into the stereotypical 'soft' vocations and boys into the stereotypical 'hard' vocations like construction, maintenance and engineering (Millman, 1985). The status of girls and women is low in both mathematics and work.

One way of redressing the imbalance between school and work mathematics and 'hard' and 'soft' work practices is to demonstrate some of the mathematical content of the 'soft' practices themselves. The analytical approach of Maths in Work was to explore and demonstrate mathematical thinking and doing within traditional women's tasks, rather than patronize their activity by applying selected 'hard' mathematics as examples of something serious girls and women can do with their 'soft' interests. The approach has also been adopted by individual teachers in their own classroom practice who have taken the very stuff of ordinary everyday life and work and used it to reveal and develop mathematics (Harris, 1989). There is no doubt that such work has proved both motivating to students and broadening in curriculum activity.

At the same time, mathematics teachers have been concerned to affirm women's status as mathematicians, both now and in the past. This has led to the writing of materials about women mathematicians in an attempt to make them more visible to students. Although women have made major contributions to the development of mathematics, their work has often gone unacknowledged. The work of several women (for example that of the women of the Pythagorean community) is remembered by the names of the men with whom they worked (Alic, 1986). Some men are less generous to their female colleagues than others. In a letter to the mathematician Grace Chisholm Young (who continued her research while producing nine children between 1897 and 1908) her husband wrote:

> I hope you enjoy this working with me. On the whole I think it is, at present at any rate, quite as it should be, seeing that we are responsible only to ourselves as to division of laurels. The work is not of a character to cause conflicting claims. I am very happy that you are getting on with the ideas. I feel partly as if I were teaching you, and setting you problems which I could not quite do myself but could enable you to. Then again I think of myself as like Klein, furnishing the steam required — the initiative, the guidance. But I feel confident too that we are rising together to new heights. You do need a good deal of criticism when you are at your best, and in your best working vein.
>
> The fact is that our papers ought to be published under our joint names, but if this were done neither of us would get the benefit of it. No. Mine the laurels now and the knowledge. Yours the knowledge only. Everything under my name now, and later when the loaves and the fishes are no more procurable in that way, everything or much under your name.
>
> There is my programme. At present you can't undertake a public career. You have your children. I can and do. (Grattan-Guiness, 1978)

It is important to make plain to students that women mathematicians have played as pivotal a role as men in the development of mathematical thought, and to this end various books and teaching packs have been written giving short biographies and associated mathematical activities. Girls are often very interested in these life histories and the struggles women have faced to do the work they wanted. It is quite clear to both students and teachers, however, that such women were and are extraordinary. We find ourselves teaching that *some* women may be mathematicians, not that *all* women can. Work on mathematics and textiles can play a part in resolving this dilemma.

> My aim is to make explicit the mathematical processes and content of traditional women's crafts, work that girls may have seen their mothers or grandmothers doing or that can be clearly seen as something that many ordinary women do or did at other times or places. I want to show girls that even while doing culturally stereotyped work they may be doing mathematics, that women traditionally have been mathematicians. At the same time I aim to make some areas of mathematics more real by putting them into an accessible context (Paechter, 1989).

Lolley and Ross (1989) were also concerned to give due status to the mathematical heritage and potential of girls through a patchwork project carefully designed to cover curriculum areas in geometry through an open-ended investigation into the geometry of polygons. The result was the development of a different model of learning in the mathematics class from a 'rather structured' approach to one that was 'so dynamic and, so self-perpetuating that ordinary teaching now lacks flavour'. The students felt satisfied in their work and directed their own learning to a great extent, learning new mathematics because they needed it. Their learning was generalized 'for example they developed a fascination for patterns, both in creating them and appreciating them' and they were able to explain their learning in everyday language (Lolley and Ross, 1989).

The use of textiles as a mathematical resource has proved to be a useful way of raising some fundamental issues in the relationship between practical and academic mathematics. An immediate outcome is the reinstating of women's work within the framework of work-related mathematics. At the same time it is possible to give status to the cultural activities of ethnic minorities by unpacking the mathematics they involve. 'Mathematics activities identified within the everyday life of societies' (Howson *et. al.*, 1985, p. 15) are the focus of Gerdes who identifies mathematics 'frozen' in the artefacts of Mozambican weavers (Gerdes, 1986). He 'defrosts' it in teaching programmes developed for the Mozambican curriculum which alert teachers to the use of familiar artefacts as resources for mathematics learning. The mathematical thinking that derives from an analysis of the structure of objects including woven mats and baskets is far from trivial.

> . . . the forms of these objects are almost never arbitrary, but generally possess many practical advantages, and are, a lot of the time, the only

possible or the optimal solutions of specific production problems . . . The traditional forms reflect accumulated experience and wisdom. They constitute an expression not only of biological and physical knowledge about the materials that are used, but also of mathematical knowledge. (Gerdes, 1986)

Of equal significance is the introduction of holistic tasks into the often fragmented mathematics curriculum. Historically there has been a tendency for work-related tasks to be presented in the form of simulations. This tokenistic approach denudes the activity of its contextual complexities, thus rendering the mathematical activity more low-level than it might otherwise be. The use of holistic design and technology tasks allows the student to be confronted with the genuine constraints involved, making such problems more real and showing their solutions to be complex. For example, Paechter (1989) describes working with students making patchwork.

> Most students find it very hard to appreciate the rigourousness of the constraints involved in such a practical piece of work. Aware of their limitations regarding accuracy of cutting and sewing, most confine themselves to using squares and rectangles at first. However, girls in particular tend to work on a very small scale and produce designs featuring 1cm^2 patches. At this stage I get out my own patchwork once more and explain again how difficult they will find it to sew patches that small. I think that it is only at this point that they really begin to grasp what the process is all about and to appreciate the reality of the exercise. Many students now treat their original design as a draft and enlarge it to make the final version. This is often the first piece of work they have done on enlargement and it is a particularly graphic way of introducing them to the effect of an enlargement on the area of a shape.

The association of mathematics with the textiles contribution to design and technology has a further effect on the balance of power within design and technology as a whole. Historically while the status of all practical subjects has been low compared with those seen as academic, the association of Craft Design and Technology with engineering has allowed it to be seen as more important than 'softer' aspects of Design and Technology[2] such as textiles. The identification of CDT as masculine and textiles a feminine has exacerbated this. While Design and Technology as a whole has been enabled to gain status due to its position as a national curriculum foundation subject, the imbalance is in danger of remaining. The association of textiles work with academic mathematical learning enhances its position and thus is one way that mathematics teachers can support equal status for all subjects contributing to design and technology.

There is richness indeed in extending the vision of mathematics beyond the limits of the mathematical topics usually selected for application to 'work' examples. If pupils become self confident in their recognition of mathematics as something they do and enjoy as part of everyday working life, then there is surely more chance of them developing the positive and confident attitudes that employers say they want. Numbing them with the diet of base skills and their rule-bound application to unreal workplace examples is unlikely to have the same effect.

A mathematics course that chooses all its examples from textile activities would of course be as unbalanced as current ones that take all their examples from traditional 'hard' interests. But there can be no doubt that a textile mathematics course could be planned that could take children from infancy to sixteen while allowing them to develop context-based design and technology tasks in an empowering way. The emancipatory potential of school mathematics is enormous.

Notes

1 The TVEI recently changed to TVEIE by the addition of the word 'extention'.

2 Design and Technology is an activity which spans the curriculum, drawing on and linking a range of subjects. By creating a new subject area, work at present undertaken in art and design, business education, craft design and technology (CDT), home economics (HE) [food and textiles] will be co-ordinated...

Design and Technology describes a way of working in which pupils investigate a need or respond to an opportunity to make or modify something. They use their knowledge and understanding to devise a method or solution, realize it practically and evaluate the end product and decisions taken during the process. (National Curriculum Council, 1990).

References

ALIC, M. (1986) *Hypatia's Heritage*, London, The Women's Press.

GERDES, P. (1986) 'How to Recognise Hidden Geometrical Thinking: a Contribution to the Development of Anthropological Mathematics', *For the Learning of Mathematics*, 6, 2 (June 1986).

GIROUX, H. A. (1985) 'Critical Pedagogy, Cultural Politics and the Discourse of Experience', *Journal of Education*, 167, 2.

GOODSON, I. (1989) *The Making of the Curriculum*, Basingstoke, Falmer Press.

GRATTAN-GUINESS, I. (1972) 'A Mathematical Union. William Henry and Grace Chisholm Young', *Annals of Science*, 19; 2. (August), 107. p. 141 Reprinted permission of Taylor and Francis Ltd. London, in Perl, T. (1978) *Maths Equals Reading*, Massachusetts, Addison Wesley p. 157.

HARRIS, M. (Ed) (1989) *Textiles in Mathematics Teaching*, Maths in Work Project, University of London Institute of Education.

KEDDIE, N. (1971) 'Classroom Knowledge', in Young, M. F. D. *Knowledge and Control*, West Drawton, Collier and Macmillan, p. 133–160.

LAYTON, D. (1972) 'Science as General Education', *Trends in Education*, 25. January.

LOLLEY, M. and ROSS, K. (1989) 'The Patchwork Quilt', in Harris, M. (Ed) *Textiles in Mathematics Teaching*, Maths in Work Project, University of London Institute of Education.

NATIONAL CURRICULUM COUNCIL (1990) *Technology, Non-Statutory Guidance: Design and Technology Capability*, York, NCC.

MILLMAN, V. (1985) 'The New Vocationalism in Secondary Schools: its influence on girls', in Whyte, J., Deem, R., Kant, L. and Cruickshank, M. *Girl Friendly Schooling*, London, Methuen.

PAECHTER, C. (1989) 'Patchwork as Medium for Tessellation', in Harris, M. (Ed) *Textiles in Mathematics Teaching*, Maths in Work Project, University of London Institute of Education.

25
Postscript: The Maths in Work Project

Mary Harris

The detailed study by the Maths in Work Project of the data of the London into Work project on skills and their contexts in the work of 16-year-old school leavers, led to the discovery that more mathematics was revealed in the contexts of questions that were not about mathematics than to questions which were (Harris, Chapter 13 this volume). Such a finding naturally had implications for further research but the project had no power to pursue them. It did however have the opportunity to work through some of the implications in its prescribed tasks of designing and producing learning materials for school, based on the research findings.

Indirect lessons were also learned during the research phase. While reporting on the data, and as a means of checking that reports would be up-to-date and realistic, visits were made to various workplaces of the type that formed the London into Work sample. These were either small companies or branches of large ones; they were not the large companies who recruit mainly 18-year-olds and graduates and whose influence is now predominant in the schools-industry movement. During the visits it was noted that a great deal of small problem solving went on where workers had to make decisions which included evaluating a number of variables, for example deciding the placement of yoghurt cartons by sell-by date, manufacturer and flavour in such a way as to maximize sales, or shelving shoes by size, style, manufacturer, colour and popularity with customers. Some of such activities had been detailed in the context data from the problem-solving questionnaires, but only when the decisions concerned a defined problem. These were not regarded as involving mathematics by the researchers, the further education teachers or the employers because they did not involve any calculating. The whole emphasis of the further education team and the employers involved in the London into Work project was on skills training for particular jobs, indeed that was the purpose of the research. Teaching associated with it therefore tended to emphasize the rules, procedures and techniques of arithmetic and it was not unusual for employers to stress that they did not want their workers to understand but to do. What learning materials existed in the mathematics and employment field generally tended to support this type of teaching by illustrating arithmetic going on at work and supplying examples to practice in class.

It was believed by the school-oriented Maths in Work project that such an approach in fact alienates people from thinking mathematically, making them dependent on their next set of instructions. It may be a good model for a docile and un-enterprising workforce but, as a model with an essentially anti-educational aim, it is not suitable for school. The teaching experience of the Project Leader of Maths in Work had been with an alternative approach based on the belief that mathematics is an active and creative process of individual minds best learned by involving pupils in activities devised to encourage them to solve problems in co-operation and discussion with fellow pupils and the teacher. During the process, mathematical relationships and structures are drawn out, discussed and argued and progress in pupils' performance and self-confidence is evident. The task of producing learning materials for Maths in Work thus became one of reconciling the activity of workplaces with the learning (as distinct from training) model of school. The materials would need to be open-ended, to invite practical, co-operative involvement and discussion, to reveal mathematical relationships and suggest ways of symbolizing and generalizing them. Promising workplace examples had already been noted in the decision-making tasks described above.

Two of the first sets of learning materials to be published internally in the Inner London Education Authority, took the work situations of laying out a counter in a chain clothes store and planning delivery routes against traffic conditions. The 'laying out a counter' activity was devised from the method then used by Marks and Spencer and was developed with their permission and assistance. It depended on the number of places available on a counter and the proportion of different sizes and colours of particular garments sold. For example, one row of a counter could hold four sweaters in the back and middle sections and two in front: more size 14s are sold than size 18s and more red than blue. Sweaters arrived at the store already packed in proportion according to previous sales data and the task was to decide how many of each size were to be placed in each position, given that parts of the counter would already be occupied by other garments.

Base boards to represent the counter were made, numbered cards representing the sweaters and to fit the spaces on the base board were prepared and customer behaviour was represented by messages on blank (on one side) playing cards with remarks such as 'customer buys one size 14 blue' or 'coach load of tourists arrives and buys three size 16 blue, four size 16 red and seven size 14 red'. Activities using these materials were structured so that layout problems could be solved firstly by simply manipulating the cards but, as the problems became more complex it would be quicker to calculate. By this time the pupils would have had plenty of experience of the practical, trial and error activity, would understand both the significance and efficiency of calculation and would be devising ways of doing it. The activities gave practise in ratio calculations under the changing circumstances of a context in which the concept of ratio was the most sensible one to develop and pursue.

The activity about delivery routes came to be called Hazards, in honour of London's traffic. Its materials were a street map with a warehouse, a station and shops marked on it and two sets of playing cards to simulate consignments of goods and traffic problems respectively. Various activities were devised in which students argued for various routes against various criteria. The activity became popular but during the trials of its various drafts it was noticed that each time the drafts returned for editing the activities had become more limited and controlled. It was felt that there was a danger of it becoming both unrealistic of work and unadventurous and constrained as learning material. Consequently, while Hazards was being trialled, another whole set of learning materials was prepared in the style preferred by Maths in Work with the intention of publishing it without any trailing at all.

While visiting warehouses during the research phase of the project it had been noticed that a variety of packages stored there were of interesting design that could profitably be introduced into school as design problems. A number of packing firms were visited and a collection of problems made (Harris, 1985a, 1985b). A main purpose of good packing design is to market a product, so marketing considerations became a criterion for the whole pack with mathematical thinking being the commodity to be marketed. Design expertise and mathematics and mathematics-learning research was consulted and a pack design drawn up. The cover bore a professionally drawn cartoon that ensured that the pack looked as little like a mathematics text book as possible, Each activity was laid out on a separate A4 sheet and designed to carry drawings and diagrams of as high a standard as could be achieved within a very limited budget. Economical and efficient language was used and the logo of the contributing firms were added if the firm concerned wished. Each activity was the subject of a different file on the word-processor and reproduced on a separate sheet; changes suggested by use could be easily inserted into the next edition of the pack or supplied separately. Mathematically the activities were mostly open-ended and with a low entry point thus allowing maximum flexibility of use by teachers who, as was already known, would want to adapt them for their own purposes. The complete pack contained twenty-four activities run off on both coloured paper and on white (thereby making them reproducible), a skill list, teachers' notes and some of the special card needed for some of the activities. One hundred copies were prepared using ordinary office materials and made available at a cost that recouped the cost of production. Within six months another 300 had to be produced and by then more, and more positive response had been received than from the formal testing of Hazards. As the result of the response, the Hazards activity was withdrawn and a completely new pack of activites including problems about the placing of bus stops, letter boxes and pedestrian crossings was prepared. As with

Wrap it Up, appropriate firms were consulted and London Transport, the Post office and the Department of Transport gave their help.[1]

It was felt that the enthusiastic use of both packs justified the approach that the Maths in Work project had taken and Wrap it Up in particular became a challenge to some of the mathematics and workplace materials then beginning to be produced. At a conference of about 250 SMILE[2] teachers, the entire conference was formed into groups to represent the visualization team of a design company with the brief of designing an up-market pack for a sports ball. Each group was asked to record the mathematics they used while doing the task, together with the mathematics they could expect to teach in school through the activity. When all the records were summarized it was found that a large part of an 'O' level syllabus had been covered, a great deal of mathematical discussion had gone on and two quite different sorts of group behaviour, distinguished here as theoretical and practical, had occurred. The theoreticians revealed their approach by standing back from the problem and dealing with it in the abstract by such remarks as 'this is simply an optimisation problem — a sphere in a cube'. In fact they neither tackled the problem, which was about concrete (not theoretical) shapes and their behaviour under particular conditions, nor did they do any mathematics. The majority tackled the problem practically with the materials provided, discovering for example that a ball inside a tetrahedron tends either to roll about or push the sides out causing additional problems of design; they produced a wide range of elegant and not-so-elegant designs as well as doing a lot of mathematics.

With its approach reasonably well established, Maths in Work attempted to move into other practical areas, intending to produce two more packs, one to be called Cardboard Engineering and another to be based on textile activities. Unfortunately it did not prove possible to find funding for the engineering project[3] but a generous and imaginative grant from the Department of Education and Science, allowed the project to survive staffing cuts in the Inner London Education Authority and transfer to the University of London Institute of Education to undertake a year's investigation of textiles activities as a resource for learning mathematics.

Part of the reason for the textiles project was to investigate gender and culture effects in work-place mathematics learning: everyone wears clothes and they are mostly made by women. Another part was that the Project Leader had always made her own clothes and was aware of at least some of the mathematics involved. The project proceeded by collecting examples of textiles activities from both domestic and industrial sources that promised to be rich resources for mathematics learning. As the collection of articles grew and as the design of the Cabbage pack[4] progressed, the idea arose of setting up an exhibition of textiles and textiles products that would demonstrate the mathematical nature of textile activity and the Common Threads[5] exhibition opened at the University of London Institute of Education in November 1987. As the direct result of demand it toured England until January 1990 and was then taken over by the British Council for an overseas tour beginning in September of the same year.

Although it had been known from the outset that textiles would provide a particularly rich seam for mathematics the effect of the Cabbage pack and Common Threads was both much wider and deeper than had been anticipated and the Project's aim of revealing the stuff of daily life as full of mathematics was vindicated. Both resources showed undeniable mathematical thinking within activities traditionally the preserve of low status groups in mathematics (women and ethnic minority women in particular) and therefore presumed to contain only low-level mathematics if any at all (Harris, 1987). The depth of this prejudice showed on occasion in in-service work with teachers. A standard Maths in Work presentation at the time, for one-off in-service events consisted of an illustrated talk followed by a practical activity such as that described for the SMILE Conference above. During the textiles project, textiles activity was substituted instead of a packaging one, partly as a way of demonstrating the traditionally limited range of activities usually taken to be mathematical and partly to note the responses of people unfamiliar with textiles work. Although the Project Leader was determined to be open-minded about these people, the majority were in fact male. The most obvious response was the widespread ignorance of simple hand-knitting and sewing technology in this group which reacted much in the way that their female pupils do when asked to take part in traditionally male activity in mathematics; they 'backed off' allowing more 'technically literate' colleagues to proceed with the task; they found a range of excuses for not doing the task or did an alternative task from within their own range of interests; on some occasions they announced that the activity was 'not mathematics' or that 'this is silly', a well-known pupil response meaning 'I don't understand this.' A not infrequent response was to enter the activity with alacrity with remarks of relief such as 'I was afraid that you were going to give us something difficult' only to find that the activity was rather more difficult than it was assumed to be and that assistance was required. On one occasion a teacher refused to do the task at all on grounds that it was women's work and that in his culture a degrading thing for a man to be seen doing. Ignorance of the basic techniques of knitting and weaving was found to be so widespread that notes had to be added to the Cabbage pack to explain for example that knitting consists of rows of interlinked loops.

A particularly interesting development was that of linking the textiles work to information technology through software programs that drive knitting machines. With such programs pupils can perform a mathematical investigation with symmetry on a micro-computer and then 'print out' their results on a scarf or other garment that they can wear. Textiles work in particular has lent itself to cross-curricular work, a finding that was of no surprise to a Project that had long since noted that outside the education system, mathematics is a cross-curriculum subject.

During the textiles phase of the Maths in Work project, direct links with the textiles industry were made and textiles factories provided some poignant examples of the under-utilization of women's intellectual skill while exploiting their labour. In one factory visited, Asian women were sewing children's track suits at high speed (they worked on piece rates) and with great accuracy. Their speed and skill depended on their ability to estimate very accurately the distance between their line of sewing and the edge of the garment so that the lines could be kept parallel to within a millimetre while stitching the tight curves of children's wear. While doing so they also tensioned the two garment pieces that they were sewing so that seams would not pucker in the finished garment. This work was classed as unskilled in comparison to the 'skilled' work of male colleagues in the same factory whose job was to pick the fluff off knitting machines.

Many of the women came from Bangladesh whence two items had been derived for the Cabbage pack. One involved symmetries obtained on woven fans and the other geometrical problems of placing points on the circumference of the embroidered circles that represent lotus flowers. Instructions for both activities were written in Bengali as well as English and teachers who used the activities with Bangladeshi girls in their classes noted the positive effects on both the girls and their colleagues as they demonstrated once again the narrowness of content and prejudices of focus within mathematics education (Harris, 1990).

Throughout the time that Maths in Work was producing these resources it noted repeatedly the different perceptions of 'low-level' mathematics both inside and outside education. The textiles project in particular demonstrated restricted thinking within mathematics education and revealed a strong male bias that restricts most mathematical activity to areas already familiar — to boys. Many of these attitudes are no more than the habits of history or the prejudices of exclusive power groups and however powerful, are without rational justification. Mathematics education has still not resolved the tension between the devalued arithmetic that is the mark of mathematics-for-work and the academic aspirations of the very small percentage of the total population who stay in the education system to study more mathematics. Definitions of 'relevance' are still dominated by the academic aspirations and the life experiences of those who have never themselves worked in a factory and the threats behind the demands of industry for basic skills are still not taken seriously (Harris, 1989). In the nine-year experience of Maths in Work, the attitudes of industry itself form some of the most negative of forces on school mathematics education and are particularly damaging at a time when the status of educators in society has been rendered so low.

'England is the worst educated country in Europe' may have a modern ring to it but the remark is from Brougham in 1820 while making a plea for state education in England and Wales for the first time. The remark that 'Britain's workforce is under-educated, under-trained and under-qualified [and that] a third of school leavers have no useful qualification to show for at least eleven years in full time education' was made 170 years later in the CBI Report 'Towards a Skills Revolution'. Between those dates, indeed since at least the sixteenth century, there have been repeated tales of falling standards in education and a constant blaming of teachers. Since there has never been a contemporary time (only a golden past in the memory of individual complainants) when it has been socially acceptable or politically expedient to remark on the good quality of education, the peak from which the standards have so regrettably been falling for so long, must indeed be mythical. The laying of blame for industry's problems at the door of education cannot be maintained with rationality and there has been plenty of time to do something about it, if that was really the cause and if society really wanted it. Evidently society prefers scape-goating to action. In spite of the generous co-operation of firms in the workplace visits of the Maths in Work Project and of the high praise with which many greeted its approach and productivity; in spite of its undoubted influence on other projects which have used its materials and ideas and in spite of generous funding from the Departments of Education and Science, and Trade and Industry, not one company or industry was prepared to support it so that it could continue its work, preferring in the main to place their financial resources in the traditional, discriminatory and increasingly discredited approach of skills-and-applications. That an alternative, more positive, more lively, more enterprising and more mathematically rich approach is available to anyone prepared to study it is ignored. Unfortunately, continued action is almost beyond the educational reach of one single-handed project.

Notes

1 Commercially produced editions of *Wrap it Up* and *Hazards* are now published by Macmillan Education.

2 The Secondary Mathematics Individual Learning Experience (SMILE) project is a project of the Inner London Education Authority. It develops and produces teacher-designed learning activities which are now used world-wide.

3 Some of the ideas researched for it and some of the activities of Wrap it Up were taken up by a joint JMB-Shell Centre (Nottingham University) project and eventually published as 'Be a Paper Engineer' by Longman.

4 The word Cabbage derives from the old French work 'coupage' and is related to the skill of being able to cut more garments from one piece of cloth than is thought possible by the employer who supplies the cutter with the cloth and the garment pattern. Cabbage has formed a major part of the earnings of impoverished garment makers, indeed sometimes their only income, and though obviously open to abuse is a highly skilled business involving a great deal of understanding of symmetry concepts.

5 The Common Threads exhibition has been described elsewhere; see References at the end of this paper.

References

HARRIS, M. (1985a) 'Wraping up Mathematics in the World of Work', London, ILEA Contact, March 8th 1985.

HARRIS, M. (1985b) 'Wrapping it Up', *Mathematics Teaching*, 113.

HARRIS, M. (1987) 'An Example of Traditional Women's Work as a Mathematics Resource, *For the Learning of Mathematics*, 7, 3 (November).

HARRIS, M. (1989) 'Basics', *Mathematics Teaching*, 128, September.

HARRIS, M. (1990) 'Art and Skill of Symmetry in Old and Modern Textile Techniques', in *Leonardo*, in press.

The *Common Threads* exhibition was reviewd in the following papers,

   Morning Star, January 26 1988

   Times Educational Supplement, March 4 1988.

and is described in the following articles,

HARRIS, M. (1988) 'Mathematics and Textiles', *Mathematics in Schools*, September.

HARRIS, M. (1988) 'Common Threads', *Mathematics Teaching*, 123, June.

Notes on Contributors

Dr. George Barr was Research Fellow in Mathematics Education at Brunel University and is now Senior Research Fellow at the City and Guilds of London Institute.

Dr. Alan Bishop is Lecturer in Mathematics Education at the Department of Education, Cambridge University.

Professor Doctor H. J. M. Bos is Senior Lecturer in the History of Mathematics at the Mathematical Institute of the University of Utrecht, Netherlands.

Dr. David Carraher is Associate Professor of Psychology at the Federal University of Pernambuco, Recife, Brazil.

Professor Ubiritan D'Ambrosio is Professor of Mathematics at the University of Campinas, Brazil.

Paul Dowling is lecturer in Mathematics Education at the University of London Institute of Education.

Pat Drake was Director of the project 'Maths in the Workplace' at Sheffield City Polytechnic before becoming Lecturer in Mathematics Education at the University of London Institute of Education.

Dr. Jeff Evans is Senior Lecturer in Statistics at the Middlesex Polytechnic.

Dr. Munir Fasheh lectures in Mathematics Education at Birzeit University, Palestine.

Annette Grey is Research Officer of the Maths and TVEI Project at the University of London Institute of Education.

Mary Harris is a Research Officer and Leader of the Maths in Work Project at the University of London Institute of Education.

Janet Holland is Research Lecturer in the Department of Policy Studies, University of London Institute of Education.

Sarah Ingham wrote her article while employed as Employee Development Manager with Courtaulds Textiles Outerwear Group.

Dr. George Joseph is Senior Lecturer in the Department of Econometrics and Social Statistics at the University of Manchester.

Eugene Maier was working at the Mathematics Learning Center of Portland State University, Oregon, when he wrote his article that is included in this volume.

David Mathews was Deputy Director of the Work Based Learning Project at the Further Education Staff College and is now a consultant in training and selection.

Jenny Maxwell wrote the article reproduced in this volume while teaching adult basic literacy and numeracy students at a community education project in Birmingham.

Dr. Richard Noss in Senior Lecturer in Mathematics Education at the University of London Institute of Education.

Carrie Paechter is Project Officer at the Centre for Educational Studies, King's College London, working on cross-curricula assessment through coursework at GCSE.

Keith Pye is Special Project Co-ordinator of the clothing and Allied Products Industry Training Board.

Dr. Rudolf Strässer is a Researcher specializing in technical and vocational education at the Institute for Didactics in Mathematics (IDM) at the university of Bielefeld, West Germany.

Sheila Sullivan was the CPVE Co-ordinator for Essex County Council before becoming Staff Development Officer at Chelmsford College of Further Education.

T. S. Wilkinson is a marine engineer currently Head of the Mathematics Department of NEI Parsons Limited.

Alison Wolf is research Lecturer and Senior Research officer in the Department of Mathematics, Statistics and Computing, University of London Institute of Education.

Index

abacus, 54, 171
Abbasid Caliphs, 48
Aborigines, 30
Abreu, G. 170, 192–4
abstraction, 36, 63, 78, 79, 82, 190–2
Abu-Kamil, 49
Abul-Wefa, 49
academism, 18, 277
accountancy, 4
accreditation gap, 150
acculturation, 37–9
Acioly, N. M. 170, 185–90
acquisition, 112, 114
Act for the Maintenance of the Navy in
    England (1540), 4
active learning, 110, 273
addition, 179–83
Adelard of Bath, translation of Euclid's
    Elements, 3, 17, 49
adult education, teaching maths in, 166–7
agricultural revolution, 3
    and sexual division of labour, 230–1
Ahmes Papyrus, 51
Alexander (356–323BC), 46
Alexander, S. 235–6
Alexandria, 46, 48, 49
Alexandrian Library, 48
Alexsandrov, A. D. 197
algebra, 49, 159
algorithms, 49, 159
Alic, M. 279
Amerindians, 30
Angolan sand drawing, 104, 106
anthropology, 5, 20, 37
    mathematical, 22, 104–10, 116, 128
Antioch, 48
Apollonius, 46

applied maths, 2, 22–3
approximations, 62, 276
Arab mathematics, 3, 45, 48–9, 50
Arabic numerals, 17, 49
Arbeitsgruppe Mathematiklehrerausbildung,
    166
Arbuthnot, 28
Archimedes, 46
architecture, 3, 17
Aristotle, 129
arithmetic
    and mathematics, 207; and political
        neutrality, 14
artificial intelligence, 88
artillery technicians, 26–7
Ascher, M. 16
Ascher, R. 16
assessment, and the accreditation gap, 150
Association of Teachers of Mathematics, 8–9
Association for Teaching Aids in Mathematics,
    see Association of Teachers of
    Mathematics
Attercliffe academy, 6
Australia, 240

Babylonian mathematics, 45, 46, 51–2, 54
Bacon, R. 3–4
Badger, M. 87
Baghdad, 48
Bailey, 124
Bait al-Hikma, 48
ballistics, maths and, 26–8
banking, 3
Banks, M. H. 147
Barnett, C. 9
Barr, G. 121, 125, 158–68
BASIC, 88

Basis for Choice, A, 256
Bath List, 124, 132
Bath Report (1981), 124–5, 138
Bauersfeld, H. 167
BBC, "Maths Counts", 143
Beattie, J. F. 72
Beechey, V. 245
behaviourism, 125
Bellone, E. 20–1
Bernal, M. 2, 118, *Black Athena*, 44
Bernstein, B. 100, 112, 115, 262
Berry, J. 80
binary system, 5
Birnbaum, B. 241
al-Biruni, 49
Bishop, A. J. 13, 29–41, 80, 107–8
Boethius, 2, 5
bookies, permutations and probability, 185–90
Bos, H. J. M. 13, 26–8
Bourbakiism, 93
Bourdieu, P. 109
Boyer, C. B. 2, 4, 47
Brahmagupta, *Brahmasputa Siddhanta*, 49
Brazil, studies, 169–201
Bromme, R. 165
Brown, S. 81
Brunel University, Mathematics Education
    Group, 125
Bruner, J. S. 32, 90
BTEC, 223, 256, 260, 262
Buxton, L. 78
Byzantine Empire, 48

Cairo, 48
calculators, 31, 65, 75–6, 93
calculus, 18
Callaghan, J. Ruskin College speech, 95, 124
candy sellers, abstraction, 190–2
canonry, 4
capitalism, patriarchy and, 231, 243–4
careers: academic, school, work and popular,
    113–16
Carraher, D. 122, 127, 169–201, 202
Carraher, T. N. 127, 169, 172, 175, 178, 181,
    182, 194, 195
Cavalcanti, C. 173
CBI: common learning objectives, 147–8;
    "Towards a Skills Revolution", 289–90

CDT, 281; *see also* design and technology
Certificate of Extended Education (CEE), 256
Cesaire, Aimé, *La Tragédie du Roi Christophe*,
    23
Charlton, K. 4
Chevallard, Y. 166–7
Chinese mathematics, 16, 44, 47, 48
Cipolla, C. 3, 4, 6–7
City and Guilds of London Institute (CGLI),
    126, 160, 222, 256, 259, 260, 262
civilization, diffusion theories of, 47–8
Clarendon Commission, 8
Clarricoates, K. 239
class relations, reproduction of, 239
classroom, Eurocentrism in the, 53–4
Closs, M. P. 30
Clothing and Allied Products Industry Training
    Board (CAPITB), 212, 221–9
clothing industry: "invisible" workers in, 237;
    and machinists' skills, 212, 220–9; maths
    and the, 226–9; sexual discrimination in,
    241
Cockburn, C. 10, 212, 239–40
Cockroft Report (1982), 94–5, 98–9, 102, 124,
    125, 132, 138–41, 146, 159, 209
cognition: in practice, 110–12; related to
    culture, 16, 203
cognitive development, "universal" stage
    theory, 34
cognitivists, 123, 127, 128–9
Cohen, P. C. 3, 5
Cole, M. 77, 80, 110, 128, 203
Colleges of Further Education, 266
colonialism, 23, 43–4, 59, 105
commercial arithmetic, 4, 5
communication skills, 74
compartmentalization, 166–7, 168
competence, 97, 98, 152–3, 156
computer based learning environments, 82–4
computer controlled machines, 74
computer games, 82
computer intraction, 80, 85
computers, 31, 65, 76, 271: at work, 14; in
    mathematical learning, 77–92; research
    and curriculum development, 1, 88–9;
    use in schools, 11
concepts (Vergnaud), 177–8
construction foreman, proportions, 183–5

constructivism, 164, 167
consultancy, in clothing industry, 225
consumer maths, 62
context, maths in, 1–14, 202, 207–8
contexts: and performance, 164–5; role in
    learning, 162–4
contextualization of maths, 93–120
control, 35, 38, 59
Cooley, M. 102–3
Cooper, B. 8
Co-operative Societies, 8
Copernican astronomy, 5
Cordoba, 48, 49
core skills, 153–5
counting tables, 171
Courtaulds Textiles, 211, 214–19, 232–3, 238
CPVE, 253, 256–65; in action, 254–7; an
    assessment, 261–63; the future, 260–61;
    origins, 259–60
Cromwell, O. 6
cross-cultural studies, 107–8
Cross-curricular Assessment Project, 278
cultural anthropology, 15
cultural context: of maths, 77–8; maths
    education in, 29–41
cultural deprivation theory, 278
cultural relativism, 10, 108,
culture: components of, 32; hierarchization of,
    18–20; related to cognition, 16; value
    components, 32, 35–6
culture conflict situation, 30, 38
culture of silence, 59, 60
Cunha, L. A. 173
Cunningham, T. W. 134
curriculum, "hidden", 130
curriculum development: research and
    computers, 88–9; in Third World
    countries, 22–3
urriculum negotiation, 253, 259

Damascus, 48
D'Ambrosio, U, 2, 13, 15–25, 30, 77, 107,
    278
Dark Ages, 44, 48
Darwin, C. 10, 23
Daune-Richard, A. M. 230
Davis, P. 82
Davy, J. 82

de la Rocha, O. 169, 206
decontextualizing, 115
deficit model, 10
dehumanization, 81, 82, 166
Descartes, R. 16
design, assessment of in clothing industry,
    225–6
Design Council, design 2000, 225–6
design and technology, 254, 281
designers, in the clothing industry, 227
deskilling, 242
Dimitriu, A. 16
Diophantus, 46
Dirac, delta function, 21
distributions, theory of, 21
doing maths, 78–80; children, 84–5
domestic labour, 230, 243
Dominicus Gomdissalinus, 17
Dorfler, W. 88
Dowling, P. 14, 93–120, 123, 124, 126, 127,
    209
Drake, P. 126, 251–2, 266–72
dual labour market theories, 242
Durkheim, E. 108

Edessa, 48
education: dualism of science and humanities,
    4; from moral instruction to moral
    rescue, 7; investment and consumption
    function, 3
Education Act (1944), 8
egalitarianism, 203
Egyptian mathematics, 16, 43, 44, 45, 46, 48,
    50–1
electronic data interchange (EDI), 219
electronics, 219
élitism, 17–18, 50, 53, 94
Ellul, J. 31
embeddedness, 79, 82, 97–9
employers: confidence and scepticism, 74–5;
    recruitment practices, 9; restructuring of
    work to male advantage, 241; views,
    73–4, 147
Employment White Paper (1988), 221
empowerment, 57–61
enculturation, 34, 37–9
endowed schools, 7
engineering, 146; needs of employers, 73–4

engineers: in the clothing industry, 228;
     training, 223
England, workplace research, 124–6
episteme, 112
Equal Opportunities Commission, 238
Equal Pay Act (1975), 238
equations, false position method of solving
     linear, 51
Eskimos, 30
essentialism, 102, 103, 110
estimation, 65, 276
ethnic minorities, 280; education difficulties 29;
     and textile maths, 288–9
ethnography, 20, 127–8
ethnomathematics, 2, 11, 13, 15–25, 30, 61,
     77–8, 107, 278; definition, 18;
     etymology of, 23–4
ethnoscience, 15, 21
Euclid 3, 5, 8, 46: *Elements*, 17
Eurocentrism, 2, 13, 23–4, 50–3:
     countering classroom, 53–4; foundations of,
     42–56
Evans, J. 121, 122, 123–31, 158–68, 202–10
Everyday Maths, 133, 143
Eves, H. 47
evolutionary theories, Eurocentrism in, 47

factories, 203–6
Fahim, H. 43
failure in school maths, 101–2
farmers, areas and lengths, 192–4
Fasheh, M. 13, 57–61
FE, core curriculum, 262
feminism, 239, 242–4
Fibonacci, 17; *Liber Abaci*, 4
Finch, B. 211
Firestone, S. 243
Fitzgerald, A. 124, 125
Fitzgibbon, C. T. 72
Fleishman, E. A. 148–9
folk mathematics, 13, 62–6, 152
For the Learning of Mathematics, 26
formalism, 78, 196
Foucault, M. 112
Freudenthal, 79
"frozen maths", 78, 95–6, 105, 128, 280–81
functionality, meaning and, 81–2

Galileo, 4, 17, 26
Gallistel, C. R. 169
Game, 240
Gardner, P. L. 36
Gardner, W. 111
Gay, J. 77, 80
GCEs, 259, 260, 276; A level maths, 74–5
Gelman, R. 169
gender: bias in vocational channelling, 278–9;
     construction of 238–41; themes for a
     theory of, 245–6
gender differences, in cognitive styles, 81
gender division, explanations for, 242–4
gender relations: in paid work, 244;
     reproduction of, 238–9
gendering: of maths, 9–10; of work, 212,
     230–52
Genghis Khan, 49
geometry, 17, 76, 85–6, 280; Arab, 49
George, 234
Gerdes, P. 30, 78–80, 95–6, 104, 105–6, 110,
     280–81
Germany, vocational education, 8
Gherardo of Cremona, 49
Gilligan, C. 81
Gillings, R. J. 51
Gimpel, J. 2, 3, 4
Giroux, H. A. 278
Glucksman, 234–5
Goodson, I. 278
government intervention, 262
Gramsci, A. 78
Grando, N. I. 194
Grattan-Guiness, I. 279
Great Debate, 95, 124, 147
Greek mathematics, 2, 16, 44, 45–7, 50, 53
Greenfield, P. M. 104, 111, 129
Grey, A. 254, 273–6
Griffiths, H. B. 69
guidance education, 259
gunnery schools, 4
Guthrie, W. 63

Hagen, 245
Hales, M. 118
Hall, 231
Hammurabi dynasty, 51
Harding, 245

Hardy, G. H. *Mathematicians Apology*, 109
Hargreaves, D. 263
Harris, M. 1–14, 121, 122, 123–31, 202–10, 254, 255, 277–83, 284–91
Harris, P. 30
Hart, K. 84, 85–6
Harun al-Rashid, 48
Harvey, J. 3
Hatano, G. 198
"Headlight Project", 90
hegemony, 57–61, 110, 271–3
Henry VIII, 5
Hersch, R. 82
Hessen, B. 28
hierarchy, gender, 212–13, 231, 236
Higginson Report (1988), 74
Hillel, J. 87
Hindu-Arabic numerals, 4, 171
historiography of maths, European bias, 43–53
history of maths, 13, 15
history of science, 20
HMI, 102; on CPVE, 261–3
holism, 16, 22–3, 54, 281
Holland, J. 212, 230–52
Holt, M. 262
homeworking, 10, 237
Horton, R. 31
Howson, A. G. 69
Howson, G. 2, 5, 6, 8, 142
Hoyles, C. 79, 82, 83, 84, 85, 87, 88, 89, 129
Hughes, M. 169
Hulaga Khan, 49
Humphries, 238
Huxley, T. 8
Huygens, C. 26

Ibn al-Haytham, 49
ibn-Qurra, Thabit, 49
ideology, 35, 59, 114, 234; male gender, 240; relation between science and, 23
ILEA, 285, 287; Schools Mathematics Inspectorate, 133–4
Inca numbers, 16
Indian mathematics, 16, 44, 45, 47, 54
indigenous cultures, 61
indigenous maths, 77
industrial revolution, 6–7, 17
industrial training boards, 220–1

industry: links with education, 146–7; links with schools, 1, 273, 284; needs of, 72–5; view of post-GCSE maths education, 71–6; women in, 211
inequality in the workplace, 212, 230–52
informal, vs. school maths, 128–9, 172–7
information technology (IT), 74, 219, 288
Ingham, S. 211–12, 214–19
intelligence, restrictive definitions of, 151–2
interests, vested, 1, 253, 262–3
internationalized maths, 37–8
intuitive maths, 50
invariants (Vergnaud), 177, 179–83, 198
investment, in training, 75
Iran, 29
Ishago bone, Lake Edward, Zaire, 53–4

Japan, 72, 75
Jensen, 245
Job Competence Model, 152–3
Job Components Inventory, 147
job segregation, by sex and family wage, 244
jobs: sex-typing of, 240; skills and task analysis, 148–9
Joint Board for Pre-Vocational Education, 256
Joint Unit for CPVE and Foundation Programmes, 259
Jollie, T. 6
Joseph, G. G. 13, 42–56
Jundishapur, 48
Justinian, 48

Keddie, N. 278
Keitel, C. 29, 79
Keohane Report (1979), 256
al-Khwarizmi, Mohamed ibn Musa, 48–9
Kline, M. 44, 47, 52
knowledge: accreditation of scientific, 21; conceptual and procedural, 198; culture-free, 30; intuitive, 198; mathematical, 43–50, 183–94; personal and collective nature of mathematical, 171–2; and power, 43
Kpelle of Liberia, 77, 80
Kronecker, L. 23
Kuhn, T. S. 22

labour force, feminization of, 245

labour process, 104, 109
Laclau, E. 103
Lancy, D. F. 16, 30, 34, 78, 80
language, mathematical dual role as signifier
    and signified, 196, 208–9
Lapps, 30
Lave, J. 87, 111, 113, 116, 126, 127, 129,
    164–5, 169, 202, 203, 208; *Adult
    Mathematics Project,* 206–7
Lawler, R. 85, 112
Lawson, J. 7
Layton, D. 278
Lazonick, 232
Lean, G. A. 30, 31
learning, work-based, 155
learning maths, children, 85–7
Lehrer, T. 93
Leonardo da Vinci, 6
Leont'ev, A. N. 110
Leron, U. 87
Levy, M. 149, 155
Lewis, D. 30, 31
Lilley, S. 4
Lindsay, R. 99–101, 111
linguistics, 117, 171, 203, 208–9
Linn, P. 118
LISP, 88
literacy, 3, 6–7, 150
logic, 16, 35–6, 50
Logo, 83, 84–5, 86–7
Lolley, M. 280
London Into Work (ILEA), 125, 132–3, 147,
    284
Loveridge, 235
Lown, 232–3
Lumsden, C. J. 20

McCormick, E. J. 148
MacFarlane-Smith, I. L. 148
machinists, in the clothing industry, 228
McIntosh, A. 126
Maier, E. 13, 62–6, 94
male gender ideology, 240–1
male violence against women, 243
al-Mamun, 48
management 156: in the clothing industry,
    224–5, 227; production in clothing
    industry, 223–4

Manpower Services Commission (MSC), 125,
    133, 153–5, 254;
    see later Training Agency
Mansfield, B. 152
al-Mansur, 48
Mark-Lawson, 231
Marx, K. 23, 104–5, 109
Marxism, 47, 243
masons, 3
mass education, 17
Masterman, L. 70
Mathematical Association, 8
mathematical knowledge 183–94: historical
    development of, 43–50
mathematical reasoning, 149
"Mathematics Outside School", xi
mathematics schools, 18th C, 6
mathematization, 79
Mathews, D. 121–2, 124, 125, 126, 145–57
maths: as a cultural phenomenon, 31–6;
    written and oral, 175–7
Maths + TVEI, 254, 273–6
maths curriculum, culturally fair, 38
maths education: cultural context, 29–41;
    history, 16–18; and values, 31, 38,
Maths in Work project, 134–43, 255, 278–9,
    284–91
Maths in the Workplace, 266–72
Maxwell, J. 14, 67–70, 81
meaning, and functionality, 81–2, 195
meanings, mathematical, 208–9
Mechanics Institutes, 8
Mellin-Olsen, S. 77, 79
Memphis, priests of, 45
Menninger, K. 171, 182
mental computations, 65
mentifacts, 19
mercantile classes, 4, 7
merchants' manuals, 4
Mesopotamian mathematics, 43, 44, 45, 50–1
Mexican weaving, 104
Middle Ages, 2–3, 17
middle class, 8, 233–4
Middleton, C. 230
military schools, 26
Millman, V. 278
Minsky, M. 112
MIST (maths incorporated in a specific task),
    97

Mitchelmore, M. 80
modelling, mathematical, 74, 159
modularization, 263
Moscow Papyrus, 51
Mouffe, C. 103
Mozambican basket makers, 29, 78, 95–6, 104, 106, 280
multiplication, 84–5
Murtaugh, M. 169, 206
mystery, 36, 38

National Assessment of Educational Progress, US, 62
National Council for Vocational Qualifications (NCVQ), 156, 261, 262
National Curriculum, 9; Statements of Attainment, 278
National Vocational Qualifications, 156, 222–6
navigation school, 6
navigation techniques, 4
NERIS, xi
Neugebauer, O. 51
neutrality, 58, 60, 67–70
"New math", 57–8, 93
Newton, I. 26; *Optiks*, 17
NFER, 259
Nicolas de Biard, 3
Noss, R. 14, 77–92
Nottingham University, Shell Centre, 99–101
number: concepts of, 16; feel for, 276; framework for, 152–3; role in work and training, 121, 145–57
number facility, 149
number system, Indian origin, 45, 49, 54
numeracy 150–1, 202, 253, 266; applications of (CBI), 147–8; low status of, 254; number or, 121, 145–57; socially defined, 150–1
numerical integration, 27
numerical part programming, 219
Nyaya-Vaisesika epistemology, 16

objectivism, 36, 38
occupational destiny, and sexual identity, 239
occupational segregation, horizontal and vertical, 235–6, 242
Omar Khayam, 49
openness, 36, 38

Oriental mathematics, 47
Osen, L. M. 53
Owen, R. 7

Paechter, C. 254, 277–83
painting, 17
paper-and-pencil maths, 65
Papert, S. 77, 81, 82, 83, 85, 86, 89, 90
Papua New Guinea, 29, 80
parabolic ballistics, 26
parish schools, 7
Parr, H. E. 94
participant observation, 109, 128
Pascal, 88
patchwork, 281
paternalism, 232–3
patriarchy, 212, 232–3, 242, 243–4; and capitalism, 231, 243–4
pattern cutters and graders, in the clothing industry, 228
pedagogy, and ethnomathematics, 18–22
Pepys, S. 6
Peters, R. 263
Piaget, J. 34, 85, 113, 169, 179
piecework payment system, 217
Pimm, D. 85
Pinchbeck, I. 231
Pinxten, R. 30, 31
place value notations, 54
Plato, 5, 16–17, 45, 48
Plato of Tivoli, 49
Plutarch, 45
politicization of maths, 11, 67–70
politics, in research, 130
post-16 curriculum, 263
Potter, K. H. 16
power: and knowledge, 43; maths and symbolic, 58–61; in science, 196
power relations, 240–1, 246
practical intelligence, 152
practical maths 76: and social control, 8
Practical Problem Solving Mathematics in the Workplace, 126
practical thinking, 129, 172, 206
practice, theories of, 164, 202–10
praxis, maths within education as, 57–61
pre-mathematics, 79
pre-vocational methodology, 259

Pringle, 240
printing, 4
private schools, until 18th C, 7
problem solving activities, 273–6
profiling, 261, 262
programmable logic controllers (plcs), 218
progress, 35, 38, 61
Ptolemy, 5; *Almagest*, 49
public schools, 8
pupils' expectations, and gender discrimination, 239
pure maths, 2, 22–3
Pye, K. 212, 220–9
Pythagoras, 45
Pythagorean community, 279
Pythagorean Theorem, 45, 105–6

QALY (Quality Adjusted Life Year), 89
quadratic ballistics, 26
quadratic equations, 45
quadrivium, 2, 17
Quaintance, M. K. 148–9
questions, open-ended, 121
quipus, 16, 54

racial discrimination, 43, 237–8
Ranson, S. 262–3
rationalism, 35, 38
Recorde, R. *Grounde of Artes*, 5, 8, 142
Records of Achievement, 261
recruitment, and sex-typing, 240
Reeder, D. 9
Rees, R. 125, 160, 165
reflection, 80, 85
reification, 18–20
relevance of maths, 13–14, 59–61, 259, 277, 278, 289
Renaissance, 3, 5–6, 17
representations (Vergnaud), 177–8
research, curriculum development and computer, 88–9
Resnick, L. 170, 196
Restoration church, 6
Revised Code, 7
Robert of Chester, 49
Robinson, O. 236–7
rod numerals, 54
Rogoff, B. 111

Roman numeral notation, 171
Romans, 17
Ross, K. 280
rote skills, 5
RSA, 256, 260
Rubery, J. 238

Saccheri, 49
Said, E. *Race and Class*, 43
Saussure, F. 117
Saxe, G. 170, 172, 190–2
scaffolding, 90; and computers, 88
Schliemann, A. 122, 127, 169, 170, 172, 175, 178, 182, 185–90
school: maths out of, 169–70; role in process of learning, 183–94
school failure, 172–3
school maths: in context, 253–5; and street maths, 173–5; versus folk maths, 62–6
schooling, 183–94; mechanism of, 21–2; role in development of mathematical knowledge, 195
schools: and the clothing industry, 226; workplace research and, 207–9
Schools Council, 256
schools-industry relations, 1, 273, 284
science, nature of, 21
Scotland, 222
Scott, A. M. 240–1
Scott, J. F. 47
Scott, J. W. 245–6
SCOTVEC, 222
Scribner, S. 111, 127, 128, 129–30, 169, 172, 203–6
set theory, 93
Seventeen + : a New Qualification (1982), 256
Sewell, B. *Use of Mathematics by Adults in Everyday Life*, 95, 96–7
sexual discrimination, 10, 211, 236, 237–8
sexual division of labour, 230–1, 234–5, 240–1, 243
sexuality, social construction of, 243
Sherman, J. 87
Siegel, A. W. 111
signification, 116, 117, 196, 208–9
Silver, H. 7
Silver, R. 126
simulations, 149

situations (Vergnaud), 178
Skemp, R. R. 159
skill transfer, *see* transfer
skills: jobs and task analysis, 148–9; or
    understanding, 121, 123, 141–2, 158–68
Skovsmose, O. 31
small companies, 1
SMILE (Secondary Maths Individual Learning
    Experience), 287, 288
Smith, A. 134
Smith, A. D. W. 148
Smith, P. 225
Smorinski, C. 15
Snell, M. 238
social bias, of maths questions, 67–70
social context, of maths, 13, 57–61
social control, 262
social practice theory, 122, 123, 129–30, 202–7
social structuring of maths, 111, 112–16
South Kensington institutions, 8
Spain, 48, 49
specialism, 277–8
Spradberry, John, 101–2, 111
standardization, 27, 28
standards, 156, 289–90
Stanworth, M. 239
state education, class-based, 7
statistics, 75
status of maths, 9, 289
status maths, in non-elitist contexts, 276–83
Sternberg, R. J. 152
STIM (specific task incorporating maths), 97
story problems, 63
Straker, N. 72
Strässer, Rudolf, 121, 125, 126, 158–68, 208
street maths, and school maths, 173–5
structural linguistics, 117
"structuring resources" (Lave), 207
Struik, D. J. 47
student-centred curriculum design, 260
subtraction, 179–83
Sulbasutras, 44
Sullivan, S. 253, 256–65
Sumerian symbols, 52
supermarket shopping, 111
supervisors, in the clothing industry, 224, 229
Sutherland, R. 88, 89
Sweden, 72

Swetz, F. J. 69
symbolic representation, 197
symbolism, mathematical, 32

task analysis, 148–9
Taunton Commission, 7
Taylor, E. G. R. 5
Taylorism, 166
teacher education, 13, 39
teachers: expectations of girls, 238–9; and role
    of computer, 89
teaching: constraints on, 165–7; maths, 87–8
technical instruction, 4–5, 8
technology, 273; of a culture, 32, 35; new in
    textile industry, 218–19
    relationship to maths education, 80–1
tertiary tripartism, 262–3
textile industry, 3, 10, 211–12, 214–19:
    women in, 231–2, 233
textiles, as a mathematical resource, 280–1,
    287–9
Thales, 45
Thatcher, M. 117
theorems in action, 197
theoretical map, 112–16
theories of practice, 202–10
Thibaut, 44
thinking at work, 130
Third World countries, 80; curriculum
    development in, 22–3; and Eurocentrism,
    42
Third World intellectuals, 59–60
Thompson, S. 18
Toledo, 48, 49
tools: maths as a set of, 93–104, 159, 166;
    maths as use of conceptual, 196–7
Towsend, C. 134, 135, 147
trade unions, and sex discrimination, 239–40
training: assessment of needs in clothing
    industry, 225; in industry, 149–50,
    220–9; investment in, 75; off-the-job,
    149–50, 266; policies, 212–13;
    prescriptions for, 147–8; vs. education,
    31, 271–2
Training Agency, 267
transfer of skills, 126–7, 155, 160, 165, 202,
    209, 266
transmission, 112, 114, 118

trigonometry, 76
trivium, 17
Turkle, S. 82–3, 87
al-Tusi, Nasir Eddin, 49
TVEI, 254, 260, 273–6

UK, decline of shipbuilding, 72–3
Ummayid Caliphs, 48, 49
understanding: instrumental and relational, 159
    levels of, 160–2; skills or, 121, 158–68;
        teaching, 165–7
unemployment (1970s), 253, 259
universality, 30, 32–3, 58, 60, 107
universities, 13th C clerical, 2
urbanization, 7
use value, 104–5
utilitarianism, 50–1, 94–104, 116, 126, 132,
    266

value-free terms, 81
values: dominant, 68; and maths education, 31,
    38; in research, 130
Van Sertima, I. 30
Vergnaud, G. 122, 177–83, 197–8
Victor, S. K. 17
video, 270–1
Vitruvius, 3
Vives, J. L. 5
vocation, in the church, 2
vocational standards, 156
vocational studies, 258
vocationalism, 253, 277
voucher system, 9
Vygotsky, L. S. 32, 88, 110

Wagner, R. K. 152
Wales, workplace research, 124–6
Walkerdine, V. 10, 112, 164, 203, 208–9
war, 4
wealth creation, 72
Weir, S. 88

Wertsch, J. V. 111
Western education, and ethnic minorities,
    29–30
Western mathematics, 13, 33–6, 37, 60
White, J. P. 263
White, L. A. 35, 36, 37; *The Evolution of
    Culture*, 31–2
White, S. H. 111
Wilder, R. L. 15
Wilkinson, T. S. 9, 14, 71–6
Williams, R. 6, 7, 8
Willis, S. 10
Wilson, E. G. 20
Wilson, H. 220
Wittgenstein, L. 114
Wolf, A. 9, 78, 121, 126, 158–68, 270
women: income differentials, 238; in industry,
    211; in the labour force, 230; married in
    part-time work, 236–7; position in the
    labour market, 235–6, 244, 245–6; status
    as mathematicians, 279–80
women's work, 212, 230, 234
work, gendering of, 230–52
work at home, 10, 237
work experience, 183–94
working class, education, 7, 8
working intelligence, 129
workplace maths, user views, 211–13
workplace research, 121–2, 123–31;
    and schools, 207–9

Young, G. C. 279
Young People Starting Work, 125
youth culture, 90
Youth Training, 266–7
Youth Training Scheme (YTS), 153, 254;
    and CAPITB 222–6; *see also* Youth Training

Zaslavsky, C. 30
zone of proximal development (Vygotsky), 88